Springer Texts in Statistics

Series Editors:
G. Casella
S. Fienberg
I. Olkin

For further volumes:
http://www.springer.com/series/417

Gareth James • Daniela Witten • Trevor Hastie
Robert Tibshirani

An Introduction to Statistical Learning

with Applications in R

 Springer

Gareth James
Department of Data Sciences and
    Operations
University of Southern California
Los Angeles, CA, USA

Daniela Witten
Department of Biostatistics
University of Washington
Seattle, WA, USA

Trevor Hastie
Department of Statistics
Stanford University
Stanford, CA, USA

Robert Tibshirani
Department of Statistics
Stanford University
Stanford, CA, USA

ISSN 1431-875X
ISBN 978-1-4614-7137-0   ISBN 978-1-4614-7138-7 (eBook)
DOI 10.1007/978-1-4614-7138-7
Springer New York Heidelberg Dordrecht London

Library of Congress Control Number: 2013936251

Printed on acid-free paper

Springer is part of Springer Science+Business Media (www.springer.com)

To our parents:

Alison and Michael James

Chiara Nappi and Edward Witten

Valerie and Patrick Hastie

Vera and Sami Tibshirani

and to our families:

Michael, Daniel, and Catherine

Tessa, Theo, and Ari

Samantha, Timothy, and Lynda

Charlie, Ryan, Julie, and Cheryl

Preface

Statistical learning refers to a set of tools for modeling and understanding complex datasets. It is a recently developed area in statistics and blends with parallel developments in computer science and, in particular, machine learning. The field encompasses many methods such as the lasso and sparse regression, classification and regression trees, and boosting and support vector machines.

With the explosion of "Big Data" problems, statistical learning has become a very hot field in many scientific areas as well as marketing, finance, and other business disciplines. People with statistical learning skills are in high demand.

One of the first books in this area—*The Elements of Statistical Learning* (ESL) (Hastie, Tibshirani, and Friedman)—was published in 2001, with a second edition in 2009. ESL has become a popular text not only in statistics but also in related fields. One of the reasons for ESL's popularity is its relatively accessible style. But ESL is intended for individuals with advanced training in the mathematical sciences. *An Introduction to Statistical Learning* (ISL) arose from the perceived need for a broader and less technical treatment of these topics. In this new book, we cover many of the same topics as ESL, but we concentrate more on the applications of the methods and less on the mathematical details. We have created labs illustrating how to implement each of the statistical learning methods using the popular statistical software package R. These labs provide the reader with valuable hands-on experience.

This book is appropriate for advanced undergraduates or master's students in statistics or related quantitative fields or for individuals in other disciplines who wish to use statistical learning tools to analyze their data. It can be used as a textbook for a course spanning one or two semesters.

We would like to thank several readers for valuable comments on preliminary drafts of this book: Pallavi Basu, Alexandra Chouldechova, Patrick Danaher, Will Fithian, Luella Fu, Sam Gross, Max Grazier G'Sell, Courtney Paulson, Xinghao Qiao, Elisa Sheng, Noah Simon, Kean Ming Tan, and Xin Lu Tan.

It's tough to make predictions, especially about the future.

-Yogi Berra

Los Angeles, USA   Gareth James
Seattle, USA   Daniela Witten
Palo Alto, USA   Trevor Hastie
Palo Alto, USA   Robert Tibshirani

Contents

1
Introduction

An Overview of Statistical Learning

Statistical learning refers to a vast set of tools for *understanding data*. These tools can be classified as *supervised* or *unsupervised*. Broadly speaking, supervised statistical learning involves building a statistical model for predicting, or estimating, an *output* based on one or more *inputs*. Problems of this nature occur in fields as diverse as business, medicine, astrophysics, and public policy. With unsupervised statistical learning, there are inputs but no supervising output; nevertheless we can learn relationships and structure from such data. To provide an illustration of some applications of statistical learning, we briefly discuss three real-world data sets that are considered in this book.

Wage Data

In this application (which we refer to as the Wage data set throughout this book), we examine a number of factors that relate to wages for a group of males from the Atlantic region of the United States. In particular, we wish to understand the association between an employee's age and education, as well as the calendar year, on his wage. Consider, for example, the left-hand panel of Figure 1.1, which displays wage versus age for each of the individuals in the data set. There is evidence that wage increases with age but then decreases again after approximately age 60. The blue line, which provides an estimate of the average wage for a given age, makes this trend clearer.

G. James et al., *An Introduction to Statistical Learning: with Applications in R*,
Springer Texts in Statistics, DOI 10.1007/978-1-4614-7138-7_1,
© Springer Science+Business Media New York 2013

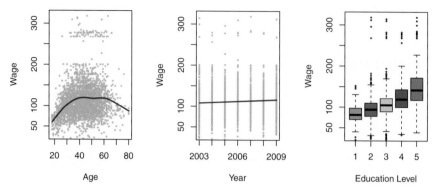

FIGURE 1.1. Wage *data, which contains income survey information for males from the central Atlantic region of the United States.* Left: wage *as a function of* age. *On average,* wage *increases with* age *until about* 60 *years of age, at which point it begins to decline.* Center: wage *as a function of* year. *There is a slow but steady increase of approximately* $10,000 *in the average* wage *between* 2003 *and* 2009. Right: *Boxplots displaying* wage *as a function of* education, *with* 1 *indicating the lowest level (no high school diploma) and* 5 *the highest level (an advanced graduate degree). On average,* wage *increases with the level of education.*

Given an employee's age, we can use this curve to *predict* his wage. However, it is also clear from Figure 1.1 that there is a significant amount of variability associated with this average value, and so age alone is unlikely to provide an accurate prediction of a particular man's wage.

We also have information regarding each employee's education level and the year in which the wage was earned. The center and right-hand panels of Figure 1.1, which display wage as a function of both year and education, indicate that both of these factors are associated with wage. Wages increase by approximately $10,000, in a roughly linear (or straight-line) fashion, between 2003 and 2009, though this rise is very slight relative to the variability in the data. Wages are also typically greater for individuals with higher education levels: men with the lowest education level (1) tend to have substantially lower wages than those with the highest education level (5). Clearly, the most accurate prediction of a given man's wage will be obtained by combining his age, his education, and the year. In Chapter 3, we discuss linear regression, which can be used to predict wage from this data set. Ideally, we should predict wage in a way that accounts for the non-linear relationship between wage and age. In Chapter 7, we discuss a class of approaches for addressing this problem.

Stock Market Data

The Wage data involves predicting a *continuous* or *quantitative* output value. This is often referred to as a *regression* problem. However, in certain cases we may instead wish to predict a non-numerical value—that is, a *categorical*

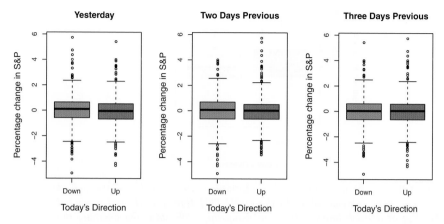

FIGURE 1.2. Left: *Boxplots of the previous day's percentage change in the S&P index for the days for which the market increased or decreased, obtained from the* Smarket *data.* Center and Right: *Same as left panel, but the percentage changes for 2 and 3 days previous are shown.*

or *qualitative* output. For example, in Chapter 4 we examine a stock market data set that contains the daily movements in the Standard & Poor's 500 (S&P) stock index over a 5-year period between 2001 and 2005. We refer to this as the Smarket data. The goal is to predict whether the index will *increase* or *decrease* on a given day using the past 5 days' percentage changes in the index. Here the statistical learning problem does not involve predicting a numerical value. Instead it involves predicting whether a given day's stock market performance will fall into the Up bucket or the Down bucket. This is known as a *classification* problem. A model that could accurately predict the direction in which the market will move would be very useful!

The left-hand panel of Figure 1.2 displays two boxplots of the previous day's percentage changes in the stock index: one for the 648 days for which the market increased on the subsequent day, and one for the 602 days for which the market decreased. The two plots look almost identical, suggesting that there is no simple strategy for using yesterday's movement in the S&P to predict today's returns. The remaining panels, which display boxplots for the percentage changes 2 and 3 days previous to today, similarly indicate little association between past and present returns. Of course, this lack of pattern is to be expected: in the presence of strong correlations between successive days' returns, one could adopt a simple trading strategy to generate profits from the market. Nevertheless, in Chapter 4, we explore these data using several different statistical learning methods. Interestingly, there are hints of some weak trends in the data that suggest that, at least for this 5-year period, it is possible to correctly predict the direction of movement in the market approximately 60% of the time (Figure 1.3).

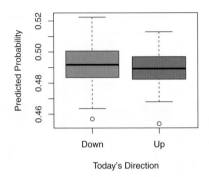

FIGURE 1.3. *We fit a quadratic discriminant analysis model to the subset of the* Smarket *data corresponding to the 2001–2004 time period, and predicted the probability of a stock market decrease using the 2005 data. On average, the predicted probability of decrease is higher for the days in which the market does decrease. Based on these results, we are able to correctly predict the direction of movement in the market 60% of the time.*

Gene Expression Data

The previous two applications illustrate data sets with both input and output variables. However, another important class of problems involves situations in which we only observe input variables, with no corresponding output. For example, in a marketing setting, we might have demographic information for a number of current or potential customers. We may wish to understand which types of customers are similar to each other by grouping individuals according to their observed characteristics. This is known as a *clustering* problem. Unlike in the previous examples, here we are not trying to predict an output variable.

We devote Chapter 10 to a discussion of statistical learning methods for problems in which no natural output variable is available. We consider the NCI60 data set, which consists of 6,830 gene expression measurements for each of 64 cancer cell lines. Instead of predicting a particular output variable, we are interested in determining whether there are groups, or clusters, among the cell lines based on their gene expression measurements. This is a difficult question to address, in part because there are thousands of gene expression measurements per cell line, making it hard to visualize the data.

The left-hand panel of Figure 1.4 addresses this problem by representing each of the 64 cell lines using just two numbers, Z_1 and Z_2. These are the first two *principal components* of the data, which summarize the $6,830$ expression measurements for each cell line down to two numbers or *dimensions*. While it is likely that this dimension reduction has resulted in

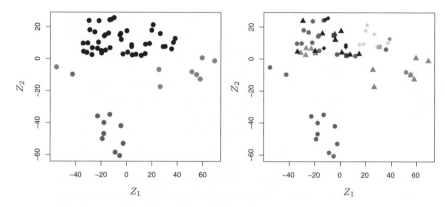

FIGURE 1.4. Left: *Representation of the* NCI60 *gene expression data set in a two-dimensional space, Z_1 and Z_2. Each point corresponds to one of the 64 cell lines. There appear to be four groups of cell lines, which we have represented using different colors.* Right: *Same as left panel except that we have represented each of the 14 different types of cancer using a different colored symbol. Cell lines corresponding to the same cancer type tend to be nearby in the two-dimensional space.*

some loss of information, it is now possible to visually examine the data for evidence of clustering. Deciding on the number of clusters is often a difficult problem. But the left-hand panel of Figure 1.4 suggests at least four groups of cell lines, which we have represented using separate colors. We can now examine the cell lines within each cluster for similarities in their types of cancer, in order to better understand the relationship between gene expression levels and cancer.

In this particular data set, it turns out that the cell lines correspond to 14 different types of cancer. (However, this information was not used to create the left-hand panel of Figure 1.4.) The right-hand panel of Figure 1.4 is identical to the left-hand panel, except that the 14 cancer types are shown using distinct colored symbols. There is clear evidence that cell lines with the same cancer type tend to be located near each other in this two-dimensional representation. In addition, even though the cancer information was not used to produce the left-hand panel, the clustering obtained does bear some resemblance to some of the actual cancer types observed in the right-hand panel. This provides some independent verification of the accuracy of our clustering analysis.

A Brief History of Statistical Learning

Though the term *statistical learning* is fairly new, many of the concepts that underlie the field were developed long ago. At the beginning of the nineteenth century, Legendre and Gauss published papers on the *method of least squares*, which implemented the earliest form of what is now known as *linear regression*. The approach was first successfully applied to problems in astronomy. Linear regression is used for predicting quantitative values, such as an individual's salary. In order to predict qualitative values, such as whether a patient survives or dies, or whether the stock market increases or decreases, Fisher proposed *linear discriminant analysis* in 1936. In the 1940s, various authors put forth an alternative approach, *logistic regression*. In the early 1970s, Nelder and Wedderburn coined the term *generalized linear models* for an entire class of statistical learning methods that include both linear and logistic regression as special cases.

By the end of the 1970s, many more techniques for learning from data were available. However, they were almost exclusively *linear* methods, because fitting *non-linear* relationships was computationally infeasible at the time. By the 1980s, computing technology had finally improved sufficiently that non-linear methods were no longer computationally prohibitive. In mid 1980s Breiman, Friedman, Olshen and Stone introduced *classification and regression trees*, and were among the first to demonstrate the power of a detailed practical implementation of a method, including cross-validation for model selection. Hastie and Tibshirani coined the term *generalized additive models* in 1986 for a class of non-linear extensions to generalized linear models, and also provided a practical software implementation.

Since that time, inspired by the advent of *machine learning* and other disciplines, statistical learning has emerged as a new subfield in statistics, focused on supervised and unsupervised modeling and prediction. In recent years, progress in statistical learning has been marked by the increasing availability of powerful and relatively user-friendly software, such as the popular and freely available R system. This has the potential to continue the transformation of the field from a set of techniques used and developed by statisticians and computer scientists to an essential toolkit for a much broader community.

This Book

The Elements of Statistical Learning (ESL) by Hastie, Tibshirani, and Friedman was first published in 2001. Since that time, it has become an important reference on the fundamentals of statistical machine learning. Its success derives from its comprehensive and detailed treatment of many important topics in statistical learning, as well as the fact that (relative to many upper-level statistics textbooks) it is accessible to a wide audience. However, the greatest factor behind the success of ESL has been its topical nature. At the time of its publication, interest in the field of statistical learning was starting to explode. ESL provided one of the first accessible and comprehensive introductions to the topic.

Since ESL was first published, the field of statistical learning has continued to flourish. The field's expansion has taken two forms. The most obvious growth has involved the development of new and improved statistical learning approaches aimed at answering a range of scientific questions across a number of fields. However, the field of statistical learning has also expanded its audience. In the 1990s, increases in computational power generated a surge of interest in the field from non-statisticians who were eager to use cutting-edge statistical tools to analyze their data. Unfortunately, the highly technical nature of these approaches meant that the user community remained primarily restricted to experts in statistics, computer science, and related fields with the training (and time) to understand and implement them.

In recent years, new and improved software packages have significantly eased the implementation burden for many statistical learning methods. At the same time, there has been growing recognition across a number of fields, from business to health care to genetics to the social sciences and beyond, that statistical learning is a powerful tool with important practical applications. As a result, the field has moved from one of primarily academic interest to a mainstream discipline, with an enormous potential audience. This trend will surely continue with the increasing availability of enormous quantities of data and the software to analyze it.

The purpose of *An Introduction to Statistical Learning* (ISL) is to facilitate the transition of statistical learning from an academic to a mainstream field. ISL is not intended to replace ESL, which is a far more comprehensive text both in terms of the number of approaches considered and the depth to which they are explored. We consider ESL to be an important companion for professionals (with graduate degrees in statistics, machine learning, or related fields) who need to understand the technical details behind statistical learning approaches. However, the community of users of statistical learning techniques has expanded to include individuals with a wider range of interests and backgrounds. Therefore, we believe that there is now a place for a less technical and more accessible version of ESL.

In teaching these topics over the years, we have discovered that they are of interest to master's and PhD students in fields as disparate as business administration, biology, and computer science, as well as to quantitatively-oriented upper-division undergraduates. It is important for this diverse group to be able to understand the models, intuitions, and strengths and weaknesses of the various approaches. But for this audience, many of the technical details behind statistical learning methods, such as optimization algorithms and theoretical properties, are not of primary interest. We believe that these students do not need a deep understanding of these aspects in order to become informed users of the various methodologies, and in order to contribute to their chosen fields through the use of statistical learning tools.

ISLR is based on the following four premises.

1. *Many statistical learning methods are relevant and useful in a wide range of academic and non-academic disciplines, beyond just the statistical sciences.* We believe that many contemporary statistical learning procedures should, and will, become as widely available and used as is currently the case for classical methods such as linear regression. As a result, rather than attempting to consider every possible approach (an impossible task), we have concentrated on presenting the methods that we believe are most widely applicable.

2. *Statistical learning should not be viewed as a series of black boxes.* No single approach will perform well in all possible applications. Without understanding all of the cogs inside the box, or the interaction between those cogs, it is impossible to select the best box. Hence, we have attempted to carefully describe the model, intuition, assumptions, and trade-offs behind each of the methods that we consider.

3. *While it is important to know what job is performed by each cog, it is not necessary to have the skills to construct the machine inside the box!* Thus, we have minimized discussion of technical details related to fitting procedures and theoretical properties. We assume that the reader is comfortable with basic mathematical concepts, but we do not assume a graduate degree in the mathematical sciences. For instance, we have almost completely avoided the use of matrix algebra, and it is possible to understand the entire book without a detailed knowledge of matrices and vectors.

4. *We presume that the reader is interested in applying statistical learning methods to real-world problems.* In order to facilitate this, as well as to motivate the techniques discussed, we have devoted a section within each chapter to R computer labs. In each lab, we walk the reader through a realistic application of the methods considered in that chapter. When we have taught this material in our courses, we have allocated roughly one-third of classroom time to working through the labs, and we have found them to be extremely useful. Many of the less computationally-oriented students who were initially intimidated by R's command level interface got the hang of things over the course of the quarter or semester. We have used R because it is freely available and is powerful enough to implement all of the methods discussed in the book. It also has optional packages that can be downloaded to implement literally thousands of additional methods. Most importantly, R is the language of choice for academic statisticians, and new approaches often become available in

R years before they are implemented in commercial packages. However, the labs in ISL are self-contained, and can be skipped if the reader wishes to use a different software package or does not wish to apply the methods discussed to real-world problems.

Who Should Read This Book?

This book is intended for anyone who is interested in using modern statistical methods for modeling and prediction from data. This group includes scientists, engineers, data analysts, or *quants*, but also less technical individuals with degrees in non-quantitative fields such as the social sciences or business. We expect that the reader will have had at least one elementary course in statistics. Background in linear regression is also useful, though not required, since we review the key concepts behind linear regression in Chapter 3. The mathematical level of this book is modest, and a detailed knowledge of matrix operations is not required. This book provides an introduction to the statistical programming language R. Previous exposure to a programming language, such as MATLAB or Python, is useful but not required.

We have successfully taught material at this level to master's and PhD students in business, computer science, biology, earth sciences, psychology, and many other areas of the physical and social sciences. This book could also be appropriate for advanced undergraduates who have already taken a course on linear regression. In the context of a more mathematically rigorous course in which ESL serves as the primary textbook, ISL could be used as a supplementary text for teaching computational aspects of the various approaches.

Notation and Simple Matrix Algebra

Choosing notation for a textbook is always a difficult task. For the most part we adopt the same notational conventions as ESL.

We will use n to represent the number of distinct data points, or observations, in our sample. We will let p denote the number of variables that are available for use in making predictions. For example, the Wage data set consists of 12 variables for 3,000 people, so we have $n = 3,000$ observations and $p = 12$ variables (such as year, age, sex, and more). Note that throughout this book, we indicate variable names using colored font: Variable Name.

In some examples, p might be quite large, such as on the order of thousands or even millions; this situation arises quite often, for example, in the analysis of modern biological data or web-based advertising data.

In general, we will let x_{ij} represent the value of the jth variable for the ith observation, where $i = 1, 2, \ldots, n$ and $j = 1, 2, \ldots, p$. Throughout this book, i will be used to index the samples or observations (from 1 to n) and j will be used to index the variables (from 1 to p). We let $\mathbf{X}$ denote a $n \times p$ matrix whose (i, j)th element is x_{ij}. That is,

$$
\mathbf{X} = \begin{pmatrix}
x_{11} & x_{12} & \cdots & x_{1p} \\
x_{21} & x_{22} & \cdots & x_{2p} \\
\vdots & \vdots & \ddots & \vdots \\
x_{n1} & x_{n2} & \cdots & x_{np}
\end{pmatrix}.
$$

For readers who are unfamiliar with matrices, it is useful to visualize $\mathbf{X}$ as a spreadsheet of numbers with n rows and p columns.

At times we will be interested in the rows of $\mathbf{X}$, which we write as $x_1, x_2, \ldots, x_n$. Here x_i is a vector of length p, containing the p variable measurements for the ith observation. That is,

$$
x_i = \begin{pmatrix}
x_{i1} \\
x_{i2} \\
\vdots \\
x_{ip}
\end{pmatrix}. \tag{1.1}
$$

(Vectors are by default represented as columns.) For example, for the Wage data, x_i is a vector of length 12, consisting of year, age, sex, and other values for the ith individual. At other times we will instead be interested in the columns of $\mathbf{X}$, which we write as $\mathbf{x}_1, \mathbf{x}_2, \ldots, \mathbf{x}_p$. Each is a vector of length n. That is,

$$
\mathbf{x}_j = \begin{pmatrix}
x_{1j} \\
x_{2j} \\
\vdots \\
x_{nj}
\end{pmatrix}.
$$

For example, for the Wage data, $\mathbf{x}_1$ contains the $n = 3{,}000$ values for year.

Using this notation, the matrix $\mathbf{X}$ can be written as

$$
\mathbf{X} = \begin{pmatrix} \mathbf{x}_1 & \mathbf{x}_2 & \cdots & \mathbf{x}_p \end{pmatrix},
$$

or

$$
\mathbf{X} = \begin{pmatrix}
x_1^T \\
x_2^T \\
\vdots \\
x_n^T
\end{pmatrix}.
$$

The ^{T} notation denotes the *transpose* of a matrix or vector. So, for example,

$$\mathbf{X}^T = \begin{pmatrix} x_{11} & x_{21} & \cdots & x_{n1} \\ x_{12} & x_{22} & \cdots & x_{n2} \\ \vdots & \vdots & & \vdots \\ x_{1p} & x_{2p} & \cdots & x_{np} \end{pmatrix},$$

while

$$x_i^T = \begin{pmatrix} x_{i1} & x_{i2} & \cdots & x_{ip} \end{pmatrix}.$$

We use y_i to denote the ith observation of the variable on which we wish to make predictions, such as wage. Hence, we write the set of all n observations in vector form as

$$\mathbf{y} = \begin{pmatrix} y_1 \\ y_2 \\ \vdots \\ y_n \end{pmatrix}.$$

Then our observed data consists of $\{(x_1, y_1), (x_2, y_2), \ldots, (x_n, y_n)\}$, where each x_i is a vector of length p. (If $p = 1$, then x_i is simply a scalar.)

In this text, a vector of length n will always be denoted in *lower case bold*; e.g.

$$\mathbf{a} = \begin{pmatrix} a_1 \\ a_2 \\ \vdots \\ a_n \end{pmatrix}.$$

However, vectors that are not of length n (such as feature vectors of length p, as in (1.1)) will be denoted in *lower case normal font*, e.g. a. Scalars will also be denoted in *lower case normal font*, e.g. a. In the rare cases in which these two uses for lower case normal font lead to ambiguity, we will clarify which use is intended. Matrices will be denoted using *bold capitals*, such as $\mathbf{A}$. Random variables will be denoted using *capital normal font*, e.g. A, regardless of their dimensions.

Occasionally we will want to indicate the dimension of a particular object. To indicate that an object is a scalar, we will use the notation $a \in \mathbb{R}$. To indicate that it is a vector of length k, we will use $a \in \mathbb{R}^k$ (or $\mathbf{a} \in \mathbb{R}^n$ if it is of length n). We will indicate that an object is a $r \times s$ matrix using $\mathbf{A} \in \mathbb{R}^{r \times s}$.

We have avoided using matrix algebra whenever possible. However, in a few instances it becomes too cumbersome to avoid it entirely. In these rare instances it is important to understand the concept of multiplying two matrices. Suppose that $\mathbf{A} \in \mathbb{R}^{r \times d}$ and $\mathbf{B} \in \mathbb{R}^{d \times s}$. Then the product of $\mathbf{A}$ and $\mathbf{B}$ is denoted $\mathbf{AB}$. The (i, j)th element of $\mathbf{AB}$ is computed by multiplying each element of the ith row of $\mathbf{A}$ by the corresponding element of the jth column of $\mathbf{B}$. That is, $(\mathbf{AB})_{ij} = \sum_{k=1}^{d} a_{ik}b_{kj}$. As an example, consider

$$\mathbf{A} = \begin{pmatrix} 1 & 2 \\ 3 & 4 \end{pmatrix} \quad \text{and} \quad \mathbf{B} = \begin{pmatrix} 5 & 6 \\ 7 & 8 \end{pmatrix}.$$

Then

$$\mathbf{AB} = \begin{pmatrix} 1 & 2 \\ 3 & 4 \end{pmatrix} \begin{pmatrix} 5 & 6 \\ 7 & 8 \end{pmatrix} = \begin{pmatrix} 1 \times 5 + 2 \times 7 & 1 \times 6 + 2 \times 8 \\ 3 \times 5 + 4 \times 7 & 3 \times 6 + 4 \times 8 \end{pmatrix} = \begin{pmatrix} 19 & 22 \\ 43 & 50 \end{pmatrix}.$$

Note that this operation produces an $r \times s$ matrix. It is only possible to compute $\mathbf{AB}$ if the number of columns of $\mathbf{A}$ is the same as the number of rows of $\mathbf{B}$.

Organization of This Book

Chapter 2 introduces the basic terminology and concepts behind statistical learning. This chapter also presents the K-*nearest neighbor* classifier, a very simple method that works surprisingly well on many problems. Chapters 3 and 4 cover classical linear methods for regression and classification. In particular, Chapter 3 reviews *linear regression*, the fundamental starting point for all regression methods. In Chapter 4 we discuss two of the most important classical classification methods, *logistic regression* and *linear discriminant analysis*.

A central problem in all statistical learning situations involves choosing the best method for a given application. Hence, in Chapter 5 we introduce *cross-validation* and the *bootstrap*, which can be used to estimate the accuracy of a number of different methods in order to choose the best one.

Much of the recent research in statistical learning has concentrated on non-linear methods. However, linear methods often have advantages over their non-linear competitors in terms of interpretability and sometimes also accuracy. Hence, in Chapter 6 we consider a host of linear methods, both classical and more modern, which offer potential improvements over standard linear regression. These include *stepwise selection, ridge regression, principal components regression, partial least squares,* and the *lasso*.

The remaining chapters move into the world of non-linear statistical learning. We first introduce in Chapter 7 a number of non-linear methods that work well for problems with a single input variable. We then show how these methods can be used to fit non-linear *additive* models for which there is more than one input. In Chapter 8, we investigate *tree*-based methods, including *bagging, boosting,* and *random forests. Support vector machines,* a set of approaches for performing both linear and non-linear classification, are discussed in Chapter 9. Finally, in Chapter 10, we consider a setting in which we have input variables but no output variable. In particular, we present *principal components analysis*, *K-means clustering*, and *hierarchical clustering*.

At the end of each chapter, we present one or more R lab sections in which we systematically work through applications of the various methods discussed in that chapter. These labs demonstrate the strengths and weaknesses of the various approaches, and also provide a useful reference for the syntax required to implement the various methods. The reader may choose to work through the labs at his or her own pace, or the labs may be the focus of group sessions as part of a classroom environment. Within each R lab, we present the results that we obtained when we performed the lab at the time of writing this book. However, new versions of R are continuously released, and over time, the packages called in the labs will be updated. Therefore, in the future, it is possible that the results shown in the lab sections may no longer correspond precisely to the results obtained by the reader who performs the labs. As necessary, we will post updates to the labs on the book website.

We use the ✦ symbol to denote sections or exercises that contain more challenging concepts. These can be easily skipped by readers who do not wish to delve as deeply into the material, or who lack the mathematical background.

Data Sets Used in Labs and Exercises

In this textbook, we illustrate statistical learning methods using applications from marketing, finance, biology, and other areas. The ISLR package available on the book website contains a number of data sets that are required in order to perform the labs and exercises associated with this book. One other data set is contained in the MASS library, and yet another is part of the base R distribution. Table 1.1 contains a summary of the data sets required to perform the labs and exercises. A couple of these data sets are also available as text files on the book website, for use in Chapter 2.

Book Website

The website for this book is located at

www.StatLearning.com

Name	Description
Auto	Gas mileage, horsepower, and other information for cars.
Boston	Housing values and other information about Boston suburbs.
Caravan	Information about individuals offered caravan insurance.
Carseats	Information about car seat sales in 400 stores.
College	Demographic characteristics, tuition, and more for USA colleges.
Default	Customer default records for a credit card company.
Hitters	Records and salaries for baseball players.
Khan	Gene expression measurements for four cancer types.
NCI60	Gene expression measurements for 64 cancer cell lines.
OJ	Sales information for Citrus Hill and Minute Maid orange juice.
Portfolio	Past values of financial assets, for use in portfolio allocation.
Smarket	Daily percentage returns for S&P 500 over a 5-year period.
USArrests	Crime statistics per 100,000 residents in 50 states of USA.
Wage	Income survey data for males in central Atlantic region of USA.
Weekly	1,089 weekly stock market returns for 21 years.

TABLE 1.1. *A list of data sets needed to perform the labs and exercises in this textbook. All data sets are available in the* ISLR *library, with the exception of* Boston *(part of* MASS*) and* USArrests *(part of the base* R *distribution).*

It contains a number of resources, including the R package associated with this book, and some additional data sets.

Acknowledgements

A few of the plots in this book were taken from ESL: Figures 6.7, 8.3, and 10.12. All other plots are new to this book.

2

Statistical Learning

2.1 What Is Statistical Learning?

In order to motivate our study of statistical learning, we begin with a simple example. Suppose that we are statistical consultants hired by a client to provide advice on how to improve sales of a particular product. The Advertising data set consists of the sales of that product in 200 different markets, along with advertising budgets for the product in each of those markets for three different media: TV, radio, and newspaper. The data are displayed in Figure 2.1. It is not possible for our client to directly increase sales of the product. On the other hand, they can control the advertising expenditure in each of the three media. Therefore, if we determine that there is an association between advertising and sales, then we can instruct our client to adjust advertising budgets, thereby indirectly increasing sales. In other words, our goal is to develop an accurate model that can be used to predict sales on the basis of the three media budgets.

In this setting, the advertising budgets are *input variables* while sales is an *output variable*. The input variables are typically denoted using the symbol X, with a subscript to distinguish them. So X_1 might be the TV budget, X_2 the radio budget, and X_3 the newspaper budget. The inputs go by different names, such as *predictors, independent variables, features,* or sometimes just *variables*. The output variable—in this case, sales—is often called the *response* or *dependent variable,* and is typically denoted using the symbol Y. Throughout this book, we will use all of these terms interchangeably.

input variable
output variable

predictor
independent variable
feature
variable
response
dependent variable

G. James et al., *An Introduction to Statistical Learning: with Applications in R*, 15
Springer Texts in Statistics, DOI 10.1007/978-1-4614-7138-7_2,
© Springer Science+Business Media New York 2013

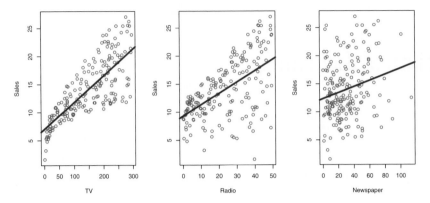

FIGURE 2.1. *The* Advertising *data set. The plot displays* sales, *in thousands of units, as a function of* TV, radio, *and* newspaper *budgets, in thousands of dollars, for* 200 *different markets. In each plot we show the simple least squares fit of* sales *to that variable, as described in Chapter 3. In other words, each blue line represents a simple model that can be used to predict* sales *using* TV, radio, *and* newspaper, *respectively.*

More generally, suppose that we observe a quantitative response Y and p different predictors, $X_1, X_2, \ldots, X_p$. We assume that there is some relationship between Y and $X = (X_1, X_2, \ldots, X_p)$, which can be written in the very general form

$$Y = f(X) + \epsilon. \tag{2.1}$$

Here f is some fixed but unknown function of $X_1, \ldots, X_p$, and ϵ is a random *error term*, which is independent of X and has mean zero. In this formula- error term
tion, f represents the *systematic* information that X provides about Y. systematic

As another example, consider the left-hand panel of Figure 2.2, a plot of income versus years of education for 30 individuals in the Income data set. The plot suggests that one might be able to predict income using years of education. However, the function f that connects the input variable to the output variable is in general unknown. In this situation one must estimate f based on the observed points. Since Income is a simulated data set, f is known and is shown by the blue curve in the right-hand panel of Figure 2.2. The vertical lines represent the error terms ϵ. We note that some of the 30 observations lie above the blue curve and some lie below it; overall, the errors have approximately mean zero.

In general, the function f may involve more than one input variable. In Figure 2.3 we plot income as a function of years of education and seniority. Here f is a two-dimensional surface that must be estimated based on the observed data.

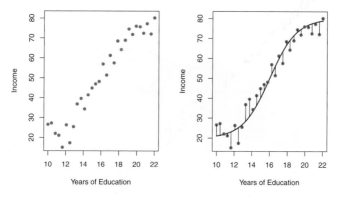

FIGURE 2.2. *The* Income *data set. Left: The red dots are the observed values of* income *(in tens of thousands of dollars) and* years *of* education *for* 30 *individuals. Right: The blue curve represents the true underlying relationship between* income *and* years *of* education, *which is generally unknown (but is known in this case because the data were simulated). The black lines represent the error associated with each observation. Note that some errors are positive (if an observation lies above the blue curve) and some are negative (if an observation lies below the curve). Overall, these errors have approximately mean zero.*

In essence, statistical learning refers to a set of approaches for estimating f. In this chapter we outline some of the key theoretical concepts that arise in estimating f, as well as tools for evaluating the estimates obtained.

2.1.1 Why Estimate f?

There are two main reasons that we may wish to estimate f: *prediction* and *inference*. We discuss each in turn.

Prediction

In many situations, a set of inputs X are readily available, but the output Y cannot be easily obtained. In this setting, since the error term averages to zero, we can predict Y using

$$\hat{Y} = \hat{f}(X), \tag{2.2}$$

where $\hat{f}$ represents our estimate for f, and $\hat{Y}$ represents the resulting prediction for Y. In this setting, $\hat{f}$ is often treated as a *black box*, in the sense that one is not typically concerned with the exact form of $\hat{f}$, provided that it yields accurate predictions for Y.

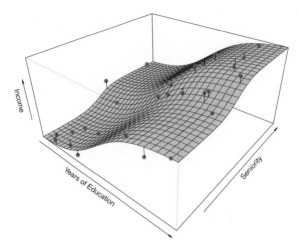

FIGURE 2.3. *The plot displays* income *as a function of* years of education *and* seniority *in the* Income *data set. The blue surface represents the true underlying relationship between* income *and* years of education *and* seniority, *which is known since the data are simulated. The red dots indicate the observed values of these quantities for* 30 *individuals.*

As an example, suppose that $X_1, \ldots, X_p$ are characteristics of a patient's blood sample that can be easily measured in a lab, and Y is a variable encoding the patient's risk for a severe adverse reaction to a particular drug. It is natural to seek to predict Y using X, since we can then avoid giving the drug in question to patients who are at high risk of an adverse reaction—that is, patients for whom the estimate of Y is high.

The accuracy of $\hat{Y}$ as a prediction for Y depends on two quantities, which we will call the *reducible error* and the *irreducible error*. In general, $\hat{f}$ will not be a perfect estimate for f, and this inaccuracy will introduce some error. This error is *reducible* because we can potentially improve the accuracy of $\hat{f}$ by using the most appropriate statistical learning technique to estimate f. However, even if it were possible to form a perfect estimate for f, so that our estimated response took the form $\hat{Y} = f(X)$, our prediction would still have some error in it! This is because Y is also a function of ϵ, which, by definition, cannot be predicted using X. Therefore, variability associated with ϵ also affects the accuracy of our predictions. This is known as the *irreducible* error, because no matter how well we estimate f, we cannot reduce the error introduced by ϵ.

Why is the irreducible error larger than zero? The quantity ϵ may contain unmeasured variables that are useful in predicting Y: since we don't measure them, f cannot use them for its prediction. The quantity ϵ may also contain unmeasurable variation. For example, the risk of an adverse reaction might vary for a given patient on a given day, depending on

reducible
error

irreducible
error

manufacturing variation in the drug itself or the patient's general feeling
of well-being on that day.

Consider a given estimate $\hat{f}$ and a set of predictors X, which yields the
prediction $\hat{Y} = \hat{f}(X)$. Assume for a moment that both $\hat{f}$ and X are fixed.
Then, it is easy to show that

$$
\begin{aligned}
E(Y - \hat{Y})^2 &= E[f(X) + \epsilon - \hat{f}(X)]^2 \\
&= \underbrace{[f(X) - \hat{f}(X)]^2}_{\text{Reducible}} + \underbrace{\text{Var}(\epsilon)}_{\text{Irreducible}} , \quad (2.3)
\end{aligned}
$$

where $E(Y - \hat{Y})^2$ represents the average, or *expected value*, of the squared
difference between the predicted and actual value of Y, and $\text{Var}(\epsilon)$ represents the *variance* associated with the error term ϵ.

expected
value

variance

The focus of this book is on techniques for estimating f with the aim of
minimizing the reducible error. It is important to keep in mind that the
irreducible error will always provide an upper bound on the accuracy of
our prediction for Y. This bound is almost always unknown in practice.

Inference

We are often interested in understanding the way that Y is affected as
$X_1, \ldots, X_p$ change. In this situation we wish to estimate f, but our goal is
not necessarily to make predictions for Y. We instead want to understand
the relationship between X and Y, or more specifically, to understand how
Y changes as a function of $X_1, \ldots, X_p$. Now $\hat{f}$ cannot be treated as a black
box, because we need to know its exact form. In this setting, one may be
interested in answering the following questions:

- *Which predictors are associated with the response?* It is often the case
  that only a small fraction of the available predictors are substantially
  associated with Y. Identifying the few *important* predictors among a
  large set of possible variables can be extremely useful, depending on
  the application.

- *What is the relationship between the response and each predictor?*
  Some predictors may have a positive relationship with Y, in the sense
  that increasing the predictor is associated with increasing values of
  Y. Other predictors may have the opposite relationship. Depending
  on the complexity of f, the relationship between the response and a
  given predictor may also depend on the values of the other predictors.

- *Can the relationship between Y and each predictor be adequately summarized using a linear equation, or is the relationship more complicated?* Historically, most methods for estimating f have taken a linear
  form. In some situations, such an assumption is reasonable or even desirable. But often the true relationship is more complicated, in which
  case a linear model may not provide an accurate representation of
  the relationship between the input and output variables.

In this book, we will see a number of examples that fall into the prediction setting, the inference setting, or a combination of the two.

For instance, consider a company that is interested in conducting a direct-marketing campaign. The goal is to identify individuals who will respond positively to a mailing, based on observations of demographic variables measured on each individual. In this case, the demographic variables serve as predictors, and response to the marketing campaign (either positive or negative) serves as the outcome. The company is not interested in obtaining a deep understanding of the relationships between each individual predictor and the response; instead, the company simply wants an accurate model to predict the response using the predictors. This is an example of modeling for prediction.

In contrast, consider the Advertising data illustrated in Figure 2.1. One may be interested in answering questions such as:

– *Which media contribute to sales?*

– *Which media generate the biggest boost in sales?* or

– *How much increase in sales is associated with a given increase in TV advertising?*

This situation falls into the inference paradigm. Another example involves modeling the brand of a product that a customer might purchase based on variables such as price, store location, discount levels, competition price, and so forth. In this situation one might really be most interested in how each of the individual variables affects the probability of purchase. For instance, *what effect will changing the price of a product have on sales?* This is an example of modeling for inference.

Finally, some modeling could be conducted both for prediction and inference. For example, in a real estate setting, one may seek to relate values of homes to inputs such as crime rate, zoning, distance from a river, air quality, schools, income level of community, size of houses, and so forth. In this case one might be interested in how the individual input variables affect the prices—that is, *how much extra will a house be worth if it has a view of the river?* This is an inference problem. Alternatively, one may simply be interested in predicting the value of a home given its characteristics: *is this house under- or over-valued?* This is a prediction problem.

Depending on whether our ultimate goal is prediction, inference, or a combination of the two, different methods for estimating f may be appropriate. For example, *linear models* allow for relatively simple and inter-   linear model
pretable inference, but may not yield as accurate predictions as some other approaches. In contrast, some of the highly non-linear approaches that we discuss in the later chapters of this book can potentially provide quite accurate predictions for Y, but this comes at the expense of a less interpretable model for which inference is more challenging.

2.1.2   How Do We Estimate f?

Throughout this book, we explore many linear and non-linear approaches
for estimating f. However, these methods generally share certain characteristics. We provide an overview of these shared characteristics in this
section. We will always assume that we have observed a set of n different
data points. For example in Figure 2.2 we observed $n = 30$ data points.
These observations are called the *training data* because we will use these
observations to train, or teach, our method how to estimate f. Let x_{ij} *training data*
represent the value of the jth predictor, or input, for observation i, where
$i = 1, 2, \ldots, n$ and $j = 1, 2, \ldots, p$. Correspondingly, let y_i represent the
response variable for the ith observation. Then our training data consist of
$\{(x_1, y_1), (x_2, y_2), \ldots, (x_n, y_n)\}$ where $x_i = (x_{i1}, x_{i2}, \ldots, x_{ip})^T$.

Our goal is to apply a statistical learning method to the training data
in order to estimate the unknown function f. In other words, we want to
find a function $\hat{f}$ such that $Y \approx \hat{f}(X)$ for any observation (X, Y). Broadly
speaking, most statistical learning methods for this task can be characterized as either *parametric* or *non-parametric*. We now briefly discuss these *parametric*
two types of approaches. *non-*
parametric

Parametric Methods

Parametric methods involve a two-step model-based approach.

1. First, we make an assumption about the functional form, or shape,
   of f. For example, one very simple assumption is that f is linear in
   X:
   $$f(X) = \beta_0 + \beta_1 X_1 + \beta_2 X_2 + \ldots + \beta_p X_p. \tag{2.4}$$
   This is a *linear model*, which will be discussed extensively in Chapter 3. Once we have assumed that f is linear, the problem of estimating f is greatly simplified. Instead of having to estimate an entirely
   arbitrary p-dimensional function $f(X)$, one only needs to estimate
   the $p + 1$ coefficients $\beta_0, \beta_1, \ldots, \beta_p$.

2. After a model has been selected, we need a procedure that uses the
   training data to *fit* or *train* the model. In the case of the linear model
   (2.4), we need to estimate the parameters $\beta_0, \beta_1, \ldots, \beta_p$. That is, we *fit*
   want to find values of these parameters such that *train*
   $$Y \approx \beta_0 + \beta_1 X_1 + \beta_2 X_2 + \ldots + \beta_p X_p.$$
   The most common approach to fitting the model (2.4) is referred to
   as *(ordinary) least squares*, which we discuss in Chapter 3. However,
   least squares is one of many possible ways to fit the linear model. In *least squares*
   Chapter 6, we discuss other approaches for estimating the parameters
   in (2.4).

The model-based approach just described is referred to as *parametric*;
it reduces the problem of estimating f down to one of estimating a set of

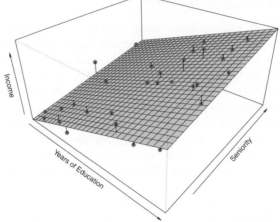

FIGURE 2.4. *A linear model fit by least squares to the* Income *data from Figure 2.3. The observations are shown in red, and the yellow plane indicates the least squares fit to the data.*

parameters. Assuming a parametric form for f simplifies the problem of estimating f because it is generally much easier to estimate a set of parameters, such as $\beta_0, \beta_1, \ldots, \beta_p$ in the linear model (2.4), than it is to fit an entirely arbitrary function f. The potential disadvantage of a parametric approach is that the model we choose will usually not match the true unknown form of f. If the chosen model is too far from the true f, then our estimate will be poor. We can try to address this problem by choosing *flexible* models that can fit many different possible functional forms for f. But in general, fitting a more flexible model requires estimating a greater number of parameters. These more complex models can lead to a phenomenon known as *overfitting* the data, which essentially means they follow the errors, or *noise*, too closely. These issues are discussed throughout this book.

flexible

overfitting

noise

Figure 2.4 shows an example of the parametric approach applied to the Income data from Figure 2.3. We have fit a linear model of the form

$$\texttt{income} \approx \beta_0 + \beta_1 \times \texttt{education} + \beta_2 \times \texttt{seniority}.$$

Since we have assumed a linear relationship between the response and the two predictors, the entire fitting problem reduces to estimating β_0, β_1, and β_2, which we do using least squares linear regression. Comparing Figure 2.3 to Figure 2.4, we can see that the linear fit given in Figure 2.4 is not quite right: the true f has some curvature that is not captured in the linear fit. However, the linear fit still appears to do a reasonable job of capturing the positive relationship between years of education and income, as well as the

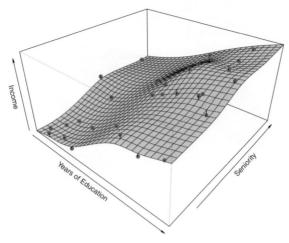

FIGURE 2.5. *A smooth thin-plate spline fit to the* Income *data from Figure 2.3 is shown in yellow; the observations are displayed in red. Splines are discussed in Chapter 7.*

slightly less positive relationship between seniority and income. It may be that with such a small number of observations, this is the best we can do.

Non-parametric Methods

Non-parametric methods do not make explicit assumptions about the functional form of f. Instead they seek an estimate of f that gets as close to the data points as possible without being too rough or wiggly. Such approaches can have a major advantage over parametric approaches: by avoiding the assumption of a particular functional form for f, they have the potential to accurately fit a wider range of possible shapes for f. Any parametric approach brings with it the possibility that the functional form used to estimate f is very different from the true f, in which case the resulting model will not fit the data well. In contrast, non-parametric approaches completely avoid this danger, since essentially no assumption about the form of f is made. But non-parametric approaches do suffer from a major disadvantage: since they do not reduce the problem of estimating f to a small number of parameters, a very large number of observations (far more than is typically needed for a parametric approach) is required in order to obtain an accurate estimate for f.

An example of a non-parametric approach to fitting the Income data is shown in Figure 2.5. A *thin-plate spline* is used to estimate f. This approach does not impose any pre-specified model on f. It instead attempts to produce an estimate for f that is as close as possible to the observed data, subject to the fit—that is, the yellow surface in Figure 2.5—being

thin-plate
spline

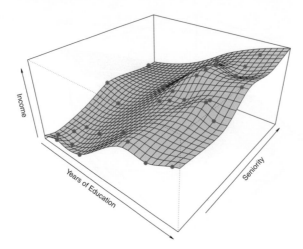

FIGURE 2.6. *A rough thin-plate spline fit to the* Income *data from Figure 2.3.* *This fit makes zero errors on the training data.*

smooth. In this case, the non-parametric fit has produced a remarkably accurate estimate of the true f shown in Figure 2.3. In order to fit a thin-plate spline, the data analyst must select a level of smoothness. Figure 2.6 shows the same thin-plate spline fit using a lower level of smoothness, allowing for a rougher fit. The resulting estimate fits the observed data perfectly! However, the spline fit shown in Figure 2.6 is far more variable than the true function f, from Figure 2.3. This is an example of overfitting the data, which we discussed previously. It is an undesirable situation because the fit obtained will not yield accurate estimates of the response on new observations that were not part of the original training data set. We discuss methods for choosing the *correct* amount of smoothness in Chapter 5. Splines are discussed in Chapter 7.

As we have seen, there are advantages and disadvantages to parametric and non-parametric methods for statistical learning. We explore both types of methods throughout this book.

2.1.3 *The Trade-Off Between Prediction Accuracy and Model Interpretability*

Of the many methods that we examine in this book, some are less flexible, or more restrictive, in the sense that they can produce just a relatively small range of shapes to estimate f. For example, linear regression is a relatively inflexible approach, because it can only generate linear functions such as the lines shown in Figure 2.1 or the plane shown in Figure 2.4.

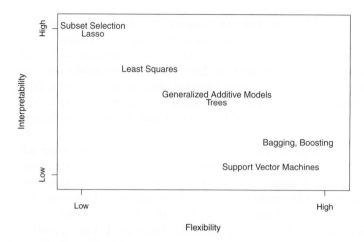

FIGURE 2.7. *A representation of the tradeoff between flexibility and inter-pretability, using different statistical learning methods. In general, as the flexibil-ity of a method increases, its interpretability decreases.*

Other methods, such as the thin plate splines shown in Figures 2.5 and 2.6, are considerably more flexible because they can generate a much wider range of possible shapes to estimate f.

One might reasonably ask the following question: *why would we ever choose to use a more restrictive method instead of a very flexible approach?* There are several reasons that we might prefer a more restrictive model. If we are mainly interested in inference, then restrictive models are much more interpretable. For instance, when inference is the goal, the linear model may be a good choice since it will be quite easy to understand the relationship between Y and $X_1, X_2, \ldots, X_p$. In contrast, very flexible approaches, such as the splines discussed in Chapter 7 and displayed in Figures 2.5 and 2.6, and the boosting methods discussed in Chapter 8, can lead to such complicated estimates of f that it is difficult to understand how any individual predictor is associated with the response.

Figure 2.7 provides an illustration of the trade-off between flexibility and interpretability for some of the methods that we cover in this book. Least squares linear regression, discussed in Chapter 3, is relatively inflexible but is quite interpretable. The *lasso*, discussed in Chapter 6, relies upon the linear model (2.4) but uses an alternative fitting procedure for estimating the coefficients $\beta_0, \beta_1, \ldots, \beta_p$. The new procedure is more restrictive in estimating the coefficients, and sets a number of them to exactly zero. Hence in this sense the lasso is a less flexible approach than linear regression. It is also more interpretable than linear regression, because in the final model the response variable will only be related to a small subset of the predictors—namely, those with nonzero coefficient estimates. *Generalized*

lasso

additive models (GAMs), discussed in Chapter 7, instead extend the linear model (2.4) to allow for certain non-linear relationships. Consequently,   generalized
GAMs are more flexible than linear regression. They are also somewhat   additive
less interpretable than linear regression, because the relationship between   model
each predictor and the response is now modeled using a curve. Finally, fully
non-linear methods such as *bagging*, *boosting*, and *support vector machines*   bagging
with non-linear kernels, discussed in Chapters 8 and 9, are highly flexible   boosting
approaches that are harder to interpret.   support
   We have established that when inference is the goal, there are clear ad-   vector
vantages to using simple and relatively inflexible statistical learning meth-   machine
ods. In some settings, however, we are only interested in prediction, and
the interpretability of the predictive model is simply not of interest. For
instance, if we seek to develop an algorithm to predict the price of a
stock, our sole requirement for the algorithm is that it predict accurately—
interpretability is not a concern. In this setting, we might expect that it
will be best to use the most flexible model available. Surprisingly, this is
not always the case! We will often obtain more accurate predictions using
a less flexible method. This phenomenon, which may seem counterintuitive
at first glance, has to do with the potential for overfitting in highly flexible
methods. We saw an example of overfitting in Figure 2.6. We will discuss
this very important concept further in Section 2.2 and throughout this
book.

2.1.4 *Supervised Versus Unsupervised Learning*

Most statistical learning problems fall into one of two categories: *supervised*   supervised
or *unsupervised*. The examples that we have discussed so far in this chap-   unsupervised
ter all fall into the supervised learning domain. For each observation of the
predictor measurement(s) x_i, $i = 1, \ldots, n$ there is an associated response
measurement y_i. We wish to fit a model that relates the response to the
predictors, with the aim of accurately predicting the response for future
observations (prediction) or better understanding the relationship between
the response and the predictors (inference). Many classical statistical learning methods such as linear regression and *logistic regression* (Chapter 4), as   logistic
well as more modern approaches such as GAM, boosting, and support vec-   regression
tor machines, operate in the supervised learning domain. The vast majority
of this book is devoted to this setting.
   In contrast, unsupervised learning describes the somewhat more challenging situation in which for every observation $i = 1, \ldots, n$, we observe
a vector of measurements x_i but no associated response y_i. It is not possible to fit a linear regression model, since there is no response variable
to predict. In this setting, we are in some sense working blind; the situation is referred to as *unsupervised* because we lack a response variable that can supervise our analysis. What sort of statistical analysis is

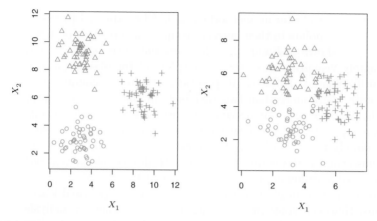

FIGURE 2.8. *A clustering data set involving three groups. Each group is shown using a different colored symbol.* Left: *The three groups are well-separated. In this setting, a clustering approach should successfully identify the three groups.* Right: *There is some overlap among the groups. Now the clustering task is more challenging.*

possible? We can seek to understand the relationships between the variables or between the observations. One statistical learning tool that we may use in this setting is *cluster analysis*, or clustering. The goal of cluster analysis is to ascertain, on the basis of $x_1, \ldots, x_n$, whether the observations fall into relatively distinct groups. For example, in a market segmentation study we might observe multiple characteristics (variables) for potential customers, such as zip code, family income, and shopping habits. We might believe that the customers fall into different groups, such as big spenders versus low spenders. If the information about each customer's spending patterns were available, then a supervised analysis would be possible. However, this information is not available—that is, we do not know whether each potential customer is a big spender or not. In this setting, we can try to cluster the customers on the basis of the variables measured, in order to identify distinct groups of potential customers. Identifying such groups can be of interest because it might be that the groups differ with respect to some property of interest, such as spending habits.

Figure 2.8 provides a simple illustration of the clustering problem. We have plotted 150 observations with measurements on two variables, X_1 and X_2. Each observation corresponds to one of three distinct groups. For illustrative purposes, we have plotted the members of each group using different colors and symbols. However, in practice the group memberships are unknown, and the goal is to determine the group to which each observation belongs. In the left-hand panel of Figure 2.8, this is a relatively easy task because the groups are well-separated. In contrast, the right-hand panel illustrates a more challenging problem in which there is some overlap

cluster
analysis

between the groups. A clustering method could not be expected to assign all of the overlapping points to their correct group (blue, green, or orange).

In the examples shown in Figure 2.8, there are only two variables, and so one can simply visually inspect the scatterplots of the observations in order to identify clusters. However, in practice, we often encounter data sets that contain many more than two variables. In this case, we cannot easily plot the observations. For instance, if there are p variables in our data set, then $p(p-1)/2$ distinct scatterplots can be made, and visual inspection is simply not a viable way to identify clusters. For this reason, automated clustering methods are important. We discuss clustering and other unsupervised learning approaches in Chapter 10.

Many problems fall naturally into the supervised or unsupervised learning paradigms. However, sometimes the question of whether an analysis should be considered supervised or unsupervised is less clear-cut. For instance, suppose that we have a set of n observations. For m of the observations, where $m < n$, we have both predictor measurements and a response measurement. For the remaining $n - m$ observations, we have predictor measurements but no response measurement. Such a scenario can arise if the predictors can be measured relatively cheaply but the corresponding responses are much more expensive to collect. We refer to this setting as a *semi-supervised learning* problem. In this setting, we wish to use a statistical learning method that can incorporate the m observations for which response measurements are available as well as the $n - m$ observations for which they are not. Although this is an interesting topic, it is beyond the scope of this book.

semi-supervised learning

2.1.5 *Regression Versus Classification Problems*

Variables can be characterized as either *quantitative* or *qualitative* (also known as *categorical*). Quantitative variables take on numerical values. Examples include a person's age, height, or income, the value of a house, and the price of a stock. In contrast, qualitative variables take on values in one of K different *classes*, or categories. Examples of qualitative variables include a person's gender (male or female), the brand of product purchased (brand A, B, or C), whether a person defaults on a debt (yes or no), or a cancer diagnosis (Acute Myelogenous Leukemia, Acute Lymphoblastic Leukemia, or No Leukemia). We tend to refer to problems with a quantitative response as *regression* problems, while those involving a qualitative response are often referred to as *classification* problems. However, the distinction is not always that crisp. Least squares linear regression (Chapter 3) is used with a quantitative response, whereas logistic regression (Chapter 4) is typically used with a qualitative (two-class, or *binary*) response. As such it is often used as a classification method. But since it estimates class probabilities, it can be thought of as a regression

quantitative
qualitative
categorical

class

regression
classification

binary

method as well. Some statistical methods, such as K-nearest neighbors (Chapters 2 and 4) and boosting (Chapter 8), can be used in the case of either quantitative or qualitative responses.

We tend to select statistical learning methods on the basis of whether the response is quantitative or qualitative; i.e. we might use linear regression when quantitative and logistic regression when qualitative. However, whether the *predictors* are qualitative or quantitative is generally considered less important. Most of the statistical learning methods discussed in this book can be applied regardless of the predictor variable type, provided that any qualitative predictors are properly *coded* before the analysis is performed. This is discussed in Chapter 3.

2.2   Assessing Model Accuracy

One of the key aims of this book is to introduce the reader to a wide range of statistical learning methods that extend far beyond the standard linear regression approach. Why is it necessary to introduce so many different statistical learning approaches, rather than just a single *best* method? *There is no free lunch in statistics:* no one method dominates all others over all possible data sets. On a particular data set, one specific method may work best, but some other method may work better on a similar but different data set. Hence it is an important task to decide for any given set of data which method produces the best results. Selecting the best approach can be one of the most challenging parts of performing statistical learning in practice.

In this section, we discuss some of the most important concepts that arise in selecting a statistical learning procedure for a specific data set. As the book progresses, we will explain how the concepts presented here can be applied in practice.

2.2.1   Measuring the Quality of Fit

In order to evaluate the performance of a statistical learning method on a given data set, we need some way to measure how well its predictions actually match the observed data. That is, we need to quantify the extent to which the predicted response value for a given observation is close to the true response value for that observation. In the regression setting, the most commonly-used measure is the *mean squared error* (MSE), given by

mean squared error

$$MSE = \frac{1}{n} \sum_{i=1}^{n} (y_i - \hat{f}(x_i))^2, \qquad (2.5)$$

where $\hat{f}(x_i)$ is the prediction that $\hat{f}$ gives for the ith observation. The MSE will be small if the predicted responses are very close to the true responses, and will be large if for some of the observations, the predicted and true responses differ substantially.

The MSE in (2.5) is computed using the training data that was used to fit the model, and so should more accurately be referred to as the *training MSE*. But in general, we do not really care how well the method works on the training data. Rather, *we are interested in the accuracy of the predictions that we obtain when we apply our method to previously unseen test data.* Why is this what we care about? Suppose that we are interested in developing an algorithm to predict a stock's price based on previous stock returns. We can train the method using stock returns from the past 6 months. But we don't really care how well our method predicts last week's stock price. We instead care about how well it will predict tomorrow's price or next month's price. On a similar note, suppose that we have clinical measurements (e.g. weight, blood pressure, height, age, family history of disease) for a number of patients, as well as information about whether each patient has diabetes. We can use these patients to train a statistical learning method to predict risk of diabetes based on clinical measurements. In practice, we want this method to accurately predict diabetes risk for *future patients* based on their clinical measurements. We are not very interested in whether or not the method accurately predicts diabetes risk for patients used to train the model, since we already know which of those patients have diabetes.

To state it more mathematically, suppose that we fit our statistical learning method on our training observations $\{(x_1, y_1), (x_2, y_2), \ldots, (x_n, y_n)\}$, and we obtain the estimate $\hat{f}$. We can then compute $\hat{f}(x_1), \hat{f}(x_2), \ldots, \hat{f}(x_n)$. If these are approximately equal to $y_1, y_2, \ldots, y_n$, then the training MSE given by (2.5) is small. However, we are really not interested in whether $\hat{f}(x_i) \approx y_i$; instead, we want to know whether $\hat{f}(x_0)$ is approximately equal to y_0, where (x_0, y_0) is a *previously unseen test observation not used to train the statistical learning method.* We want to choose the method that gives the lowest *test MSE*, as opposed to the lowest training MSE. In other words, if we had a large number of test observations, we could compute

$$\text{Ave}(y_0 - \hat{f}(x_0))^2, \tag{2.6}$$

the average squared prediction error for these test observations (x_0, y_0). We'd like to select the model for which the average of this quantity—the test MSE—is as small as possible.

How can we go about trying to select a method that minimizes the test MSE? In some settings, we may have a test data set available—that is, we may have access to a set of observations that were not used to train the statistical learning method. We can then simply evaluate (2.6) on the test observations, and select the learning method for which the test MSE is

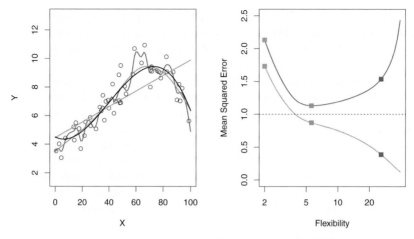

FIGURE 2.9. Left: *Data simulated from f, shown in black. Three estimates of f are shown: the linear regression line (orange curve), and two smoothing spline fits (blue and green curves).* Right: *Training MSE (grey curve), test MSE (red curve), and minimum possible test MSE over all methods (dashed line). Squares represent the training and test MSEs for the three fits shown in the left-hand panel.*

smallest. But what if no test observations are available? In that case, one might imagine simply selecting a statistical learning method that minimizes the training MSE (2.5). This seems like it might be a sensible approach, since the training MSE and the test MSE appear to be closely related. Unfortunately, there is a fundamental problem with this strategy: there is no guarantee that the method with the lowest training MSE will also have the lowest test MSE. Roughly speaking, the problem is that many statistical methods specifically estimate coefficients so as to minimize the training set MSE. For these methods, the training set MSE can be quite small, but the test MSE is often much larger.

Figure 2.9 illustrates this phenomenon on a simple example. In the left-hand panel of Figure 2.9, we have generated observations from (2.1) with the true f given by the black curve. The orange, blue and green curves illustrate three possible estimates for f obtained using methods with increasing levels of flexibility. The orange line is the linear regression fit, which is relatively inflexible. The blue and green curves were produced using *smoothing splines*, discussed in Chapter 7, with different levels of smoothness. It is clear that as the level of flexibility increases, the curves fit the observed data more closely. The green curve is the most flexible and matches the data very well; however, we observe that it fits the true f (shown in black) poorly because it is too wiggly. By adjusting the level of flexibility of the smoothing spline fit, we can produce many different fits to this data.

We now move on to the right-hand panel of Figure 2.9. The grey curve displays the average training MSE as a function of flexibility, or more formally the *degrees of freedom*, for a number of smoothing splines. The degrees of freedom is a quantity that summarizes the flexibility of a curve; it is discussed more fully in Chapter 7. The orange, blue and green squares indicate the MSEs associated with the corresponding curves in the left-hand panel. A more restricted and hence smoother curve has fewer degrees of freedom than a wiggly curve—note that in Figure 2.9, linear regression is at the most restrictive end, with two degrees of freedom. The training MSE declines monotonically as flexibility increases. In this example the true f is non-linear, and so the orange linear fit is not flexible enough to estimate f well. The green curve has the lowest training MSE of all three methods, since it corresponds to the most flexible of the three curves fit in the left-hand panel.

degrees of freedom

In this example, we know the true function f, and so we can also compute the test MSE over a very large test set, as a function of flexibility. (Of course, in general f is unknown, so this will not be possible.) The test MSE is displayed using the red curve in the right-hand panel of Figure 2.9. As with the training MSE, the test MSE initially declines as the level of flexibility increases. However, at some point the test MSE levels off and then starts to increase again. Consequently, the orange and green curves both have high test MSE. The blue curve minimizes the test MSE, which should not be surprising given that visually it appears to estimate f the best in the left-hand panel of Figure 2.9. The horizontal dashed line indicates $\text{Var}(\epsilon)$, the irreducible error in (2.3), which corresponds to the lowest achievable test MSE among all possible methods. Hence, the smoothing spline represented by the blue curve is close to optimal.

In the right-hand panel of Figure 2.9, as the flexibility of the statistical learning method increases, we observe a monotone decrease in the training MSE and a *U-shape* in the test MSE. This is a fundamental property of statistical learning that holds regardless of the particular data set at hand and regardless of the statistical method being used. As model flexibility increases, training MSE will decrease, but the test MSE may not. When a given method yields a small training MSE but a large test MSE, we are said to be *overfitting* the data. This happens because our statistical learning procedure is working too hard to find patterns in the training data, and may be picking up some patterns that are just caused by random chance rather than by true properties of the unknown function f. When we overfit the training data, the test MSE will be very large because the supposed patterns that the method found in the training data simply don't exist in the test data. Note that regardless of whether or not overfitting has occurred, we almost always expect the training MSE to be smaller than the test MSE because most statistical learning methods either directly or indirectly seek to minimize the training MSE. Overfitting refers specifically to the case in which a less flexible model would have yielded a smaller test MSE.

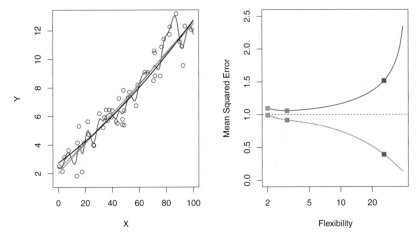

FIGURE 2.10. *Details are as in Figure 2.9, using a different true f that is much closer to linear. In this setting, linear regression provides a very good fit to the data.*

Figure 2.10 provides another example in which the true f is approximately linear. Again we observe that the training MSE decreases monotonically as the model flexibility increases, and that there is a U-shape in the test MSE. However, because the truth is close to linear, the test MSE only decreases slightly before increasing again, so that the orange least squares fit is substantially better than the highly flexible green curve. Finally, Figure 2.11 displays an example in which f is highly non-linear. The training and test MSE curves still exhibit the same general patterns, but now there is a rapid decrease in both curves before the test MSE starts to increase slowly.

In practice, one can usually compute the training MSE with relative ease, but estimating test MSE is considerably more difficult because usually no test data are available. As the previous three examples illustrate, the flexibility level corresponding to the model with the minimal test MSE can vary considerably among data sets. Throughout this book, we discuss a variety of approaches that can be used in practice to estimate this minimum point. One important method is *cross-validation* (Chapter 5), which is a method for estimating test MSE using the training data.

cross-validation

2.2.2 The Bias-Variance Trade-Off

The U-shape observed in the test MSE curves (Figures 2.9–2.11) turns out to be the result of two competing properties of statistical learning methods. Though the mathematical proof is beyond the scope of this book, it is possible to show that the expected test MSE, for a given value x_0, can

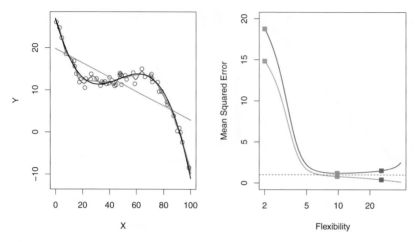

FIGURE 2.11. *Details are as in Figure 2.9, using a different f that is far from linear. In this setting, linear regression provides a very poor fit to the data.*

always be decomposed into the sum of three fundamental quantities: the *variance* of $\hat{f}(x_0)$, the squared *bias* of $\hat{f}(x_0)$ and the variance of the error terms ϵ. That is,

variance
bias

$$E\left(y_0 - \hat{f}(x_0)\right)^2 = \mathrm{Var}(\hat{f}(x_0)) + [\mathrm{Bias}(\hat{f}(x_0))]^2 + \mathrm{Var}(\epsilon). \qquad (2.7)$$

Here the notation $E\left(y_0 - \hat{f}(x_0)\right)^2$ defines the *expected test MSE*, and refers to the average test MSE that we would obtain if we repeatedly estimated f using a large number of training sets, and tested each at x_0. The overall expected test MSE can be computed by averaging $E\left(y_0 - \hat{f}(x_0)\right)^2$ over all possible values of x_0 in the test set.

expected
test MSE

Equation 2.7 tells us that in order to minimize the expected test error, we need to select a statistical learning method that simultaneously achieves *low variance* and *low bias*. Note that variance is inherently a nonnegative quantity, and squared bias is also nonnegative. Hence, we see that the expected test MSE can never lie below $\mathrm{Var}(\epsilon)$, the irreducible error from (2.3).

What do we mean by the *variance* and *bias* of a statistical learning method? *Variance* refers to the amount by which $\hat{f}$ would change if we estimated it using a different training data set. Since the training data are used to fit the statistical learning method, different training data sets will result in a different $\hat{f}$. But ideally the estimate for f should not vary too much between training sets. However, if a method has high variance then small changes in the training data can result in large changes in $\hat{f}$. In general, more flexible statistical methods have higher variance. Consider the green and orange curves in Figure 2.9. The flexible green curve is following the observations very closely. It has high variance because changing any one of these data points may cause the estimate $\hat{f}$ to change considerably. In contrast, the orange least squares line is relatively inflexible and has low variance, because moving any single observation will likely cause only a small shift in the position of the line.

On the other hand, *bias* refers to the error that is introduced by approximating a real-life problem, which may be extremely complicated, by a much simpler model. For example, linear regression assumes that there is a linear relationship between Y and $X_1, X_2, \ldots, X_p$. It is unlikely that any real-life problem truly has such a simple linear relationship, and so performing linear regression will undoubtedly result in some bias in the estimate of f. In Figure 2.11, the true f is substantially non-linear, so no matter how many training observations we are given, it will not be possible to produce an accurate estimate using linear regression. In other words, linear regression results in high bias in this example. However, in Figure 2.10 the true f is very close to linear, and so given enough data, it should be possible for linear regression to produce an accurate estimate. Generally, more flexible methods result in less bias.

As a general rule, as we use more flexible methods, the variance will increase and the bias will decrease. The relative rate of change of these two quantities determines whether the test MSE increases or decreases. As we increase the flexibility of a class of methods, the bias tends to initially decrease faster than the variance increases. Consequently, the expected test MSE declines. However, at some point increasing flexibility has little impact on the bias but starts to significantly increase the variance. When this happens the test MSE increases. Note that we observed this pattern of decreasing test MSE followed by increasing test MSE in the right-hand panels of Figures 2.9–2.11.

The three plots in Figure 2.12 illustrate Equation 2.7 for the examples in Figures 2.9–2.11. In each case the blue solid curve represents the squared bias, for different levels of flexibility, while the orange curve corresponds to the variance. The horizontal dashed line represents $\text{Var}(\epsilon)$, the irreducible error. Finally, the red curve, corresponding to the test set MSE, is the sum of these three quantities. In all three cases, the variance increases and the bias decreases as the method's flexibility increases. However, the flexibility level corresponding to the optimal test MSE differs considerably among the three data sets, because the squared bias and variance change at different rates in each of the data sets. In the left-hand panel of Figure 2.12, the bias initially decreases rapidly, resulting in an initial sharp decrease in the expected test MSE. On the other hand, in the center panel of Figure 2.12 the true f is close to linear, so there is only a small decrease in bias as flexibility increases, and the test MSE only declines slightly before increasing rapidly as the variance increases. Finally, in the right-hand panel of Figure 2.12, as flexibility increases, there is a dramatic decline in bias because

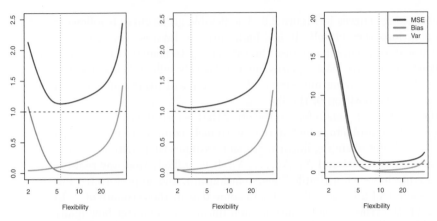

FIGURE 2.12. *Squared bias (blue curve), variance (orange curve), Var(ε) (dashed line), and test MSE (red curve) for the three data sets in Figures 2.9–2.11. The vertical dotted line indicates the flexibility level corresponding to the smallest test MSE.*

the true f is very non-linear. There is also very little increase in variance as flexibility increases. Consequently, the test MSE declines substantially before experiencing a small increase as model flexibility increases.

The relationship between bias, variance, and test set MSE given in Equation 2.7 and displayed in Figure 2.12 is referred to as the *bias-variance trade-off*. Good test set performance of a statistical learning method requires low variance as well as low squared bias. This is referred to as a
trade-off because it is easy to obtain a method with extremely low bias but high variance (for instance, by drawing a curve that passes through every single training observation) or a method with very low variance but high bias (by fitting a horizontal line to the data). The challenge lies in finding a method for which both the variance and the squared bias are low. This trade-off is one of the most important recurring themes in this book.

In a real-life situation in which f is unobserved, it is generally not pos-sible to explicitly compute the test MSE, bias, or variance for a statistical learning method. Nevertheless, one should always keep the bias-variance trade-off in mind. In this book we explore methods that are extremely flexible and hence can essentially eliminate bias. However, this does not guarantee that they will outperform a much simpler method such as linear regression. To take an extreme example, suppose that the true f is linear. In this situation linear regression will have no bias, making it very hard for a more flexible method to compete. In contrast, if the true f is highly non-linear and we have an ample number of training observations, then we may do better using a highly flexible approach, as in Figure 2.11. In Chapter 5 we discuss cross-validation, which is a way to estimate the test MSE using the training data.

bias-variance
trade-off

2.2.3   The Classification Setting

Thus far, our discussion of model accuracy has been focused on the regression setting. But many of the concepts that we have encountered, such as the bias-variance trade-off, transfer over to the classification setting with only some modifications due to the fact that y_i is no longer numerical. Suppose that we seek to estimate f on the basis of training observations $\{(x_1, y_1), \ldots, (x_n, y_n)\}$, where now $y_1, \ldots, y_n$ are qualitative. The most common approach for quantifying the accuracy of our estimate $\hat{f}$ is the training *error rate*, the proportion of mistakes that are made if we apply our estimate $\hat{f}$ to the training observations:

$$\frac{1}{n} \sum_{i=1}^{n} I(y_i \neq \hat{y}_i). \tag{2.8}$$

error rate

Here $\hat{y}_i$ is the predicted class label for the ith observation using $\hat{f}$. And $I(y_i \neq \hat{y}_i)$ is an *indicator variable* that equals 1 if $y_i \neq \hat{y}_i$ and zero if $y_i = \hat{y}_i$. If $I(y_i \neq \hat{y}_i) = 0$ then the ith observation was classified correctly by our classification method; otherwise it was misclassified. Hence Equation 2.8 computes the fraction of incorrect classifications.

indicator variable

Equation 2.8 is referred to as the *training error* rate because it is computed based on the data that was used to train our classifier. As in the regression setting, we are most interested in the error rates that result from applying our classifier to test observations that were not used in training. The *test error* rate associated with a set of test observations of the form (x_0, y_0) is given by

training error

test error

$$\text{Ave}\left(I(y_0 \neq \hat{y}_0)\right), \tag{2.9}$$

where $\hat{y}_0$ is the predicted class label that results from applying the classifier to the test observation with predictor x_0. A *good* classifier is one for which the test error (2.9) is smallest.

The Bayes Classifier

It is possible to show (though the proof is outside of the scope of this book) that the test error rate given in (2.9) is minimized, on average, by a very simple classifier that *assigns each observation to the most likely class, given its predictor values.* In other words, we should simply assign a test observation with predictor vector x_0 to the class j for which

$$\Pr(Y = j | X = x_0) \tag{2.10}$$

is largest. Note that (2.10) is a *conditional probability*: it is the probability that $Y = j$, given the observed predictor vector x_0. This very simple classifier is called the *Bayes classifier*. In a two-class problem where there are only two possible response values, say *class 1* or *class 2*, the Bayes classifier

conditional probability

Bayes classifier

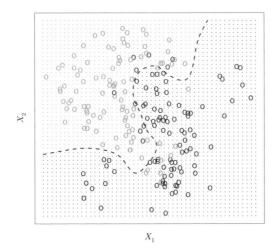

FIGURE 2.13. *A simulated data set consisting of* 100 *observations in each of two groups, indicated in blue and in orange. The purple dashed line represents the Bayes decision boundary. The orange background grid indicates the region in which a test observation will be assigned to the orange class, and the blue background grid indicates the region in which a test observation will be assigned to the blue class.*

corresponds to predicting class one if $\Pr(Y = 1|X = x_0) > 0.5$, and class two otherwise.

Figure 2.13 provides an example using a simulated data set in a two-dimensional space consisting of predictors X_1 and X_2. The orange and blue circles correspond to training observations that belong to two different classes. For each value of X_1 and X_2, there is a different probability of the response being orange or blue. Since this is simulated data, we know how the data were generated and we can calculate the conditional probabilities for each value of X_1 and X_2. The orange shaded region reflects the set of points for which $\Pr(Y = \text{orange}|X)$ is greater than 50%, while the blue shaded region indicates the set of points for which the probability is below 50%. The purple dashed line represents the points where the probability is exactly 50%. This is called the *Bayes decision boundary*. The Bayes classifier's prediction is determined by the Bayes decision boundary; an observation that falls on the orange side of the boundary will be assigned to the orange class, and similarly an observation on the blue side of the boundary will be assigned to the blue class.

The Bayes classifier produces the lowest possible test error rate, called the *Bayes error rate*. Since the Bayes classifier will always choose the class for which (2.10) is largest, the error rate at $X = x_0$ will be $1 - \max_j \Pr(Y = j|X = x_0)$. In general, the overall Bayes error rate is given by

$$1 - E\left(\max_j \Pr(Y = j|X)\right),\qquad(2.11)$$

Bayes
decision
boundary

Bayes error
rate

where the expectation averages the probability over all possible values of X. For our simulated data, the Bayes error rate is 0.1304. It is greater than zero, because the classes overlap in the true population so $\max_j \Pr(Y = j|X = x_0) < 1$ for some values of x_0. The Bayes error rate is analogous to the irreducible error, discussed earlier.

K-Nearest Neighbors

In theory we would always like to predict qualitative responses using the Bayes classifier. But for real data, we do not know the conditional distribution of Y given X, and so computing the Bayes classifier is impossible. Therefore, the Bayes classifier serves as an unattainable gold standard against which to compare other methods. Many approaches attempt to estimate the conditional distribution of Y given X, and then classify a given observation to the class with highest *estimated* probability. One such method is the *K-nearest neighbors* (KNN) classifier. Given a positive integer K and a test observation x_0, the KNN classifier first identifies the K points in the training data that are closest to x_0, represented by $\mathcal{N}_0$. It then estimates the conditional probability for class j as the fraction of points in $\mathcal{N}_0$ whose response values equal j:

K-nearest neighbors

$$\Pr(Y = j|X = x_0) = \frac{1}{K} \sum_{i \in \mathcal{N}_0} I(y_i = j). \qquad (2.12)$$

Finally, KNN applies Bayes rule and classifies the test observation x_0 to the class with the largest probability.

Figure 2.14 provides an illustrative example of the KNN approach. In the left-hand panel, we have plotted a small training data set consisting of six blue and six orange observations. Our goal is to make a prediction for the point labeled by the black cross. Suppose that we choose $K = 3$. Then KNN will first identify the three observations that are closest to the cross. This neighborhood is shown as a circle. It consists of two blue points and one orange point, resulting in estimated probabilities of 2/3 for the blue class and 1/3 for the orange class. Hence KNN will predict that the black cross belongs to the blue class. In the right-hand panel of Figure 2.14 we have applied the KNN approach with $K = 3$ at all of the possible values for X_1 and X_2, and have drawn in the corresponding KNN decision boundary.

Despite the fact that it is a very simple approach, KNN can often produce classifiers that are surprisingly close to the optimal Bayes classifier. Figure 2.15 displays the KNN decision boundary, using $K = 10$, when applied to the larger simulated data set from Figure 2.13. Notice that even though the true distribution is not known by the KNN classifier, the KNN decision boundary is very close to that of the Bayes classifier. The test error rate using KNN is 0.1363, which is close to the Bayes error rate of 0.1304.

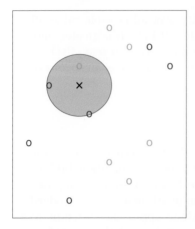 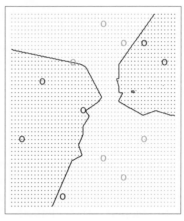

FIGURE 2.14. *The KNN approach, using* $K = 3$, *is illustrated in a simple situation with six blue observations and six orange observations. Left: a test observation at which a predicted class label is desired is shown as a black cross. The three closest points to the test observation are identified, and it is predicted that the test observation belongs to the most commonly-occurring class, in this case blue. Right: The KNN decision boundary for this example is shown in black. The blue grid indicates the region in which a test observation will be assigned to the blue class, and the orange grid indicates the region in which it will be assigned to the orange class.*

The choice of K has a drastic effect on the KNN classifier obtained. Figure 2.16 displays two KNN fits to the simulated data from Figure 2.13, using $K = 1$ and $K = 100$. When $K = 1$, the decision boundary is overly flexible and finds patterns in the data that don't correspond to the Bayes decision boundary. This corresponds to a classifier that has low bias but very high variance. As K grows, the method becomes less flexible and produces a decision boundary that is close to linear. This corresponds to a low-variance but high-bias classifier. On this simulated data set, neither $K = 1$ nor $K = 100$ give good predictions: they have test error rates of 0.1695 and 0.1925, respectively.

Just as in the regression setting, there is not a strong relationship between the training error rate and the test error rate. With $K = 1$, the KNN training error rate is 0, but the test error rate may be quite high. In general, as we use more flexible classification methods, the training error rate will decline but the test error rate may not. In Figure 2.17, we have plotted the KNN test and training errors as a function of $1/K$. As $1/K$ increases, the method becomes more flexible. As in the regression setting, the training error rate consistently declines as the flexibility increases. However, the test error exhibits a characteristic U-shape, declining at first (with a minimum at approximately $K = 10$) before increasing again when the method becomes excessively flexible and overfits.

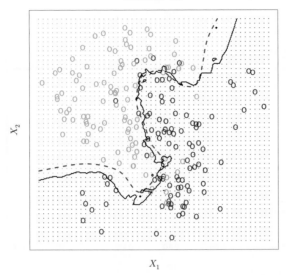

FIGURE 2.15. *The black curve indicates the KNN decision boundary on the data from Figure 2.13, using $K = 10$. The Bayes decision boundary is shown as a purple dashed line. The KNN and Bayes decision boundaries are very similar.*

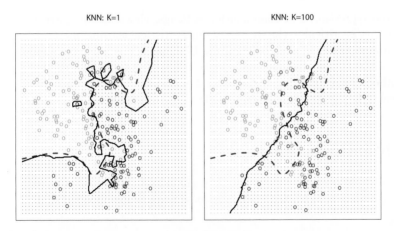

FIGURE 2.16. *A comparison of the KNN decision boundaries (solid black curves) obtained using $K = 1$ and $K = 100$ on the data from Figure 2.13. With $K = 1$, the decision boundary is overly flexible, while with $K = 100$ it is not sufficiently flexible. The Bayes decision boundary is shown as a purple dashed line.*

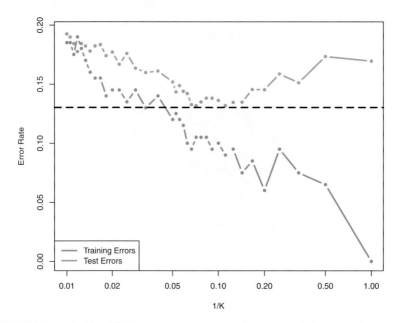

FIGURE 2.17. *The KNN training error rate (blue, 200 observations) and test error rate (orange, 5,000 observations) on the data from Figure 2.13, as the level of flexibility (assessed using $1/K$) increases, or equivalently as the number of neighbors K decreases. The black dashed line indicates the Bayes error rate. The jumpiness of the curves is due to the small size of the training data set.*

In both the regression and classification settings, choosing the correct level of flexibility is critical to the success of any statistical learning method. The bias-variance tradeoff, and the resulting U-shape in the test error, can make this a difficult task. In Chapter 5, we return to this topic and discuss various methods for estimating test error rates and thereby choosing the optimal level of flexibility for a given statistical learning method.

2.3   Lab: Introduction to R

In this lab, we will introduce some simple R commands. The best way to learn a new language is to try out the commands. R can be downloaded from

http://cran.r-project.org/

2.3.1   Basic Commands

R uses *functions* to perform operations. To run a function called funcname, we type funcname(input1, input2), where the inputs (or *arguments*) input1

function

argument

and input2 tell R how to run the function. A function can have any number
of inputs. For example, to create a vector of numbers, we use the function
c() (for *concatenate*). Any numbers inside the parentheses are joined together. The following command instructs R to join together the numbers
1, 3, 2, and 5, and to save them as a *vector* named x. When we type x, it
gives us back the vector.

c()

vector

```
> x <- c(1,3,2,5)
> x
[1] 1 3 2 5
```

Note that the > is not part of the command; rather, it is printed by R to
indicate that it is ready for another command to be entered. We can also
save things using = rather than <-:

```
> x = c(1,6,2)
> x
[1] 1 6 2
> y = c(1,4,3)
```

Hitting the *up* arrow multiple times will display the previous commands,
which can then be edited. This is useful since one often wishes to repeat
a similar command. In addition, typing ?funcname will always cause R to
open a new help file window with additional information about the function
funcname.

We can tell R to add two sets of numbers together. It will then add the
first number from x to the first number from y, and so on. However, x and
y should be the same length. We can check their length using the length()
function.

length()

```
> length(x)
[1] 3
> length(y)
[1] 3
> x+y
[1]  2 10  5
```

The ls() function allows us to look at a list of all of the objects, such
as data and functions, that we have saved so far. The rm() function can be
used to delete any that we don't want.

ls()

rm()

```
> ls()
[1] "x" "y"
> rm(x,y)
> ls()
character(0)
```

It's also possible to remove all objects at once:

```
> rm(list=ls())
```

The matrix() function can be used to create a matrix of numbers. Before we use the matrix() function, we can learn more about it:

matrix()

```
> ?matrix
```

The help file reveals that the matrix() function takes a number of inputs, but for now we focus on the first three: the data (the entries in the matrix), the number of rows, and the number of columns. First, we create a simple matrix.

```
> x=matrix(data=c(1,2,3,4), nrow=2, ncol=2)
> x
     [,1] [,2]
[1,]    1    3
[2,]    2    4
```

Note that we could just as well omit typing data=, nrow=, and ncol= in the matrix() command above: that is, we could just type

```
> x=matrix(c(1,2,3,4),2,2)
```

and this would have the same effect. However, it can sometimes be useful to specify the names of the arguments passed in, since otherwise R will assume that the function arguments are passed into the function in the same order that is given in the function's help file. As this example illustrates, by default R creates matrices by successively filling in columns. Alternatively, the byrow=TRUE option can be used to populate the matrix in order of the rows.

```
> matrix(c(1,2,3,4),2,2,byrow=TRUE)
     [,1] [,2]
[1,]    1    2
[2,]    3    4
```

Notice that in the above command we did not assign the matrix to a value such as x. In this case the matrix is printed to the screen but is not saved for future calculations. The sqrt() function returns the square root of each element of a vector or matrix. The command x^2 raises each element of x to the power 2; any powers are possible, including fractional or negative powers.

sqrt()

```
> sqrt(x)
     [,1] [,2]
[1,] 1.00 1.73
[2,] 1.41 2.00
> x^2
     [,1] [,2]
[1,]    1    9
[2,]    4   16
```

The rnorm() function generates a vector of random normal variables, with first argument n the sample size. Each time we call this function, we will get a different answer. Here we create two correlated sets of numbers, x and y, and use the cor() function to compute the correlation between them.

rnorm()

cor()

```
> x=rnorm(50)
> y=x+rnorm(50,mean=50,sd=.1)
> cor(x,y)
[1] 0.995
```

By default, rnorm() creates standard normal random variables with a mean
of 0 and a standard deviation of 1. However, the mean and standard deviation can be altered using the mean and sd arguments, as illustrated above.
Sometimes we want our code to reproduce the exact same set of random
numbers; we can use the set.seed() function to do this. The set.seed()
function takes an (arbitrary) integer argument.

set.seed()

```
> set.seed(1303)
> rnorm(50)
[1] -1.1440  1.3421  2.1854  0.5364  0.0632  0.5022 -0.0004
. . .
```

We use set.seed() throughout the labs whenever we perform calculations
involving random quantities. In general this should allow the user to reproduce our results. However, it should be noted that as new versions of
R become available it is possible that some small discrepancies may form
between the book and the output from R.

The mean() and var() functions can be used to compute the mean and
variance of a vector of numbers. Applying sqrt() to the output of var()
will give the standard deviation. Or we can simply use the sd() function.

mean()
var()
sd()

```
> set.seed(3)
> y=rnorm(100)
> mean(y)
[1] 0.0110
> var(y)
[1] 0.7329
> sqrt(var(y))
[1] 0.8561
> sd(y)
[1] 0.8561
```

2.3.2 Graphics

The plot() function is the primary way to plot data in R. For instance,
plot(x,y) produces a scatterplot of the numbers in x versus the numbers
in y. There are many additional options that can be passed in to the plot()
function. For example, passing in the argument xlab will result in a label
on the x-axis. To find out more information about the plot() function,
type ?plot.

plot()

```
> x=rnorm(100)
> y=rnorm(100)
> plot(x,y)
> plot(x,y,xlab="this is the x-axis",ylab="this is the y-axis",
       main="Plot of X vs Y")
```

We will often want to save the output of an R plot. The command that we use to do this will depend on the file type that we would like to create. For instance, to create a pdf, we use the pdf() function, and to create a jpeg, we use the jpeg() function.

pdf()
jpeg()

```
> pdf ("Figure.pdf")
> plot(x,y,col="green")
> dev.off()
null device
          1
```

The function dev.off() indicates to R that we are done creating the plot. Alternatively, we can simply copy the plot window and paste it into an appropriate file type, such as a Word document.

dev.off()

The function seq() can be used to create a sequence of numbers. For instance, seq(a,b) makes a vector of integers between a and b. There are many other options: for instance, seq(0,1,length=10) makes a sequence of 10 numbers that are equally spaced between 0 and 1. Typing 3:11 is a shorthand for seq(3,11) for integer arguments.

seq()

```
> x=seq(1,10)
> x
 [1]  1  2  3  4  5  6  7  8  9 10
> x=1:10
> x
 [1]  1  2  3  4  5  6  7  8  9 10
> x=seq(-pi,pi,length=50)
```

We will now create some more sophisticated plots. The contour() function produces a *contour plot* in order to represent three-dimensional data; it is like a topographical map. It takes three arguments:

contour()
contour plot

1. A vector of the x values (the first dimension),

2. A vector of the y values (the second dimension), and

3. A matrix whose elements correspond to the z value (the third dimension) for each pair of (x,y) coordinates.

As with the plot() function, there are many other inputs that can be used to fine-tune the output of the contour() function. To learn more about these, take a look at the help file by typing ?contour.

```
> y=x
> f=outer(x,y,function(x,y)cos(y)/(1+x^2))
> contour(x,y,f)
> contour(x,y,f,nlevels=45,add=T)
> fa=(f-t(f))/2
> contour(x,y,fa,nlevels=15)
```

The image() function works the same way as contour(), except that it produces a color-coded plot whose colors depend on the z value. This is

image()

known as a *heatmap*, and is sometimes used to plot temperature in weather
forecasts. Alternatively, persp() can be used to produce a three-dimensional
plot. The arguments theta and phi control the angles at which the plot is
viewed.

heatmap

persp()

```
> image(x,y,fa)
> persp(x,y,fa)
> persp(x,y,fa,theta=30)
> persp(x,y,fa,theta=30,phi=20)
> persp(x,y,fa,theta=30,phi=70)
> persp(x,y,fa,theta=30,phi=40)
```

2.3.3 Indexing Data

We often wish to examine part of a set of data. Suppose that our data is
stored in the matrix A.

```
> A=matrix(1:16,4,4)
> A
     [,1] [,2] [,3] [,4]
[1,]    1    5    9   13
[2,]    2    6   10   14
[3,]    3    7   11   15
[4,]    4    8   12   16
```

Then, typing

```
> A[2,3]
[1] 10
```

will select the element corresponding to the second row and the third column. The first number after the open-bracket symbol [always refers to
the row, and the second number always refers to the column. We can also
select multiple rows and columns at a time, by providing vectors as the
indices.

```
> A[c(1,3),c(2,4)]
     [,1] [,2]
[1,]    5   13
[2,]    7   15
> A[1:3,2:4]
     [,1] [,2] [,3]
[1,]    5    9   13
[2,]    6   10   14
[3,]    7   11   15
> A[1:2,]
     [,1] [,2] [,3] [,4]
[1,]    1    5    9   13
[2,]    2    6   10   14
> A[,1:2]
     [,1] [,2]
[1,]    1    5
[2,]    2    6
```

```
[3,]      3    7
[4,]      4    8
```

The last two examples include either no index for the columns or no index
for the rows. These indicate that R should include all columns or all rows,
respectively. R treats a single row or column of a matrix as a vector.

```
> A[1,]
[1]  1   5   9 13
```

The use of a negative sign – in the index tells R to keep all rows or columns
except those indicated in the index.

```
> A[-c(1,3),]
      [,1] [,2] [,3] [,4]
[1,]    2    6   10   14
[2,]    4    8   12   16
> A[-c(1,3),-c(1,3,4)]
[1] 6 8
```

The dim() function outputs the number of rows followed by the number of
columns of a given matrix.

dim()

```
> dim(A)
[1] 4 4
```

2.3.4 Loading Data

For most analyses, the first step involves importing a data set into R. The
read.table() function is one of the primary ways to do this. The help file
contains details about how to use this function. We can use the function
write.table() to export data.

read.table()

write.
table()

   Before attempting to load a data set, we must make sure that R knows
to search for the data in the proper directory. For example on a Windows
system one could select the directory using the Change dir... option under
the File menu. However, the details of how to do this depend on the operating system (e.g. Windows, Mac, Unix) that is being used, and so we
do not give further details here. We begin by loading in the Auto data set.
This data is part of the ISLR library (we discuss libraries in Chapter 3) but
to illustrate the read.table() function we load it now from a text file. The
following command will load the Auto.data file into R and store it as an
object called Auto, in a format referred to as a *data frame*. (The text file
can be obtained from this book's website.) Once the data has been loaded,
the fix() function can be used to view it in a spreadsheet like window.
However, the window must be closed before further R commands can be
entered.

data frame

```
> Auto=read.table("Auto.data")
> fix(Auto)
```

Note that Auto.data is simply a text file, which you could alternatively open on your computer using a standard text editor. It is often a good idea to view a data set using a text editor or other software such as Excel before loading it into R.

This particular data set has not been loaded correctly, because R has assumed that the variable names are part of the data and so has included them in the first row. The data set also includes a number of missing observations, indicated by a question mark ?. Missing values are a common occurrence in real data sets. Using the option header=T (or header=TRUE) in the read.table() function tells R that the first line of the file contains the variable names, and using the option na.strings tells R that any time it sees a particular character or set of characters (such as a question mark), it should be treated as a missing element of the data matrix.

```
> Auto=read.table("Auto.data",header=T,na.strings="?")
> fix(Auto)
```

Excel is a common-format data storage program. An easy way to load such data into R is to save it as a csv (comma separated value) file and then use the read.csv() function to load it in.

```
> Auto=read.csv("Auto.csv",header=T,na.strings="?")
> fix(Auto)
> dim(Auto)
[1] 397 9
> Auto[1:4,]
```

The dim() function tells us that the data has 397 observations, or rows, and nine variables, or columns. There are various ways to deal with the missing data. In this case, only five of the rows contain missing observations, and so we choose to use the na.omit() function to simply remove these rows.

`dim()`

`na.omit()`

```
> Auto=na.omit(Auto)
> dim(Auto)
[1] 392   9
```

Once the data are loaded correctly, we can use names() to check the variable names.

`names()`

```
> names(Auto)
[1] "mpg"          "cylinders"    "displacement" "horsepower"
[5] "weight"       "acceleration" "year"         "origin"
[9] "name"
```

2.3.5 Additional Graphical and Numerical Summaries

We can use the plot() function to produce *scatterplots* of the quantitative variables. However, simply typing the variable names will produce an error message, because R does not know to look in the Auto data set for those variables.

scatterplot

```
> plot(cylinders, mpg)
Error in plot(cylinders, mpg) : object 'cylinders' not found
```

To refer to a variable, we must type the data set and the variable name joined with a $ symbol. Alternatively, we can use the `attach()` function in order to tell R to make the variables in this data frame available by name.

`attach()`

```
> plot(Auto$cylinders, Auto$mpg)
> attach(Auto)
> plot(cylinders, mpg)
```

The `cylinders` variable is stored as a numeric vector, so R has treated it as quantitative. However, since there are only a small number of possible values for `cylinders`, one may prefer to treat it as a qualitative variable. The `as.factor()` function converts quantitative variables into qualitative variables.

`as.factor()`

```
> cylinders=as.factor(cylinders)
```

If the variable plotted on the x-axis is categorial, then *boxplots* will automatically be produced by the `plot()` function. As usual, a number of options can be specified in order to customize the plots.

boxplot

```
> plot(cylinders, mpg)
> plot(cylinders, mpg, col="red")
> plot(cylinders, mpg, col="red", varwidth=T)
> plot(cylinders, mpg, col="red", varwidth=T,horizontal=T)
> plot(cylinders, mpg, col="red", varwidth=T, xlab="cylinders",
    ylab="MPG")
```

The `hist()` function can be used to plot a *histogram*. Note that `col=2` has the same effect as `col="red"`.

`hist()`

histogram

```
> hist(mpg)
> hist(mpg,col=2)
> hist(mpg,col=2,breaks=15)
```

The `pairs()` function creates a *scatterplot matrix* i.e. a scatterplot for every pair of variables for any given data set. We can also produce scatterplots for just a subset of the variables.

scatterplot matrix

```
> pairs(Auto)
> pairs(~ mpg + displacement + horsepower + weight +
    acceleration, Auto)
```

In conjunction with the `plot()` function, `identify()` provides a useful interactive method for identifying the value for a particular variable for points on a plot. We pass in three arguments to `identify()`: the x-axis variable, the y-axis variable, and the variable whose values we would like to see printed for each point. Then clicking on a given point in the plot will cause R to print the value of the variable of interest. Right-clicking on the plot will exit the `identify()` function (control-click on a Mac). The numbers printed under the `identify()` function correspond to the rows for the selected points.

`identify()`

```
> plot(horsepower,mpg)
> identify(horsepower,mpg,name)
```

The summary() function produces a numerical summary of each variable in a particular data set.

summary()

```
> summary(Auto)
      mpg              cylinders        displacement
 Min.   : 9.00    Min.   :3.000    Min.   : 68.0
 1st Qu.:17.00    1st Qu.:4.000    1st Qu.:105.0
 Median :22.75    Median :4.000    Median :151.0
 Mean   :23.45    Mean   :5.472    Mean   :194.4
 3rd Qu.:29.00    3rd Qu.:8.000    3rd Qu.:275.8
 Max.   :46.60    Max.   :8.000    Max.   :455.0

   horsepower          weight         acceleration
 Min.   : 46.0    Min.   :1613    Min.   : 8.00
 1st Qu.: 75.0    1st Qu.:2225    1st Qu.:13.78
 Median : 93.5    Median :2804    Median :15.50
 Mean   :104.5    Mean   :2978    Mean   :15.54
 3rd Qu.:126.0    3rd Qu.:3615    3rd Qu.:17.02
 Max.   :230.0    Max.   :5140    Max.   :24.80

      year             origin                     name
 Min.   :70.00    Min.   :1.000    amc matador      :  5
 1st Qu.:73.00    1st Qu.:1.000    ford pinto       :  5
 Median :76.00    Median :1.000    toyota corolla   :  5
 Mean   :75.98    Mean   :1.577    amc gremlin      :  4
 3rd Qu.:79.00    3rd Qu.:2.000    amc hornet       :  4
 Max.   :82.00    Max.   :3.000    chevrolet chevette:  4
                                   (Other)          :365
```

For qualitative variables such as name, R will list the number of observations that fall in each category. We can also produce a summary of just a single variable.

```
> summary(mpg)
   Min. 1st Qu.  Median    Mean 3rd Qu.    Max.
   9.00   17.00   22.75   23.45   29.00   46.60
```

Once we have finished using R, we type q() in order to shut it down, or quit. When exiting R, we have the option to save the current *workspace* so that all objects (such as data sets) that we have created in this R session will be available next time. Before exiting R, we may want to save a record of all of the commands that we typed in the most recent session; this can be accomplished using the savehistory() function. Next time we enter R, we can load that history using the loadhistory() function.

q()
workspace

savehistory()
loadhistory()

2.4   Exercises

Conceptual

1. For each of parts (a) through (d), indicate whether we would generally expect the performance of a flexible statistical learning method to be better or worse than an inflexible method. Justify your answer.

   (a) The sample size n is extremely large, and the number of predictors p is small.

   (b) The number of predictors p is extremely large, and the number of observations n is small.

   (c) The relationship between the predictors and response is highly non-linear.

   (d) The variance of the error terms, i.e. $\sigma^2 = \text{Var}(\epsilon)$, is extremely high.

2. Explain whether each scenario is a classification or regression problem, and indicate whether we are most interested in inference or prediction. Finally, provide n and p.

   (a) We collect a set of data on the top 500 firms in the US. For each firm we record profit, number of employees, industry and the CEO salary. We are interested in understanding which factors affect CEO salary.

   (b) We are considering launching a new product and wish to know whether it will be a *success* or a *failure*. We collect data on 20 similar products that were previously launched. For each product we have recorded whether it was a success or failure, price charged for the product, marketing budget, competition price, and ten other variables.

   (c) We are interested in predicting the % change in the USD/Euro exchange rate in relation to the weekly changes in the world stock markets. Hence we collect weekly data for all of 2012. For each week we record the % change in the USD/Euro, the % change in the US market, the % change in the British market, and the % change in the German market.

3. We now revisit the bias-variance decomposition.

   (a) Provide a sketch of typical (squared) bias, variance, training error, test error, and Bayes (or irreducible) error curves, on a single plot, as we go from less flexible statistical learning methods towards more flexible approaches. The x-axis should represent the amount of flexibility in the method, and the y-axis should represent the values for each curve. There should be five curves. Make sure to label each one.

(b) Explain why each of the five curves has the shape displayed in part (a).

4. You will now think of some real-life applications for statistical learning.

   (a) Describe three real-life applications in which *classification* might be useful. Describe the response, as well as the predictors. Is the goal of each application inference or prediction? Explain your answer.

   (b) Describe three real-life applications in which *regression* might be useful. Describe the response, as well as the predictors. Is the goal of each application inference or prediction? Explain your answer.

   (c) Describe three real-life applications in which *cluster analysis* might be useful.

5. What are the advantages and disadvantages of a very flexible (versus a less flexible) approach for regression or classification? Under what circumstances might a more flexible approach be preferred to a less flexible approach? When might a less flexible approach be preferred?

6. Describe the differences between a parametric and a non-parametric statistical learning approach. What are the advantages of a parametric approach to regression or classification (as opposed to a non-parametric approach)? What are its disadvantages?

7. The table below provides a training data set containing six observations, three predictors, and one qualitative response variable.

Obs.	X_1	X_2	X_3	Y
1	0	3	0	Red
2	2	0	0	Red
3	0	1	3	Red
4	0	1	2	Green
5	−1	0	1	Green
6	1	1	1	Red

Suppose we wish to use this data set to make a prediction for Y when $X_1 = X_2 = X_3 = 0$ using K-nearest neighbors.

   (a) Compute the Euclidean distance between each observation and the test point, $X_1 = X_2 = X_3 = 0$.

(b) What is our prediction with $K = 1$? Why?

(c) What is our prediction with $K = 3$? Why?

(d) If the Bayes decision boundary in this problem is highly non-linear, then would we expect the *best* value for K to be large or small? Why?

Applied

8. This exercise relates to the College data set, which can be found in the file College.csv. It contains a number of variables for 777 different universities and colleges in the US. The variables are

   - Private : Public/private indicator
   - Apps : Number of applications received
   - Accept : Number of applicants accepted
   - Enroll : Number of new students enrolled
   - Top10perc : New students from top 10 % of high school class
   - Top25perc : New students from top 25 % of high school class
   - F.Undergrad : Number of full-time undergraduates
   - P.Undergrad : Number of part-time undergraduates
   - Outstate : Out-of-state tuition
   - Room.Board : Room and board costs
   - Books : Estimated book costs
   - Personal : Estimated personal spending
   - PhD : Percent of faculty with Ph.D.'s
   - Terminal : Percent of faculty with terminal degree
   - S.F.Ratio : Student/faculty ratio
   - perc.alumni : Percent of alumni who donate
   - Expend : Instructional expenditure per student
   - Grad.Rate : Graduation rate

   Before reading the data into R, it can be viewed in Excel or a text editor.

   (a) Use the read.csv() function to read the data into R. Call the loaded data college. Make sure that you have the directory set to the correct location for the data.

   (b) Look at the data using the fix() function. You should notice that the first column is just the name of each university. We don't really want R to treat this as data. However, it may be handy to have these names for later. Try the following commands:

```
> rownames(college)=college[,1]
> fix(college)
```

You should see that there is now a `row.names` column with the name of each university recorded. This means that R has given each row a name corresponding to the appropriate university. R will not try to perform calculations on the row names. However, we still need to eliminate the first column in the data where the names are stored. Try

```
> college=college[,-1]
> fix(college)
```

Now you should see that the first data column is `Private`. Note that another column labeled `row.names` now appears before the `Private` column. However, this is not a data column but rather the name that R is giving to each row.

(c)   i. Use the `summary()` function to produce a numerical summary of the variables in the data set.

ii. Use the `pairs()` function to produce a scatterplot matrix of the first ten columns or variables of the data. Recall that you can reference the first ten columns of a matrix `A` using `A[,1:10]`.

iii. Use the `plot()` function to produce side-by-side boxplots of `Outstate` versus `Private`.

iv. Create a new qualitative variable, called `Elite`, by *binning* the `Top10perc` variable. We are going to divide universities into two groups based on whether or not the proportion of students coming from the top 10 % of their high school classes exceeds 50 %.

```
> Elite=rep("No",nrow(college))
> Elite[college$Top10perc >50]="Yes"
> Elite=as.factor(Elite)
> college=data.frame(college,Elite)
```

Use the `summary()` function to see how many elite universities there are. Now use the `plot()` function to produce side-by-side boxplots of `Outstate` versus `Elite`.

v. Use the `hist()` function to produce some histograms with differing numbers of bins for a few of the quantitative variables. You may find the command `par(mfrow=c(2,2))` useful: it will divide the print window into four regions so that four plots can be made simultaneously. Modifying the arguments to this function will divide the screen in other ways.

vi. Continue exploring the data, and provide a brief summary of what you discover.

9. This exercise involves the `Auto` data set studied in the lab. Make sure that the missing values have been removed from the data.

   (a) Which of the predictors are quantitative, and which are qualitative?

   (b) What is the *range* of each quantitative predictor? You can answer this using the `range()` function.

   range()

   (c) What is the mean and standard deviation of each quantitative predictor?

   (d) Now remove the 10th through 85th observations. What is the range, mean, and standard deviation of each predictor in the subset of the data that remains?

   (e) Using the full data set, investigate the predictors graphically, using scatterplots or other tools of your choice. Create some plots highlighting the relationships among the predictors. Comment on your findings.

   (f) Suppose that we wish to predict gas mileage (`mpg`) on the basis of the other variables. Do your plots suggest that any of the other variables might be useful in predicting `mpg`? Justify your answer.

10. This exercise involves the `Boston` housing data set.

    (a) To begin, load in the `Boston` data set. The `Boston` data set is part of the `MASS` *library* in R.

    ```
    > library(MASS)
    ```

    Now the data set is contained in the object `Boston`.

    ```
    > Boston
    ```

    Read about the data set:

    ```
    > ?Boston
    ```

    How many rows are in this data set? How many columns? What do the rows and columns represent?

    (b) Make some pairwise scatterplots of the predictors (columns) in this data set. Describe your findings.

    (c) Are any of the predictors associated with per capita crime rate? If so, explain the relationship.

    (d) Do any of the suburbs of Boston appear to have particularly high crime rates? Tax rates? Pupil-teacher ratios? Comment on the range of each predictor.

    (e) How many of the suburbs in this data set bound the Charles river?

(f) What is the median pupil-teacher ratio among the towns in this data set?

(g) Which suburb of Boston has lowest median value of owner-occupied homes? What are the values of the other predictors for that suburb, and how do those values compare to the overall ranges for those predictors? Comment on your findings.

(h) In this data set, how many of the suburbs average more than seven rooms per dwelling? More than eight rooms per dwelling? Comment on the suburbs that average more than eight rooms per dwelling.

3
Linear Regression

This chapter is about *linear regression*, a very simple approach for supervised learning. In particular, linear regression is a useful tool for predicting a quantitative response. Linear regression has been around for a long time and is the topic of innumerable textbooks. Though it may seem somewhat dull compared to some of the more modern statistical learning approaches described in later chapters of this book, linear regression is still a useful and widely used statistical learning method. Moreover, it serves as a good jumping-off point for newer approaches: as we will see in later chapters, many fancy statistical learning approaches can be seen as generalizations or extensions of linear regression. Consequently, the importance of having a good understanding of linear regression before studying more complex learning methods cannot be overstated. In this chapter, we review some of the key ideas underlying the linear regression model, as well as the least squares approach that is most commonly used to fit this model.

Recall the `Advertising` data from Chapter 2. Figure 2.1 displays `sales` (in thousands of units) for a particular product as a function of advertising budgets (in thousands of dollars) for `TV`, `radio`, and `newspaper` media. Suppose that in our role as statistical consultants we are asked to suggest, on the basis of this data, a marketing plan for next year that will result in high product sales. What information would be useful in order to provide such a recommendation? Here are a few important questions that we might seek to address:

1. *Is there a relationship between advertising budget and sales?*
   Our first goal should be to determine whether the data provide

G. James et al., *An Introduction to Statistical Learning: with Applications in R*,   59
Springer Texts in Statistics, DOI 10.1007/978-1-4614-7138-7_3,
© Springer Science+Business Media New York 2013

evidence of an association between advertising expenditure and sales. If the evidence is weak, then one might argue that no money should be spent on advertising!

2. *How strong is the relationship between advertising budget and sales?*
   Assuming that there is a relationship between advertising and sales, we would like to know the strength of this relationship. In other words, given a certain advertising budget, can we predict sales with a high level of accuracy? This would be a strong relationship. Or is a prediction of sales based on advertising expenditure only slightly better than a random guess? This would be a weak relationship.

3. *Which media contribute to sales?*
   Do all three media—TV, radio, and newspaper—contribute to sales, or do just one or two of the media contribute? To answer this question, we must find a way to separate out the individual effects of each medium when we have spent money on all three media.

4. *How accurately can we estimate the effect of each medium on sales?*
   For every dollar spent on advertising in a particular medium, by what amount will sales increase? How accurately can we predict this amount of increase?

5. *How accurately can we predict future sales?*
   For any given level of television, radio, or newspaper advertising, what is our prediction for sales, and what is the accuracy of this prediction?

6. *Is the relationship linear?*
   If there is approximately a straight-line relationship between advertising expenditure in the various media and sales, then linear regression is an appropriate tool. If not, then it may still be possible to transform the predictor or the response so that linear regression can be used.

7. *Is there synergy among the advertising media?*
   Perhaps spending $50,000 on television advertising and $50,000 on radio advertising results in more sales than allocating $100,000 to either television or radio individually. In marketing, this is known as a *synergy* effect, while in statistics it is called an *interaction* effect.

   synergy
   interaction

It turns out that linear regression can be used to answer each of these questions. We will first discuss all of these questions in a general context, and then return to them in this specific context in Section 3.4.

3.1 Simple Linear Regression

simple linear regression

Simple linear regression lives up to its name: it is a very straightforward approach for predicting a quantitative response Y on the basis of a single predictor variable X. It assumes that there is approximately a linear relationship between X and Y. Mathematically, we can write this linear relationship as

$$Y \approx \beta_0 + \beta_1 X. \qquad (3.1)$$

You might read "$\approx$" as "*is approximately modeled as*". We will sometimes describe (3.1) by saying that we are *regressing Y on X* (or Y *onto* X). For example, X may represent TV advertising and Y may represent sales. Then we can regress sales onto TV by fitting the model

$$\text{sales} \approx \beta_0 + \beta_1 \times \text{TV}.$$

intercept
slope
coefficient
parameter

In Equation 3.1, β_0 and β_1 are two unknown constants that represent the *intercept* and *slope* terms in the linear model. Together, β_0 and β_1 are known as the model *coefficients* or *parameters*. Once we have used our training data to produce estimates $\hat{\beta}_0$ and $\hat{\beta}_1$ for the model coefficients, we can predict future sales on the basis of a particular value of TV advertising by computing

$$\hat{y} = \hat{\beta}_0 + \hat{\beta}_1 x, \qquad (3.2)$$

where $\hat{y}$ indicates a prediction of Y on the basis of $X = x$. Here we use a *hat* symbol, $\hat{}$, to denote the estimated value for an unknown parameter or coefficient, or to denote the predicted value of the response.

3.1.1 Estimating the Coefficients

In practice, β_0 and β_1 are unknown. So before we can use (3.1) to make predictions, we must use data to estimate the coefficients. Let

$$(x_1, y_1), \ (x_2, y_2), \ldots, \ (x_n, y_n)$$

represent n observation pairs, each of which consists of a measurement of X and a measurement of Y. In the Advertising example, this data set consists of the TV advertising budget and product sales in $n = 200$ different markets. (Recall that the data are displayed in Figure 2.1.) Our goal is to obtain coefficient estimates $\hat{\beta}_0$ and $\hat{\beta}_1$ such that the linear model (3.1) fits the available data well—that is, so that $y_i \approx \hat{\beta}_0 + \hat{\beta}_1 x_i$ for $i = 1, \ldots, n$. In other words, we want to find an intercept $\hat{\beta}_0$ and a slope $\hat{\beta}_1$ such that the resulting line is as close as possible to the $n = 200$ data points. There are a number of ways of measuring *closeness*. However, by far the most common approach involves minimizing the *least squares* criterion, and we take that approach in this chapter. Alternative approaches will be considered in Chapter 6.

least squares

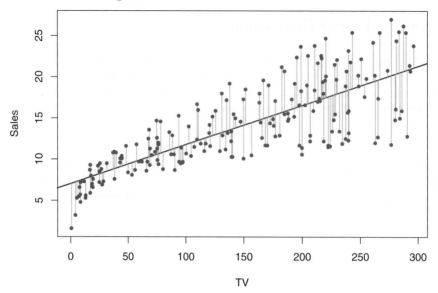

FIGURE 3.1. *For the* Advertising *data, the least squares fit for the regression of* sales *onto* TV *is shown. The fit is found by minimizing the sum of squared errors. Each grey line segment represents an error, and the fit makes a compromise by averaging their squares. In this case a linear fit captures the essence of the relationship, although it is somewhat deficient in the left of the plot.*

Let $\hat{y}_i = \hat{\beta}_0 + \hat{\beta}_1 x_i$ be the prediction for Y based on the ith value of X. Then $e_i = y_i - \hat{y}_i$ represents the ith *residual*—this is the difference between the ith observed response value and the ith response value that is predicted by our linear model. We define the *residual sum of squares* (RSS) as

residual

residual sum of squares

$$\text{RSS} = e_1^2 + e_2^2 + \cdots + e_n^2,$$

or equivalently as

$$\text{RSS} = (y_1 - \hat{\beta}_0 - \hat{\beta}_1 x_1)^2 + (y_2 - \hat{\beta}_0 - \hat{\beta}_1 x_2)^2 + \ldots + (y_n - \hat{\beta}_0 - \hat{\beta}_1 x_n)^2. \quad (3.3)$$

The least squares approach chooses $\hat{\beta}_0$ and $\hat{\beta}_1$ to minimize the RSS. Using some calculus, one can show that the minimizers are

$$\hat{\beta}_1 = \frac{\sum_{i=1}^n (x_i - \bar{x})(y_i - \bar{y})}{\sum_{i=1}^n (x_i - \bar{x})^2},$$

$$\hat{\beta}_0 = \bar{y} - \hat{\beta}_1 \bar{x}, \quad (3.4)$$

where $\bar{y} \equiv \frac{1}{n} \sum_{i=1}^n y_i$ and $\bar{x} \equiv \frac{1}{n} \sum_{i=1}^n x_i$ are the sample means. In other words, (3.4) defines the *least squares coefficient estimates* for simple linear regression.

Figure 3.1 displays the simple linear regression fit to the Advertising data, where $\hat{\beta}_0 = 7.03$ and $\hat{\beta}_1 = 0.0475$. In other words, according to

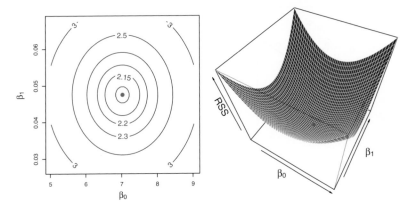

FIGURE 3.2. *Contour and three-dimensional plots of the RSS on the Advertising data, using* sales *as the response and* TV *as the predictor. The red dots correspond to the least squares estimates* $\hat{\beta}_0$ *and* $\hat{\beta}_1$*, given by (3.4).*

this approximation, an additional \$1,000 spent on TV advertising is associated with selling approximately 47.5 additional units of the product. In Figure 3.2, we have computed RSS for a number of values of β_0 and β_1, using the advertising data with sales as the response and TV as the predictor. In each plot, the red dot represents the pair of least squares estimates $(\hat{\beta}_0, \hat{\beta}_1)$ given by (3.4). These values clearly minimize the RSS.

3.1.2 Assessing the Accuracy of the Coefficient Estimates

Recall from (2.1) that we assume that the *true* relationship between X and Y takes the form $Y = f(X) + \epsilon$ for some unknown function f, where ϵ is a mean-zero random error term. If f is to be approximated by a linear function, then we can write this relationship as

$$Y = \beta_0 + \beta_1 X + \epsilon. \tag{3.5}$$

Here β_0 is the intercept term—that is, the expected value of Y when $X = 0$, and β_1 is the slope—the average increase in Y associated with a one-unit increase in X. The error term is a catch-all for what we miss with this simple model: the true relationship is probably not linear, there may be other variables that cause variation in Y, and there may be measurement error. We typically assume that the error term is independent of X.

The model given by (3.5) defines the *population regression line*, which is the best linear approximation to the true relationship between X and Y.[1] The least squares regression coefficient estimates (3.4) characterize the *least squares line* (3.2). The left-hand panel of Figure 3.3 displays these

population regression line

least squares line

---

[1] The assumption of linearity is often a useful working model. However, despite what many textbooks might tell us, we seldom believe that the true relationship is linear.

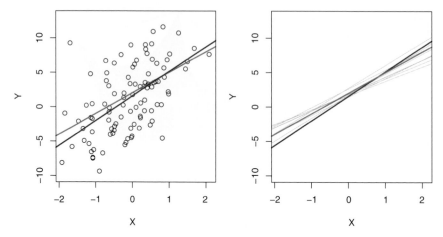

FIGURE 3.3. *A simulated data set.* Left: *The red line represents the true relationship, $f(X) = 2 + 3X$, which is known as the population regression line. The blue line is the least squares line; it is the least squares estimate for $f(X)$ based on the observed data, shown in black.* Right: *The population regression line is again shown in red, and the least squares line in dark blue. In light blue, ten least squares lines are shown, each computed on the basis of a separate random set of observations. Each least squares line is different, but on average, the least squares lines are quite close to the population regression line.*

two lines in a simple simulated example. We created 100 random Xs, and generated 100 corresponding Ys from the model

$$Y = 2 + 3X + \epsilon, \tag{3.6}$$

where ϵ was generated from a normal distribution with mean zero. The red line in the left-hand panel of Figure 3.3 displays the *true* relationship, $f(X) = 2 + 3X$, while the blue line is the least squares estimate based on the observed data. The true relationship is generally not known for real data, but the least squares line can always be computed using the coefficient estimates given in (3.4). In other words, in real applications, we have access to a set of observations from which we can compute the least squares line; however, the population regression line is unobserved. In the right-hand panel of Figure 3.3 we have generated ten different data sets from the model given by (3.6) and plotted the corresponding ten least squares lines. Notice that different data sets generated from the same true model result in slightly different least squares lines, but the unobserved population regression line does not change.

At first glance, the difference between the population regression line and the least squares line may seem subtle and confusing. We only have one data set, and so what does it mean that two different lines describe the relationship between the predictor and the response? Fundamentally, the concept of these two lines is a natural extension of the standard statistical approach of using information from a sample to estimate characteristics of a large population. For example, suppose that we are interested in knowing the population mean μ of some random variable Y. Unfortunately, μ is unknown, but we do have access to n observations from Y, which we can write as $y_1, \ldots, y_n$, and which we can use to estimate μ. A reasonable estimate is $\hat{\mu} = \bar{y}$, where $\bar{y} = \frac{1}{n} \sum_{i=1}^{n} y_i$ is the sample mean. The sample mean and the population mean are different, but in general the sample mean will provide a good estimate of the population mean. In the same way, the unknown coefficients β_0 and β_1 in linear regression define the population regression line. We seek to estimate these unknown coefficients using $\hat{\beta}_0$ and $\hat{\beta}_1$ given in (3.4). These coefficient estimates define the least squares line.

The analogy between linear regression and estimation of the mean of a random variable is an apt one based on the concept of *bias*. If we use the sample mean $\hat{\mu}$ to estimate μ, this estimate is *unbiased*, in the sense that on average, we expect $\hat{\mu}$ to equal μ. What exactly does this mean? It means that on the basis of one particular set of observations $y_1, \ldots, y_n$, $\hat{\mu}$ might overestimate μ, and on the basis of another set of observations, $\hat{\mu}$ might underestimate μ. But if we could average a huge number of estimates of μ obtained from a huge number of sets of observations, then this average would *exactly* equal μ. Hence, an unbiased estimator does not *systematically* over- or under-estimate the true parameter. The property of unbiasedness holds for the least squares coefficient estimates given by (3.4) as well: if we estimate β_0 and β_1 on the basis of a particular data set, then our estimates won't be exactly equal to β_0 and β_1. But if we could average the estimates obtained over a huge number of data sets, then the average of these estimates would be spot on! In fact, we can see from the right-hand panel of Figure 3.3 that the average of many least squares lines, each estimated from a separate data set, is pretty close to the true population regression line.

We continue the analogy with the estimation of the population mean μ of a random variable Y. A natural question is as follows: how accurate is the sample mean $\hat{\mu}$ as an estimate of μ? We have established that the average of $\hat{\mu}$'s over many data sets will be very close to μ, but that a single estimate $\hat{\mu}$ may be a substantial underestimate or overestimate of μ. How far off will that single estimate of $\hat{\mu}$ be? In general, we answer this question by computing the *standard error* of $\hat{\mu}$, written as $\text{SE}(\hat{\mu})$. We have the well-known formula

bias

unbiased

standard error

$$\text{Var}(\hat{\mu}) = \text{SE}(\hat{\mu})^2 = \frac{\sigma^2}{n}, \tag{3.7}$$

where σ is the standard deviation of each of the realizations y_i of Y.[2] Roughly speaking, the standard error tells us the average amount that this estimate $\hat{\mu}$ differs from the actual value of μ. Equation 3.7 also tells us how this deviation shrinks with n—the more observations we have, the smaller the standard error of $\hat{\mu}$. In a similar vein, we can wonder how close $\hat{\beta}_0$ and $\hat{\beta}_1$ are to the true values β_0 and β_1. To compute the standard errors associated with $\hat{\beta}_0$ and $\hat{\beta}_1$, we use the following formulas:

$$\text{SE}(\hat{\beta}_0)^2 = \sigma^2 \left[\frac{1}{n} + \frac{\bar{x}^2}{\sum_{i=1}^{n}(x_i - \bar{x})^2} \right], \quad \text{SE}(\hat{\beta}_1)^2 = \frac{\sigma^2}{\sum_{i=1}^{n}(x_i - \bar{x})^2}, \quad (3.8)$$

where $\sigma^2 = \text{Var}(\epsilon)$. For these formulas to be strictly valid, we need to assume that the errors ϵ_i for each observation are uncorrelated with common variance σ^2. This is clearly not true in Figure 3.1, but the formula still turns out to be a good approximation. Notice in the formula that $\text{SE}(\hat{\beta}_1)$ is smaller when the x_i are more spread out; intuitively we have more *leverage* to estimate a slope when this is the case. We also see that $\text{SE}(\hat{\beta}_0)$ would be the same as $\text{SE}(\hat{\mu})$ if $\bar{x}$ were zero (in which case $\hat{\beta}_0$ would be equal to $\bar{y}$). In general, σ^2 is not known, but can be estimated from the data. The estimate of σ is known as the *residual standard error*, and is given by the formula $\text{RSE} = \sqrt{\text{RSS}/(n-2)}$. Strictly speaking, when σ^2 is estimated from the data we should write $\widehat{\text{SE}}(\hat{\beta}_1)$ to indicate that an estimate has been made, but for simplicity of notation we will drop this extra "hat".

residual standard error

Standard errors can be used to compute *confidence intervals*. A 95 % confidence interval is defined as a range of values such that with 95 % probability, the range will contain the true unknown value of the parameter. The range is defined in terms of lower and upper limits computed from the sample of data. For linear regression, the 95 % confidence interval for β_1 approximately takes the form

confidence interval

$$\hat{\beta}_1 \pm 2 \cdot \text{SE}(\hat{\beta}_1). \quad (3.9)$$

That is, there is approximately a 95 % chance that the interval

$$\left[\hat{\beta}_1 - 2 \cdot \text{SE}(\hat{\beta}_1), \ \hat{\beta}_1 + 2 \cdot \text{SE}(\hat{\beta}_1) \right] \quad (3.10)$$

will contain the true value of β_1.[3] Similarly, a confidence interval for β_0 approximately takes the form

$$\hat{\beta}_0 \pm 2 \cdot \text{SE}(\hat{\beta}_0). \quad (3.11)$$

---

[2] This formula holds provided that the n observations are uncorrelated.

[3] *Approximately* for several reasons. Equation 3.10 relies on the assumption that the errors are Gaussian. Also, the factor of 2 in front of the $\text{SE}(\hat{\beta}_1)$ term will vary slightly depending on the number of observations n in the linear regression. To be precise, rather than the number 2, (3.10) should contain the 97.5 % quantile of a t-distribution with $n-2$ degrees of freedom. Details of how to compute the 95 % confidence interval precisely in R will be provided later in this chapter.

In the case of the advertising data, the 95 % confidence interval for β_0 is $[6.130, 7.935]$ and the 95 % confidence interval for β_1 is $[0.042, 0.053]$. Therefore, we can conclude that in the absence of any advertising, sales will, on average, fall somewhere between 6,130 and 7,940 units. Furthermore, for each \$1,000 increase in television advertising, there will be an average increase in sales of between 42 and 53 units.

Standard errors can also be used to perform *hypothesis tests* on the coefficients. The most common hypothesis test involves testing the *null hypothesis* of

hypothesis test

null hypothesis

$$H_0 : \text{There is no relationship between } X \text{ and } Y \qquad (3.12)$$

versus the *alternative hypothesis*

alternative hypothesis

$$H_a : \text{There is some relationship between } X \text{ and } Y. \qquad (3.13)$$

Mathematically, this corresponds to testing

$$H_0 : \beta_1 = 0$$

versus

$$H_a : \beta_1 \neq 0,$$

since if $\beta_1 = 0$ then the model (3.5) reduces to $Y = \beta_0 + \epsilon$, and X is not associated with Y. To test the null hypothesis, we need to determine whether $\hat{\beta}_1$, our estimate for β_1, is sufficiently far from zero that we can be confident that β_1 is non-zero. How far is far enough? This of course depends on the accuracy of $\hat{\beta}_1$—that is, it depends on $\text{SE}(\hat{\beta}_1)$. If $\text{SE}(\hat{\beta}_1)$ is small, then even relatively small values of $\hat{\beta}_1$ may provide strong evidence that $\beta_1 \neq 0$, and hence that there is a relationship between X and Y. In contrast, if $\text{SE}(\hat{\beta}_1)$ is large, then $\hat{\beta}_1$ must be large in absolute value in order for us to reject the null hypothesis. In practice, we compute a *t-statistic*, given by

t-statistic

$$t = \frac{\hat{\beta}_1 - 0}{\text{SE}(\hat{\beta}_1)}, \qquad (3.14)$$

which measures the number of standard deviations that $\hat{\beta}_1$ is away from 0. If there really is no relationship between X and Y, then we expect that (3.14) will have a t-distribution with $n - 2$ degrees of freedom. The t-distribution has a bell shape and for values of n greater than approximately 30 it is quite similar to the normal distribution. Consequently, it is a simple matter to compute the probability of observing any number equal to $|t|$ or larger in absolute value, assuming $\beta_1 = 0$. We call this probability the *p-value*. Roughly speaking, we interpret the p-value as follows: a small p-value indicates that it is unlikely to observe such a substantial association between the predictor and the response due to chance, in the absence of any real association between the predictor and the response. Hence, if we see a small p-value,

p-value

then we can infer that there is an association between the predictor and the response. We *reject the null hypothesis*—that is, we declare a relationship to exist between X and Y—if the p-value is small enough. Typical p-value cutoffs for rejecting the null hypothesis are 5 or 1%. When $n = 30$, these correspond to t-statistics (3.14) of around 2 and 2.75, respectively.

	Coefficient	Std. error	t-statistic	p-value
Intercept	7.0325	0.4578	15.36	< 0.0001
TV	0.0475	0.0027	17.67	< 0.0001

TABLE 3.1. *For the* Advertising *data, coefficients of the least squares model for the regression of number of units sold on TV advertising budget. An increase of $1,000 in the TV advertising budget is associated with an increase in sales by around 50 units (Recall that the* sales *variable is in thousands of units, and the* TV *variable is in thousands of dollars).*

Table 3.1 provides details of the least squares model for the regression of number of units sold on TV advertising budget for the Advertising data. Notice that the coefficients for $\hat{\beta}_0$ and $\hat{\beta}_1$ are very large relative to their standard errors, so the t-statistics are also large; the probabilities of seeing such values if H_0 is true are virtually zero. Hence we can conclude that $\beta_0 \neq 0$ and $\beta_1 \neq 0$.[4]

3.1.3 *Assessing the Accuracy of the Model*

Once we have rejected the null hypothesis (3.12) in favor of the alternative hypothesis (3.13), it is natural to want to quantify *the extent to which the model fits the data*. The quality of a linear regression fit is typically assessed using two related quantities: the *residual standard error* (RSE) and the R^2 statistic.

Table 3.2 displays the RSE, the R^2 statistic, and the F-statistic (to be described in Section 3.2.2) for the linear regression of number of units sold on TV advertising budget.

Residual Standard Error

Recall from the model (3.5) that associated with each observation is an error term ϵ. Due to the presence of these error terms, even if we knew the true regression line (i.e. even if β_0 and β_1 were known), we would not be able to perfectly predict Y from X. The RSE is an estimate of the standard

---

[4]In Table 3.1, a small p-value for the intercept indicates that we can reject the null hypothesis that $\beta_0 = 0$, and a small p-value for TV indicates that we can reject the null hypothesis that $\beta_1 = 0$. Rejecting the latter null hypothesis allows us to conclude that there is a relationship between TV and sales. Rejecting the former allows us to conclude that in the absence of TV expenditure, sales are non-zero.

Quantity	Value
Residual standard error	3.26
R^2	0.612
F-statistic	312.1

TABLE 3.2. *For the* Advertising *data, more information about the least squares model for the regression of number of units sold on TV advertising budget.*

deviation of ϵ. Roughly speaking, it is the average amount that the response will deviate from the true regression line. It is computed using the formula

$$\text{RSE} = \sqrt{\frac{1}{n-2}\text{RSS}} = \sqrt{\frac{1}{n-2}\sum_{i=1}^{n}(y_i - \hat{y}_i)^2}. \tag{3.15}$$

Note that RSS was defined in Section 3.1.1, and is given by the formula

$$\text{RSS} = \sum_{i=1}^{n}(y_i - \hat{y}_i)^2. \tag{3.16}$$

In the case of the advertising data, we see from the linear regression output in Table 3.2 that the RSE is 3.26. In other words, actual sales in each market deviate from the true regression line by approximately 3,260 units, on average. Another way to think about this is that even if the model were correct and the true values of the unknown coefficients β_0 and β_1 were known exactly, any prediction of sales on the basis of TV advertising would still be off by about 3,260 units on average. Of course, whether or not 3,260 units is an acceptable prediction error depends on the problem context. In the advertising data set, the mean value of sales over all markets is approximately 14,000 units, and so the percentage error is $3,260/14,000 = 23\%$.

The RSE is considered a measure of the *lack of fit* of the model (3.5) to the data. If the predictions obtained using the model are very close to the true outcome values—that is, if $\hat{y}_i \approx y_i$ for $i = 1, \ldots, n$—then (3.15) will be small, and we can conclude that the model fits the data very well. On the other hand, if $\hat{y}_i$ is very far from y_i for one or more observations, then the RSE may be quite large, indicating that the model doesn't fit the data well.

R^2 Statistic

The RSE provides an absolute measure of lack of fit of the model (3.5) to the data. But since it is measured in the units of Y, it is not always clear what constitutes a good RSE. The R^2 statistic provides an alternative measure of fit. It takes the form of a *proportion*—the proportion of variance explained—and so it always takes on a value between 0 and 1, and is independent of the scale of Y.

To calculate R^2, we use the formula

$$R^2 = \frac{\text{TSS} - \text{RSS}}{\text{TSS}} = 1 - \frac{\text{RSS}}{\text{TSS}} \qquad (3.17)$$

where $\text{TSS} = \sum(y_i - \bar{y})^2$ is the *total sum of squares*, and RSS is defined in (3.16). TSS measures the total variance in the response Y, and can be thought of as the amount of variability inherent in the response before the regression is performed. In contrast, RSS measures the amount of variability that is left unexplained after performing the regression. Hence, $\text{TSS} - \text{RSS}$ measures the amount of variability in the response that is explained (or removed) by performing the regression, and R^2 measures the *proportion of variability in Y that can be explained using X*. An R^2 statistic that is close to 1 indicates that a large proportion of the variability in the response has been explained by the regression. A number near 0 indicates that the regression did not explain much of the variability in the response; this might occur because the linear model is wrong, or the inherent error σ^2 is high, or both. In Table 3.2, the R^2 was 0.61, and so just under two-thirds of the variability in sales is explained by a linear regression on TV.

The R^2 statistic (3.17) has an interpretational advantage over the RSE (3.15), since unlike the RSE, it always lies between 0 and 1. However, it can still be challenging to determine what is a *good* R^2 value, and in general, this will depend on the application. For instance, in certain problems in physics, we may know that the data truly comes from a linear model with a small residual error. In this case, we would expect to see an R^2 value that is extremely close to 1, and a substantially smaller R^2 value might indicate a serious problem with the experiment in which the data were generated. On the other hand, in typical applications in biology, psychology, marketing, and other domains, the linear model (3.5) is at best an extremely rough approximation to the data, and residual errors due to other unmeasured factors are often very large. In this setting, we would expect only a very small proportion of the variance in the response to be explained by the predictor, and an R^2 value well below 0.1 might be more realistic!

The R^2 statistic is a measure of the linear relationship between X and Y. Recall that *correlation*, defined as

$$\text{Cor}(X, Y) = \frac{\sum_{i=1}^{n}(x_i - \bar{x})(y_i - \bar{y})}{\sqrt{\sum_{i=1}^{n}(x_i - \bar{x})^2}\sqrt{\sum_{i=1}^{n}(y_i - \bar{y})^2}}, \qquad (3.18)$$

is also a measure of the linear relationship between X and Y.[5] This suggests that we might be able to use $r = \text{Cor}(X, Y)$ instead of R^2 in order to assess the fit of the linear model. In fact, it can be shown that in the simple linear regression setting, $R^2 = r^2$. In other words, the squared correlation

---

[5] We note that in fact, the right-hand side of (3.18) is the sample correlation; thus, it would be more correct to write $\widehat{\text{Cor}}(X, Y)$; however, we omit the "hat" for ease of notation.

and the R^2 statistic are identical. However, in the next section we will discuss the multiple linear regression problem, in which we use several predictors simultaneously to predict the response. The concept of correlation between the predictors and the response does not extend automatically to this setting, since correlation quantifies the association between a single pair of variables rather than between a larger number of variables. We will see that R^2 fills this role.

3.2 Multiple Linear Regression

Simple linear regression is a useful approach for predicting a response on the basis of a single predictor variable. However, in practice we often have more than one predictor. For example, in the Advertising data, we have examined the relationship between sales and TV advertising. We also have data for the amount of money spent advertising on the radio and in newspapers, and we may want to know whether either of these two media is associated with sales. How can we extend our analysis of the advertising data in order to accommodate these two additional predictors?

One option is to run three separate simple linear regressions, each of which uses a different advertising medium as a predictor. For instance, we can fit a simple linear regression to predict sales on the basis of the amount spent on radio advertisements. Results are shown in Table 3.3 (top table). We find that a $1,000 increase in spending on radio advertising is associated with an increase in sales by around 203 units. Table 3.3 (bottom table) contains the least squares coefficients for a simple linear regression of sales onto newspaper advertising budget. A $1,000 increase in newspaper advertising budget is associated with an increase in sales by approximately 55 units.

However, the approach of fitting a separate simple linear regression model for each predictor is not entirely satisfactory. First of all, it is unclear how to make a single prediction of sales given levels of the three advertising media budgets, since each of the budgets is associated with a separate regression equation. Second, each of the three regression equations ignores the other two media in forming estimates for the regression coefficients. We will see shortly that if the media budgets are correlated with each other in the 200 markets that constitute our data set, then this can lead to very misleading estimates of the individual media effects on sales.

Instead of fitting a separate simple linear regression model for each predictor, a better approach is to extend the simple linear regression model (3.5) so that it can directly accommodate multiple predictors. We can do this by giving each predictor a separate slope coefficient in a single model. In general, suppose that we have p distinct predictors. Then the multiple linear regression model takes the form

$$Y = \beta_0 + \beta_1 X_1 + \beta_2 X_2 + \cdots + \beta_p X_p + \epsilon, \qquad (3.19)$$

Simple regression of sales on radio

	Coefficient	Std. error	t-statistic	p-value
Intercept	9.312	0.563	16.54	< 0.0001
radio	0.203	0.020	9.92	< 0.0001

Simple regression of sales on newspaper

	Coefficient	Std. error	t-statistic	p-value
Intercept	12.351	0.621	19.88	< 0.0001
newspaper	0.055	0.017	3.30	0.00115

TABLE 3.3. *More simple linear regression models for the* Advertising *data. Coefficients of the simple linear regression model for number of units sold on* Top: *radio advertising budget and* Bottom: *newspaper advertising budget. A $1,000 increase in spending on radio advertising is associated with an average increase in sales by around 203 units, while the same increase in spending on newspaper advertising is associated with an average increase in sales by around 55 units (Note that the* sales *variable is in thousands of units, and the* radio *and* newspaper *variables are in thousands of dollars).*

where X_j represents the jth predictor and β_j quantifies the association between that variable and the response. We interpret β_j as the *average* effect on Y of a one unit increase in X_j, *holding all other predictors fixed*. In the advertising example, (3.19) becomes

$$\texttt{sales} = \beta_0 + \beta_1 \times \texttt{TV} + \beta_2 \times \texttt{radio} + \beta_3 \times \texttt{newspaper} + \epsilon. \qquad (3.20)$$

3.2.1 Estimating the Regression Coefficients

As was the case in the simple linear regression setting, the regression coefficients $\beta_0, \beta_1, \ldots, \beta_p$ in (3.19) are unknown, and must be estimated. Given estimates $\hat{\beta}_0, \hat{\beta}_1, \ldots, \hat{\beta}_p$, we can make predictions using the formula

$$\hat{y} = \hat{\beta}_0 + \hat{\beta}_1 x_1 + \hat{\beta}_2 x_2 + \cdots + \hat{\beta}_p x_p. \qquad (3.21)$$

The parameters are estimated using the same least squares approach that we saw in the context of simple linear regression. We choose $\beta_0, \beta_1, \ldots, \beta_p$ to minimize the sum of squared residuals

$$
\begin{aligned}
\text{RSS} &= \sum_{i=1}^{n} (y_i - \hat{y}_i)^2 \\
&= \sum_{i=1}^{n} (y_i - \hat{\beta}_0 - \hat{\beta}_1 x_{i1} - \hat{\beta}_2 x_{i2} - \cdots - \hat{\beta}_p x_{ip})^2. \qquad (3.22)
\end{aligned}
$$

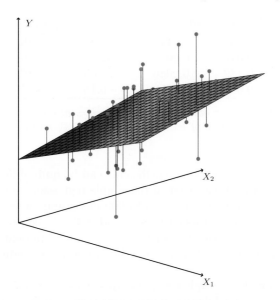

FIGURE 3.4. *In a three-dimensional setting, with two predictors and one response, the least squares regression line becomes a plane. The plane is chosen to minimize the sum of the squared vertical distances between each observation (shown in red) and the plane.*

The values $\hat{\beta}_0, \hat{\beta}_1, \ldots, \hat{\beta}_p$ that minimize (3.22) are the multiple least squares regression coefficient estimates. Unlike the simple linear regression estimates given in (3.4), the multiple regression coefficient estimates have somewhat complicated forms that are most easily represented using matrix algebra. For this reason, we do not provide them here. Any statistical software package can be used to compute these coefficient estimates, and later in this chapter we will show how this can be done in R. Figure 3.4 illustrates an example of the least squares fit to a toy data set with $p = 2$ predictors.

Table 3.4 displays the multiple regression coefficient estimates when TV, radio, and newspaper advertising budgets are used to predict product sales using the Advertising data. We interpret these results as follows: for a given amount of TV and newspaper advertising, spending an additional $1,000 on radio advertising leads to an increase in sales by approximately 189 units. Comparing these coefficient estimates to those displayed in Tables 3.1 and 3.3, we notice that the multiple regression coefficient estimates for TV and radio are pretty similar to the simple linear regression coefficient estimates. However, while the newspaper regression coefficient estimate in Table 3.3 was significantly non-zero, the coefficient estimate for newspaper in the multiple regression model is close to zero, and the corresponding p-value is no longer significant, with a value around 0.86. This illustrates

	Coefficient	Std. error	t-statistic	p-value
Intercept	2.939	0.3119	9.42	< 0.0001
TV	0.046	0.0014	32.81	< 0.0001
radio	0.189	0.0086	21.89	< 0.0001
newspaper	−0.001	0.0059	−0.18	0.8599

TABLE 3.4. *For the* Advertising *data, least squares coefficient estimates of the multiple linear regression of number of units sold on radio, TV, and newspaper advertising budgets.*

that the simple and multiple regression coefficients can be quite different. This difference stems from the fact that in the simple regression case, the slope term represents the average effect of a $1,000 increase in newspaper advertising, ignoring other predictors such as TV and radio. In contrast, in the multiple regression setting, the coefficient for newspaper represents the average effect of increasing newspaper spending by $1,000 while holding TV and radio fixed.

Does it make sense for the multiple regression to suggest no relationship between sales and newspaper while the simple linear regression implies the opposite? In fact it does. Consider the correlation matrix for the three predictor variables and response variable, displayed in Table 3.5. Notice that the correlation between radio and newspaper is 0.35. This reveals a tendency to spend more on newspaper advertising in markets where more is spent on radio advertising. Now suppose that the multiple regression is correct and newspaper advertising has no direct impact on sales, but radio advertising does increase sales. Then in markets where we spend more on radio our sales will tend to be higher, and as our correlation matrix shows, we also tend to spend more on newspaper advertising in those same markets. Hence, in a simple linear regression which only examines sales versus newspaper, we will observe that higher values of newspaper tend to be associated with higher values of sales, even though newspaper advertising does not actually affect sales. So newspaper sales are a surrogate for radio advertising; newspaper gets "credit" for the effect of radio on sales.

This slightly counterintuitive result is very common in many real life situations. Consider an absurd example to illustrate the point. Running a regression of shark attacks versus ice cream sales for data collected at a given beach community over a period of time would show a positive relationship, similar to that seen between sales and newspaper. Of course no one (yet) has suggested that ice creams should be banned at beaches to reduce shark attacks. In reality, higher temperatures cause more people to visit the beach, which in turn results in more ice cream sales and more shark attacks. A multiple regression of attacks versus ice cream sales and temperature reveals that, as intuition implies, the former predictor is no longer significant after adjusting for temperature.

	TV	radio	newspaper	sales
TV	1.0000	0.0548	0.0567	0.7822
radio		1.0000	0.3541	0.5762
newspaper			1.0000	0.2283
sales				1.0000

TABLE 3.5. *Correlation matrix for* TV, radio, newspaper, *and* sales *for the* Advertising *data.*

3.2.2 Some Important Questions

When we perform multiple linear regression, we usually are interested in answering a few important questions.

1. *Is at least one of the predictors $X_1, X_2, \ldots, X_p$ useful in predicting the response?*

2. *Do all the predictors help to explain Y, or is only a subset of the predictors useful?*

3. *How well does the model fit the data?*

4. *Given a set of predictor values, what response value should we predict, and how accurate is our prediction?*

We now address each of these questions in turn.

One: Is There a Relationship Between the Response and Predictors?

Recall that in the simple linear regression setting, in order to determine whether there is a relationship between the response and the predictor we can simply check whether $\beta_1 = 0$. In the multiple regression setting with p predictors, we need to ask whether all of the regression coefficients are zero, i.e. whether $\beta_1 = \beta_2 = \cdots = \beta_p = 0$. As in the simple linear regression setting, we use a hypothesis test to answer this question. We test the null hypothesis,

$$H_0 : \beta_1 = \beta_2 = \cdots = \beta_p = 0$$

versus the alternative

$$H_a : \text{ at least one } \beta_j \text{ is non-zero.}$$

This hypothesis test is performed by computing the *F-statistic*,

F-statistic

$$F = \frac{(\text{TSS} - \text{RSS})/p}{\text{RSS}/(n - p - 1)}, \tag{3.23}$$

Quantity	Value
Residual standard error	1.69
R^2	0.897
F-statistic	570

TABLE 3.6. *More information about the least squares model for the regression of number of units sold on TV, newspaper, and radio advertising budgets in the* Advertising *data. Other information about this model was displayed in Table 3.4.*

where, as with simple linear regression, TSS $= \sum(y_i - \bar{y})^2$ and RSS $= \sum(y_i - \hat{y}_i)^2$. If the linear model assumptions are correct, one can show that

$$E\{\text{RSS}/(n - p - 1)\} = \sigma^2$$

and that, provided H_0 is true,

$$E\{(\text{TSS} - \text{RSS})/p\} = \sigma^2.$$

Hence, when there is no relationship between the response and predictors, one would expect the F-statistic to take on a value close to 1. On the other hand, if H_a is true, then $E\{(\text{TSS} - \text{RSS})/p\} > \sigma^2$, so we expect F to be greater than 1.

The F-statistic for the multiple linear regression model obtained by regressing sales onto radio, TV, and newspaper is shown in Table 3.6. In this example the F-statistic is 570. Since this is far larger than 1, it provides compelling evidence against the null hypothesis H_0. In other words, the large F-statistic suggests that at least one of the advertising media must be related to sales. However, what if the F-statistic had been closer to 1? How large does the F-statistic need to be before we can reject H_0 and conclude that there is a relationship? It turns out that the answer depends on the values of n and p. When n is large, an F-statistic that is just a little larger than 1 might still provide evidence against H_0. In contrast, a larger F-statistic is needed to reject H_0 if n is small. When H_0 is true and the errors ϵ_i have a normal distribution, the F-statistic follows an F-distribution.[6] For any given value of n and p, any statistical software package can be used to compute the p-value associated with the F-statistic using this distribution. Based on this p-value, we can determine whether or not to reject H_0. For the advertising data, the p-value associated with the F-statistic in Table 3.6 is essentially zero, so we have extremely strong evidence that at least one of the media is associated with increased sales.

In (3.23) we are testing H_0 that all the coefficients are zero. Sometimes we want to test that a particular subset of q of the coefficients are zero. This corresponds to a null hypothesis

$$H_0: \quad \beta_{p-q+1} = \beta_{p-q+2} = \ldots = \beta_p = 0,$$

---

[6]Even if the errors are not normally-distributed, the F-statistic approximately follows an F-distribution provided that the sample size n is large.

where for convenience we have put the variables chosen for omission at the end of the list. In this case we fit a second model that uses all the variables *except* those last q. Suppose that the residual sum of squares for that model is RSS_0. Then the appropriate F-statistic is

$$F = \frac{(RSS_0 - RSS)/q}{RSS/(n - p - 1)}. \qquad (3.24)$$

Notice that in Table 3.4, for each individual predictor a t-statistic and a p-value were reported. These provide information about whether each individual predictor is related to the response, after adjusting for the other predictors. It turns out that each of these are exactly equivalent[7] to the F-test that omits that single variable from the model, leaving all the others in—i.e. $q=1$ in (3.24). So it reports the *partial effect* of adding that variable to the model. For instance, as we discussed earlier, these p-values indicate that TV and radio are related to sales, but that there is no evidence that newspaper is associated with sales, in the presence of these two.

Given these individual p-values for each variable, why do we need to look at the overall F-statistic? After all, it seems likely that if any one of the p-values for the individual variables is very small, then *at least one of the predictors is related to the response*. However, this logic is flawed, especially when the number of predictors p is large.

For instance, consider an example in which $p = 100$ and $H_0 : \beta_1 = \beta_2 = \ldots = \beta_p = 0$ is true, so no variable is truly associated with the response. In this situation, about 5 % of the p-values associated with each variable (of the type shown in Table 3.4) will be below 0.05 by chance. In other words, we expect to see approximately five *small* p-values even in the absence of any true association between the predictors and the response. In fact, we are almost guaranteed that we will observe at least one p-value below 0.05 by chance! Hence, if we use the individual t-statistics and associated p-values in order to decide whether or not there is any association between the variables and the response, there is a very high chance that we will incorrectly conclude that there is a relationship. However, the F-statistic does not suffer from this problem because it adjusts for the number of predictors. Hence, if H_0 is true, there is only a 5 % chance that the F-statistic will result in a p-value below 0.05, regardless of the number of predictors or the number of observations.

The approach of using an F-statistic to test for any association between the predictors and the response works when p is relatively small, and certainly small compared to n. However, sometimes we have a very large number of variables. If $p > n$ then there are more coefficients β_j to estimate than observations from which to estimate them. In this case we cannot even fit the multiple linear regression model using least squares, so the

---

[7]The square of each t-statistic is the corresponding F-statistic.

F-statistic cannot be used, and neither can most of the other concepts that we have seen so far in this chapter. When p is large, some of the approaches discussed in the next section, such as *forward selection*, can be used. This *high-dimensional* setting is discussed in greater detail in Chapter 6.

high-dimensional

Two: Deciding on Important Variables

As discussed in the previous section, the first step in a multiple regression analysis is to compute the F-statistic and to examine the associated p-value. If we conclude on the basis of that p-value that at least one of the predictors is related to the response, then it is natural to wonder *which* are the guilty ones! We could look at the individual p-values as in Table 3.4, but as discussed, if p is large we are likely to make some false discoveries.

It is possible that all of the predictors are associated with the response, but it is more often the case that the response is only related to a subset of the predictors. The task of determining which predictors are associated with the response, in order to fit a single model involving only those predictors, is referred to as *variable selection*. The variable selection problem is studied extensively in Chapter 6, and so here we will provide only a brief outline of some classical approaches.

variable selection

Ideally, we would like to perform variable selection by trying out a lot of different models, each containing a different subset of the predictors. For instance, if $p = 2$, then we can consider four models: (1) a model containing no variables, (2) a model containing X_1 only, (3) a model containing X_2 only, and (4) a model containing both X_1 and X_2. We can then select the *best* model out of all of the models that we have considered. How do we determine which model is best? Various statistics can be used to judge the quality of a model. These include *Mallow's C_p, Akaike information criterion* (AIC), *Bayesian information criterion* (BIC), and *adjusted R^2*. These are discussed in more detail in Chapter 6. We can also determine which model is best by plotting various model outputs, such as the residuals, in order to search for patterns.

Mallow's C_p

Akaike information criterion

Bayesian information criterion

adjusted R^2

Unfortunately, there are a total of 2^p models that contain subsets of p variables. This means that even for moderate p, trying out every possible subset of the predictors is infeasible. For instance, we saw that if $p = 2$, then there are $2^2 = 4$ models to consider. But if $p = 30$, then we must consider $2^{30} = 1,073,741,824$ models! This is not practical. Therefore, unless p is very small, we cannot consider all 2^p models, and instead we need an automated and efficient approach to choose a smaller set of models to consider. There are three classical approaches for this task:

- *Forward selection.* We begin with the *null model*—a model that contains an intercept but no predictors. We then fit p simple linear regressions and add to the null model the variable that results in the lowest RSS. We then add to that model the variable that results

forward selection

null model

in the lowest RSS for the new two-variable model. This approach is continued until some stopping rule is satisfied.

- *Backward selection.* We start with all variables in the model, and remove the variable with the largest p-value—that is, the variable that is the least statistically significant. The new $(p - 1)$-variable model is fit, and the variable with the largest p-value is removed. This procedure continues until a stopping rule is reached. For instance, we may stop when all remaining variables have a p-value below some threshold.

  <small>backward selection</small>

- *Mixed selection.* This is a combination of forward and backward selection. We start with no variables in the model, and as with forward selection, we add the variable that provides the best fit. We continue to add variables one-by-one. Of course, as we noted with the Advertising example, the p-values for variables can become larger as new predictors are added to the model. Hence, if at any point the p-value for one of the variables in the model rises above a certain threshold, then we remove that variable from the model. We continue to perform these forward and backward steps until all variables in the model have a sufficiently low p-value, and all variables outside the model would have a large p-value if added to the model.

  <small>mixed selection</small>

Backward selection cannot be used if $p > n$, while forward selection can always be used. Forward selection is a greedy approach, and might include variables early that later become redundant. Mixed selection can remedy this.

Three: Model Fit

Two of the most common numerical measures of model fit are the RSE and R^2, the fraction of variance explained. These quantities are computed and interpreted in the same fashion as for simple linear regression.

Recall that in simple regression, R^2 is the square of the correlation of the response and the variable. In multiple linear regression, it turns out that it equals $\text{Cor}(Y, \hat{Y})^2$, the square of the correlation between the response and the fitted linear model; in fact one property of the fitted linear model is that it maximizes this correlation among all possible linear models.

An R^2 value close to 1 indicates that the model explains a large portion of the variance in the response variable. As an example, we saw in Table 3.6 that for the Advertising data, the model that uses all three advertising media to predict sales has an R^2 of 0.8972. On the other hand, the model that uses only TV and radio to predict sales has an R^2 value of 0.89719. In other words, there is a *small* increase in R^2 if we include newspaper advertising in the model that already contains TV and radio advertising, even though we saw earlier that the p-value for newspaper advertising in Table 3.4 is not significant. It turns out that R^2 will always increase when more variables are added to the model, even if those variables are only weakly associated with the response. This is due to the fact that adding another variable to the least squares equations must allow us to fit the training data (though not necessarily the testing data) more accurately. Thus, the R^2 statistic, which is also computed on the training data, must increase. The fact that adding newspaper advertising to the model containing only TV and radio advertising leads to just a tiny increase in R^2 provides additional evidence that newspaper can be dropped from the model. Essentially, newspaper provides no real improvement in the model fit to the training samples, and its inclusion will likely lead to poor results on independent test samples due to overfitting.

In contrast, the model containing only TV as a predictor had an R^2 of 0.61 (Table 3.2). Adding radio to the model leads to a substantial improvement in R^2. This implies that a model that uses TV and radio expenditures to predict sales is substantially better than one that uses only TV advertising. We could further quantify this improvement by looking at the p-value for the radio coefficient in a model that contains only TV and radio as predictors.

The model that contains only TV and radio as predictors has an RSE of 1.681, and the model that also contains newspaper as a predictor has an RSE of 1.686 (Table 3.6). In contrast, the model that contains only TV has an RSE of 3.26 (Table 3.2). This corroborates our previous conclusion that a model that uses TV and radio expenditures to predict sales is much more accurate (on the training data) than one that only uses TV spending. Furthermore, given that TV and radio expenditures are used as predictors, there is no point in also using newspaper spending as a predictor in the model. The observant reader may wonder how RSE can increase when newspaper is added to the model given that RSS must decrease. In general RSE is defined as

$$\text{RSE} = \sqrt{\frac{1}{n-p-1}\text{RSS}}, \tag{3.25}$$

which simplifies to (3.15) for a simple linear regression. Thus, models with more variables can have higher RSE if the decrease in RSS is small relative to the increase in p.

In addition to looking at the RSE and R^2 statistics just discussed, it can be useful to plot the data. Graphical summaries can reveal problems with a model that are not visible from numerical statistics. For example, Figure 3.5 displays a three-dimensional plot of TV and radio versus sales. We see that some observations lie above and some observations lie below the least squares regression plane. In particular, the linear model seems to overestimate sales for instances in which most of the advertising money was spent exclusively on either TV or radio. It underestimates sales for instances where the budget was split between the two media. This pronounced non-linear pattern cannot be modeled accurately using linear re-

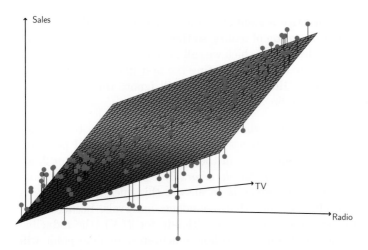

FIGURE 3.5. *For the* Advertising *data, a linear regression fit to* sales *using* TV *and* radio *as predictors. From the pattern of the residuals, we can see that there is a pronounced non-linear relationship in the data. The positive residuals (those visible above the surface), tend to lie along the 45-degree line, where TV and Radio budgets are split evenly. The negative residuals (most not visible), tend to lie away from this line, where budgets are more lopsided.*

gression. It suggests a *synergy* or *interaction* effect between the advertising media, whereby combining the media together results in a bigger boost to sales than using any single medium. In Section 3.3.2, we will discuss extending the linear model to accommodate such synergistic effects through the use of interaction terms.

Four: Predictions

Once we have fit the multiple regression model, it is straightforward to apply (3.21) in order to predict the response Y on the basis of a set of values for the predictors $X_1, X_2, \ldots, X_p$. However, there are three sorts of uncertainty associated with this prediction.

1. The coefficient estimates $\hat{\beta}_0, \hat{\beta}_1, \ldots, \hat{\beta}_p$ are estimates for $\beta_0, \beta_1, \ldots, \beta_p$. That is, the *least squares plane*

$$\hat{Y} = \hat{\beta}_0 + \hat{\beta}_1 X_1 + \cdots + \hat{\beta}_p X_p$$

is only an estimate for the *true population regression plane*

$$f(X) = \beta_0 + \beta_1 X_1 + \cdots + \beta_p X_p.$$

The inaccuracy in the coefficient estimates is related to the *reducible error* from Chapter 2. We can compute a *confidence interval* in order to determine how close $\hat{Y}$ will be to $f(X)$.

2. Of course, in practice assuming a linear model for $f(X)$ is almost always an approximation of reality, so there is an additional source of potentially reducible error which we call *model bias*. So when we use a linear model, we are in fact estimating the best linear approximation to the true surface. However, here we will ignore this discrepancy, and operate as if the linear model were correct.

3. Even if we knew $f(X)$—that is, even if we knew the true values for $\beta_0, \beta_1, \ldots, \beta_p$—the response value cannot be predicted perfectly because of the random error ϵ in the model (3.21). In Chapter 2, we referred to this as the *irreducible error*. How much will Y vary from $\hat{Y}$? We use *prediction intervals* to answer this question. Prediction intervals are always wider than confidence intervals, because they incorporate both the error in the estimate for $f(X)$ (the reducible error) and the uncertainty as to how much an individual point will differ from the population regression plane (the irreducible error).

We use a *confidence interval* to quantify the uncertainty surrounding the *average* sales over a large number of cities. For example, given that $100,000 is spent on TV advertising and $20,000 is spent on radio advertising in each city, the 95 % confidence interval is [10,985, 11,528]. We interpret this to mean that 95 % of intervals of this form will contain the true value of $f(X)$.[8] On the other hand, a *prediction interval* can be used to quantify the uncertainty surrounding sales for a *particular* city. Given that $100,000 is spent on TV advertising and $20,000 is spent on radio advertising in that city the 95 % prediction interval is [7,930, 14,580]. We interpret this to mean that 95 % of intervals of this form will contain the true value of Y for this city. Note that both intervals are centered at 11,256, but that the prediction interval is substantially wider than the confidence interval, reflecting the increased uncertainty about sales for a given city in comparison to the average sales over many locations.

confidence interval

prediction interval

3.3   Other Considerations in the Regression Model

3.3.1   Qualitative Predictors

In our discussion so far, we have assumed that all variables in our linear regression model are *quantitative*. But in practice, this is not necessarily the case; often some predictors are *qualitative*.

---

[8] In other words, if we collect a large number of data sets like the Advertising data set, and we construct a confidence interval for the average sales on the basis of each data set (given $100,000 in TV and $20,000 in radio advertising), then 95 % of these confidence intervals will contain the true value of average sales.

For example, the Credit data set displayed in Figure 3.6 records balance (average credit card debt for a number of individuals) as well as several quantitative predictors: age, cards (number of credit cards), education (years of education), income (in thousands of dollars), limit (credit limit), and rating (credit rating). Each panel of Figure 3.6 is a scatterplot for a pair of variables whose identities are given by the corresponding row and column labels. For example, the scatterplot directly to the right of the word "Balance" depicts balance versus age, while the plot directly to the right of "Age" corresponds to age versus cards. In addition to these quantitative variables, we also have four qualitative variables: gender, student (student status), status (marital status), and ethnicity (Caucasian, African American or Asian).

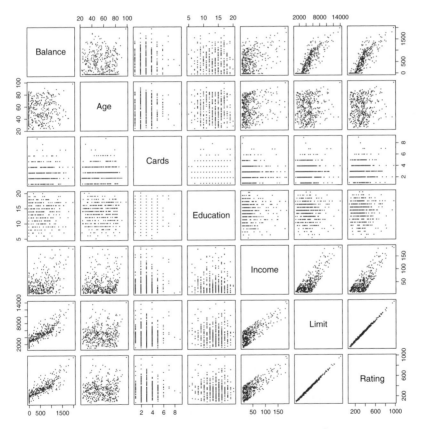

FIGURE 3.6. *The* Credit *data set contains information about* balance, age, cards, education, income, limit, *and* rating *for a number of potential customers.*

	Coefficient	Std. error	t-statistic	p-value
Intercept	509.80	33.13	15.389	< 0.0001
gender[Female]	19.73	46.05	0.429	0.6690

TABLE 3.7. *Least squares coefficient estimates associated with the regression of* balance *onto* gender *in the* Credit *data set. The linear model is given in (3.27). That is,* gender *is encoded as a dummy variable, as in (3.26).*

Predictors with Only Two Levels

Suppose that we wish to investigate differences in credit card balance between males and females, ignoring the other variables for the moment. If a qualitative predictor (also known as a *factor*) only has two *levels*, or possible values, then incorporating it into a regression model is very simple. We simply create an indicator or *dummy variable* that takes on two possible numerical values. For example, based on the gender variable, we can create a new variable that takes the form

$$x_i = \begin{cases} 1 & \text{if } i\text{th person is female} \\ 0 & \text{if } i\text{th person is male,} \end{cases} \tag{3.26}$$

and use this variable as a predictor in the regression equation. This results in the model

$$y_i = \beta_0 + \beta_1 x_i + \epsilon_i = \begin{cases} \beta_0 + \beta_1 + \epsilon_i & \text{if } i\text{th person is female} \\ \beta_0 + \epsilon_i & \text{if } i\text{th person is male.} \end{cases} \tag{3.27}$$

Now β_0 can be interpreted as the average credit card balance among males, $\beta_0 + \beta_1$ as the average credit card balance among females, and β_1 as the average difference in credit card balance between females and males.

Table 3.7 displays the coefficient estimates and other information associated with the model (3.27). The average credit card debt for males is estimated to be \$509.80, whereas females are estimated to carry \$19.73 in additional debt for a total of \$509.80 + \$19.73 = \$529.53. However, we notice that the p-value for the dummy variable is very high. This indicates that there is no statistical evidence of a difference in average credit card balance between the genders.

The decision to code females as 1 and males as 0 in (3.27) is arbitrary, and has no effect on the regression fit, but does alter the interpretation of the coefficients. If we had coded males as 1 and females as 0, then the estimates for β_0 and β_1 would have been 529.53 and -19.73, respectively, leading once again to a prediction of credit card debt of \$529.53 − \$19.73 = \$509.80 for males and a prediction of \$529.53 for females. Alternatively, instead of a 0/1 coding scheme, we could create a dummy variable

$$x_i = \begin{cases} 1 & \text{if } i\text{th person is female} \\ -1 & \text{if } i\text{th person is male} \end{cases}$$

and use this variable in the regression equation. This results in the model

$$y_i = \beta_0 + \beta_1 x_i + \epsilon_i = \begin{cases} \beta_0 + \beta_1 + \epsilon_i & \text{if } i\text{th person is female} \\ \beta_0 - \beta_1 + \epsilon_i & \text{if } i\text{th person is male.} \end{cases}$$

Now β_0 can be interpreted as the overall average credit card balance (ignoring the gender effect), and β_1 is the amount that females are above the average and males are below the average. In this example, the estimate for β_0 would be \$519.665, halfway between the male and female averages of \$509.80 and \$529.53. The estimate for β_1 would be \$9.865, which is half of \$19.73, the average difference between females and males. It is important to note that the final predictions for the credit balances of males and females will be identical regardless of the coding scheme used. The only difference is in the way that the coefficients are interpreted.

Qualitative Predictors with More than Two Levels

When a qualitative predictor has more than two levels, a single dummy variable cannot represent all possible values. In this situation, we can create additional dummy variables. For example, for the `ethnicity` variable we create two dummy variables. The first could be

$$x_{i1} = \begin{cases} 1 & \text{if } i\text{th person is Asian} \\ 0 & \text{if } i\text{th person is not Asian,} \end{cases} \tag{3.28}$$

and the second could be

$$x_{i2} = \begin{cases} 1 & \text{if } i\text{th person is Caucasian} \\ 0 & \text{if } i\text{th person is not Caucasian.} \end{cases} \tag{3.29}$$

Then both of these variables can be used in the regression equation, in order to obtain the model

$$y_i = \beta_0 + \beta_1 x_{i1} + \beta_2 x_{i2} + \epsilon_i = \begin{cases} \beta_0 + \beta_1 + \epsilon_i & \text{if } i\text{th person is Asian} \\ \beta_0 + \beta_2 + \epsilon_i & \text{if } i\text{th person is Caucasian} \\ \beta_0 + \epsilon_i & \text{if } i\text{th person is African American.} \end{cases} \tag{3.30}$$

Now β_0 can be interpreted as the average credit card balance for African Americans, β_1 can be interpreted as the difference in the average balance between the Asian and African American categories, and β_2 can be interpreted as the difference in the average balance between the Caucasian and

	Coefficient	Std. error	t-statistic	p-value
Intercept	531.00	46.32	11.464	< 0.0001
ethnicity[Asian]	−18.69	65.02	−0.287	0.7740
ethnicity[Caucasian]	−12.50	56.68	−0.221	0.8260

TABLE 3.8. *Least squares coefficient estimates associated with the regression of* balance *onto* ethnicity *in the* Credit *data set. The linear model is given in (3.30). That is, ethnicity is encoded via two dummy variables (3.28) and (3.29).*

African American categories. There will always be one fewer dummy variable than the number of levels. The level with no dummy variable—African American in this example—is known as the *baseline*.

From Table 3.8, we see that the estimated balance for the baseline, African American, is \$531.00. It is estimated that the Asian category will have \$18.69 less debt than the African American category, and that the Caucasian category will have \$12.50 less debt than the African American category. However, the p-values associated with the coefficient estimates for the two dummy variables are very large, suggesting no statistical evidence of a real difference in credit card balance between the ethnicities. Once again, the level selected as the baseline category is arbitrary, and the final predictions for each group will be the same regardless of this choice. However, the coefficients and their p-values do depend on the choice of dummy variable coding. Rather than rely on the individual coefficients, we can use an F-test to test $H_0 : \beta_1 = \beta_2 = 0$; this does not depend on the coding. This F-test has a p-value of 0.96, indicating that we cannot reject the null hypothesis that there is no relationship between balance and ethnicity.

Using this dummy variable approach presents no difficulties when incorporating both quantitative and qualitative predictors. For example, to regress balance on both a quantitative variable such as income and a qualitative variable such as student, we must simply create a dummy variable for student and then fit a multiple regression model using income and the dummy variable as predictors for credit card balance.

There are many different ways of coding qualitative variables besides the dummy variable approach taken here. All of these approaches lead to equivalent model fits, but the coefficients are different and have different interpretations, and are designed to measure particular *contrasts*. This topic is beyond the scope of the book, and so we will not pursue it further.

3.3.2 Extensions of the Linear Model

The standard linear regression model (3.19) provides interpretable results and works quite well on many real-world problems. However, it makes several highly restrictive assumptions that are often violated in practice. Two of the most important assumptions state that the relationship between the predictors and response are *additive* and *linear*. The additive assumption means that the effect of changes in a predictor X_j on the response Y is independent of the values of the other predictors. The linear assumption states that the change in the response Y due to a one-unit change in X_j is constant, regardless of the value of X_j. In this book, we examine a number of sophisticated methods that relax these two assumptions. Here, we briefly examine some common classical approaches for extending the linear model.

Removing the Additive Assumption

In our previous analysis of the Advertising data, we concluded that both TV and radio seem to be associated with sales. The linear models that formed the basis for this conclusion assumed that the effect on sales of increasing one advertising medium is independent of the amount spent on the other media. For example, the linear model (3.20) states that the average effect on sales of a one-unit increase in TV is always β_1, regardless of the amount spent on radio.

However, this simple model may be incorrect. Suppose that spending money on radio advertising actually increases the effectiveness of TV advertising, so that the slope term for TV should increase as radio increases. In this situation, given a fixed budget of \$100,000, spending half on radio and half on TV may increase sales more than allocating the entire amount to either TV or to radio. In marketing, this is known as a *synergy* effect, and in statistics it is referred to as an *interaction* effect. Figure 3.5 suggests that such an effect may be present in the advertising data. Notice that when levels of either TV or radio are low, then the true sales are lower than predicted by the linear model. But when advertising is split between the two media, then the model tends to underestimate sales.

Consider the standard linear regression model with two variables,

$$Y = \beta_0 + \beta_1 X_1 + \beta_2 X_2 + \epsilon.$$

According to this model, if we increase X_1 by one unit, then Y will increase by an average of β_1 units. Notice that the presence of X_2 does not alter this statement—that is, regardless of the value of X_2, a one-unit increase in X_1 will lead to a β_1-unit increase in Y. One way of extending this model to allow for interaction effects is to include a third predictor, called an *interaction term*, which is constructed by computing the product of X_1 and X_2. This results in the model

$$Y = \beta_0 + \beta_1 X_1 + \beta_2 X_2 + \beta_3 X_1 X_2 + \epsilon. \tag{3.31}$$

How does inclusion of this interaction term relax the additive assumption? Notice that (3.31) can be rewritten as

$$\begin{aligned} Y &= \beta_0 + (\beta_1 + \beta_3 X_2) X_1 + \beta_2 X_2 + \epsilon \\ &= \beta_0 + \tilde{\beta}_1 X_1 + \beta_2 X_2 + \epsilon \end{aligned} \tag{3.32}$$

	Coefficient	Std. error	t-statistic	p-value
Intercept	6.7502	0.248	27.23	< 0.0001
TV	0.0191	0.002	12.70	< 0.0001
radio	0.0289	0.009	3.24	0.0014
TV×radio	0.0011	0.000	20.73	< 0.0001

TABLE 3.9. *For the* Advertising *data, least squares coefficient estimates associated with the regression of* sales *onto* TV *and* radio, *with an interaction term, as in (3.33).*

where $\tilde{\beta}_1 = \beta_1 + \beta_3 X_2$. Since $\tilde{\beta}_1$ changes with X_2, the effect of X_1 on Y is no longer constant: adjusting X_2 will change the impact of X_1 on Y.

For example, suppose that we are interested in studying the productivity of a factory. We wish to predict the number of units produced on the basis of the number of production lines and the total number of workers. It seems likely that the effect of increasing the number of production lines will depend on the number of workers, since if no workers are available to operate the lines, then increasing the number of lines will not increase production. This suggests that it would be appropriate to include an interaction term between lines and workers in a linear model to predict units. Suppose that when we fit the model, we obtain

$$
\begin{aligned}
\text{units} &\approx 1.2 + 3.4 \times \text{lines} + 0.22 \times \text{workers} + 1.4 \times (\text{lines} \times \text{workers}) \\
&= 1.2 + (3.4 + 1.4 \times \text{workers}) \times \text{lines} + 0.22 \times \text{workers}.
\end{aligned}
$$

In other words, adding an additional line will increase the number of units produced by $3.4 + 1.4 \times$ workers. Hence the more workers we have, the stronger will be the effect of lines.

We now return to the Advertising example. A linear model that uses radio, TV, and an interaction between the two to predict sales takes the form

$$
\begin{aligned}
\text{sales} &= \beta_0 + \beta_1 \times \text{TV} + \beta_2 \times \text{radio} + \beta_3 \times (\text{radio} \times \text{TV}) + \epsilon \\
&= \beta_0 + (\beta_1 + \beta_3 \times \text{radio}) \times \text{TV} + \beta_2 \times \text{radio} + \epsilon. \qquad (3.33)
\end{aligned}
$$

We can interpret β_3 as the increase in the effectiveness of TV advertising for a one unit increase in radio advertising (or vice-versa). The coefficients that result from fitting the model (3.33) are given in Table 3.9.

The results in Table 3.9 strongly suggest that the model that includes the interaction term is superior to the model that contains only *main effects*. The p-value for the interaction term, TV×radio, is extremely low, indicating that there is strong evidence for $H_a : \beta_3 \neq 0$. In other words, it is clear that the true relationship is not additive. The R^2 for the model (3.33) is 96.8 %, compared to only 89.7 % for the model that predicts sales using TV and radio without an interaction term. This means that $(96.8 - 89.7)/(100 - 89.7) = 69\%$ of the variability in sales that remains after fitting the additive model has been explained by the interaction term. The coefficient

main effect

estimates in Table 3.9 suggest that an increase in TV advertising of \$1,000 is associated with increased sales of $(\hat{\beta}_1 + \hat{\beta}_3 \times \texttt{radio}) \times 1{,}000 = 19 + 1.1 \times \texttt{radio}$ units. And an increase in radio advertising of \$1,000 will be associated with an increase in sales of $(\hat{\beta}_2 + \hat{\beta}_3 \times \texttt{TV}) \times 1{,}000 = 29 + 1.1 \times \texttt{TV}$ units.

In this example, the p-values associated with TV, radio, and the interaction term all are statistically significant (Table 3.9), and so it is obvious that all three variables should be included in the model. However, it is sometimes the case that an interaction term has a very small p-value, but the associated main effects (in this case, TV and radio) do not. The *hierarchical principle* states that *if we include an interaction in a model, we should also include the main effects, even if the p-values associated with their coefficients are not significant.* In other words, if the interaction between X_1 and X_2 seems important, then we should include both X_1 and X_2 in the model even if their coefficient estimates have large p-values. The rationale for this principle is that if $X_1 \times X_2$ is related to the response, then whether or not the coefficients of X_1 or X_2 are exactly zero is of little interest. Also $X_1 \times X_2$ is typically correlated with X_1 and X_2, and so leaving them out tends to alter the meaning of the interaction.

In the previous example, we considered an interaction between TV and radio, both of which are quantitative variables. However, the concept of interactions applies just as well to qualitative variables, or to a combination of quantitative and qualitative variables. In fact, an interaction between a qualitative variable and a quantitative variable has a particularly nice interpretation. Consider the Credit data set from Section 3.3.1, and suppose that we wish to predict balance using the income (quantitative) and student (qualitative) variables. In the absence of an interaction term, the model takes the form

$$
\begin{aligned}
\texttt{balance}_i &\approx \beta_0 + \beta_1 \times \texttt{income}_i +
\begin{cases}
\beta_2 & \text{if } i\text{th person is a student} \\
0 & \text{if } i\text{th person is not a student}
\end{cases} \\[2mm]
&= \beta_1 \times \texttt{income}_i +
\begin{cases}
\beta_0 + \beta_2 & \text{if } i\text{th person is a student} \\
\beta_0 & \text{if } i\text{th person is not a student.}
\end{cases}
\end{aligned}
$$

(3.34)

Notice that this amounts to fitting two parallel lines to the data, one for students and one for non-students. The lines for students and non-students have different intercepts, $\beta_0 + \beta_2$ versus β_0, but the same slope, β_1. This is illustrated in the left-hand panel of Figure 3.7. The fact that the lines are parallel means that the average effect on balance of a one-unit increase in income does not depend on whether or not the individual is a student. This represents a potentially serious limitation of the model, since in fact a change in income may have a very different effect on the credit card balance of a student versus a non-student.

This limitation can be addressed by adding an interaction variable, created by multiplying income with the dummy variable for student. Our

hierarchical principle

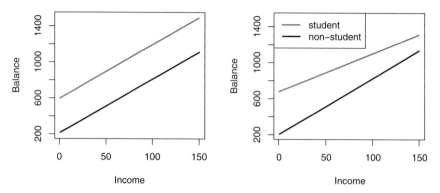

FIGURE 3.7. *For the* Credit *data, the least squares lines are shown for prediction of* balance *from* income *for students and non-students. Left: The model (3.34) was fit. There is no interaction between* income *and* student. *Right: The model (3.35) was fit. There is an interaction term between* income *and* student.

model now becomes

$$
\begin{aligned}
\texttt{balance}_i \;\approx\; & \beta_0 + \beta_1 \times \texttt{income}_i +
\begin{cases}
\beta_2 + \beta_3 \times \texttt{income}_i & \text{if student} \\
0 & \text{if not student}
\end{cases} \\[2mm]
=\; &
\begin{cases}
(\beta_0 + \beta_2) + (\beta_1 + \beta_3) \times \texttt{income}_i & \text{if student} \\
\beta_0 + \beta_1 \times \texttt{income}_i & \text{if not student}
\end{cases}
\end{aligned}
\tag{3.35}
$$

Once again, we have two different regression lines for the students and the non-students. But now those regression lines have different intercepts, $\beta_0 + \beta_2$ versus β_0, as well as different slopes, $\beta_1 + \beta_3$ versus β_1. This allows for the possibility that changes in income may affect the credit card balances of students and non-students differently. The right-hand panel of Figure 3.7 shows the estimated relationships between income and balance for students and non-students in the model (3.35). We note that the slope for students is lower than the slope for non-students. This suggests that increases in income are associated with smaller increases in credit card balance among students as compared to non-students.

Non-linear Relationships

As discussed previously, the linear regression model (3.19) assumes a linear relationship between the response and predictors. But in some cases, the true relationship between the response and the predictors may be non-linear. Here we present a very simple way to directly extend the linear model to accommodate non-linear relationships, using *polynomial regression*. In later chapters, we will present more complex approaches for performing non-linear fits in more general settings.

polynomial regression

Consider Figure 3.8, in which the mpg (gas mileage in miles per gallon) versus horsepower is shown for a number of cars in the Auto data set. The

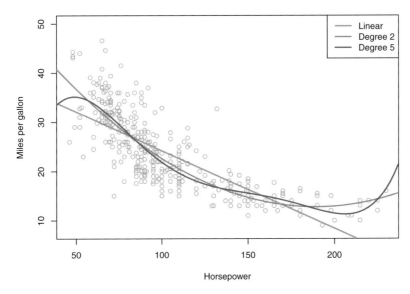

FIGURE 3.8. *The* `Auto` *data set. For a number of cars,* `mpg` *and* `horsepower` *are shown. The linear regression fit is shown in orange. The linear regression fit for a model that includes* `horsepower`^{2} *is shown as a blue curve. The linear regression fit for a model that includes all polynomials of* `horsepower` *up to fifth-degree is shown in green.*

orange line represents the linear regression fit. There is a pronounced relationship between `mpg` and `horsepower`, but it seems clear that this relationship is in fact non-linear: the data suggest a curved relationship. A simple approach for incorporating non-linear associations in a linear model is to include transformed versions of the predictors in the model. For example, the points in Figure 3.8 seem to have a *quadratic* shape, suggesting that a model of the form

quadratic

$$\texttt{mpg} = \beta_0 + \beta_1 \times \texttt{horsepower} + \beta_2 \times \texttt{horsepower}^2 + \epsilon \qquad (3.36)$$

may provide a better fit. Equation 3.36 involves predicting `mpg` using a non-linear function of `horsepower`. *But it is still a linear model!* That is, (3.36) is simply a multiple linear regression model with $X_1 = $ `horsepower` and $X_2 = $ `horsepower`^{2}. So we can use standard linear regression software to estimate $\beta_0, \beta_1,$ and β_2 in order to produce a non-linear fit. The blue curve in Figure 3.8 shows the resulting quadratic fit to the data. The quadratic fit appears to be substantially better than the fit obtained when just the linear term is included. The R^2 of the quadratic fit is 0.688, compared to 0.606 for the linear fit, and the p-value in Table 3.10 for the quadratic term is highly significant.

If including `horsepower`^{2} led to such a big improvement in the model, why not include `horsepower`^{3}, `horsepower`^{4}, or even `horsepower`^{5}? The green curve

	Coefficient	Std. error	t-statistic	p-value
Intercept	56.9001	1.8004	31.6	< 0.0001
horsepower	−0.4662	0.0311	−15.0	< 0.0001
horsepower^2	0.0012	0.0001	10.1	< 0.0001

TABLE 3.10. *For the* Auto *data set, least squares coefficient estimates associated with the regression of* mpg *onto* horsepower *and* horsepower^2.

in Figure 3.8 displays the fit that results from including all polynomials up to fifth degree in the model (3.36). The resulting fit seems unnecessarily wiggly—that is, it is unclear that including the additional terms really has led to a better fit to the data.

The approach that we have just described for extending the linear model to accommodate non-linear relationships is known as *polynomial regression*, since we have included polynomial functions of the predictors in the regression model. We further explore this approach and other non-linear extensions of the linear model in Chapter 7.

3.3.3 Potential Problems

When we fit a linear regression model to a particular data set, many problems may occur. Most common among these are the following:

1. *Non-linearity of the response-predictor relationships.*

2. *Correlation of error terms.*

3. *Non-constant variance of error terms.*

4. *Outliers.*

5. *High-leverage points.*

6. *Collinearity.*

In practice, identifying and overcoming these problems is as much an art as a science. Many pages in countless books have been written on this topic. Since the linear regression model is not our primary focus here, we will provide only a brief summary of some key points.

1. Non-linearity of the Data

The linear regression model assumes that there is a straight-line relationship between the predictors and the response. If the true relationship is far from linear, then virtually all of the conclusions that we draw from the fit are suspect. In addition, the prediction accuracy of the model can be significantly reduced.

Residual plots are a useful graphical tool for identifying non-linearity. Given a simple linear regression model, we can plot the residuals, $e_i = y_i - \hat{y}_i$, versus the predictor x_i. In the case of a multiple regression model,

residual plot

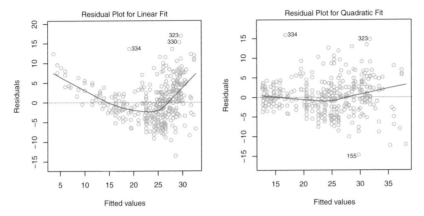

FIGURE 3.9. *Plots of residuals versus predicted (or fitted) values for the* Auto *data set. In each plot, the red line is a smooth fit to the residuals, intended to make it easier to identify a trend.* Left: *A linear regression of* mpg *on* horsepower. *A strong pattern in the residuals indicates non-linearity in the data.* Right: *A linear regression of* mpg *on* horsepower *and* horsepower^2. *There is little pattern in the residuals.*

since there are multiple predictors, we instead plot the residuals versus the predicted (or *fitted*) values $\hat{y}_i$. Ideally, the residual plot will show no discernible pattern. The presence of a pattern may indicate a problem with some aspect of the linear model.

The left panel of Figure 3.9 displays a residual plot from the linear regression of mpg onto horsepower on the Auto data set that was illustrated in Figure 3.8. The red line is a smooth fit to the residuals, which is displayed in order to make it easier to identify any trends. The residuals exhibit a clear U-shape, which provides a strong indication of non-linearity in the data. In contrast, the right-hand panel of Figure 3.9 displays the residual plot that results from the model (3.36), which contains a quadratic term. There appears to be little pattern in the residuals, suggesting that the quadratic term improves the fit to the data.

If the residual plot indicates that there are non-linear associations in the data, then a simple approach is to use non-linear transformations of the predictors, such as $\log X$, $\sqrt{X}$, and X^2, in the regression model. In the later chapters of this book, we will discuss other more advanced non-linear approaches for addressing this issue.

2. Correlation of Error Terms

An important assumption of the linear regression model is that the error terms, $\epsilon_1, \epsilon_2, \ldots, \epsilon_n$, are uncorrelated. What does this mean? For instance, if the errors are uncorrelated, then the fact that ϵ_i is positive provides little or no information about the sign of ϵ_{i+1}. The standard errors that are computed for the estimated regression coefficients or the fitted values are based on the assumption of uncorrelated error terms. If in fact there is correlation among the error terms, then the estimated standard errors will tend to underestimate the true standard errors. As a result, confidence and prediction intervals will be narrower than they should be. For example, a 95 % confidence interval may in reality have a much lower probability than 0.95 of containing the true value of the parameter. In addition, p-values associated with the model will be lower than they should be; this could cause us to erroneously conclude that a parameter is statistically significant. In short, if the error terms are correlated, we may have an unwarranted sense of confidence in our model.

As an extreme example, suppose we accidentally doubled our data, leading to observations and error terms identical in pairs. If we ignored this, our standard error calculations would be as if we had a sample of size $2n$, when in fact we have only n samples. Our estimated parameters would be the same for the $2n$ samples as for the n samples, but the confidence intervals would be narrower by a factor of $\sqrt{2}$!

Why might correlations among the error terms occur? Such correlations frequently occur in the context of *time series* data, which consists of observations for which measurements are obtained at discrete points in time. In many cases, observations that are obtained at adjacent time points will have positively correlated errors. In order to determine if this is the case for a given data set, we can plot the residuals from our model as a function of time. If the errors are uncorrelated, then there should be no discernible pattern. On the other hand, if the error terms are positively correlated, then we may see *tracking* in the residuals—that is, adjacent residuals may have similar values. Figure 3.10 provides an illustration. In the top panel, we see the residuals from a linear regression fit to data generated with uncorrelated errors. There is no evidence of a time-related trend in the residuals. In contrast, the residuals in the bottom panel are from a data set in which adjacent errors had a correlation of 0.9. Now there is a clear pattern in the residuals—adjacent residuals tend to take on similar values. Finally, the center panel illustrates a more moderate case in which the residuals had a correlation of 0.5. There is still evidence of tracking, but the pattern is less clear.

Many methods have been developed to properly take account of correlations in the error terms in time series data. Correlation among the error terms can also occur outside of time series data. For instance, consider a study in which individuals' heights are predicted from their weights. The assumption of uncorrelated errors could be violated if some of the individuals in the study are members of the same family, or eat the same diet, or have been exposed to the same environmental factors. In general, the assumption of uncorrelated errors is extremely important for linear regression as well as for other statistical methods, and good experimental design is crucial in order to mitigate the risk of such correlations.

time series

tracking

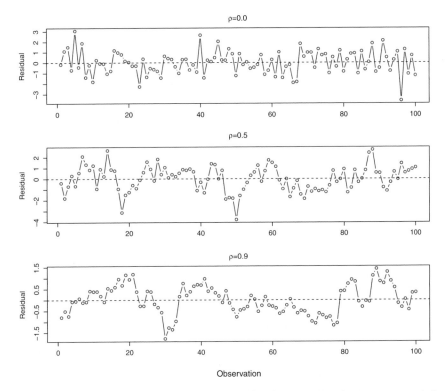

FIGURE 3.10. *Plots of residuals from simulated time series data sets generated with differing levels of correlation ρ between error terms for adjacent time points.*

3. Non-constant Variance of Error Terms

Another important assumption of the linear regression model is that the error terms have a constant variance, $\text{Var}(\epsilon_i) = \sigma^2$. The standard errors, confidence intervals, and hypothesis tests associated with the linear model rely upon this assumption.

Unfortunately, it is often the case that the variances of the error terms are non-constant. For instance, the variances of the error terms may increase with the value of the response. One can identify non-constant variances in the errors, or *heteroscedasticity*, from the presence of a *funnel shape* in the residual plot. An example is shown in the left-hand panel of Figure 3.11, in which the magnitude of the residuals tends to increase with the fitted values. When faced with this problem, one possible solution is to transform the response Y using a concave function such as $\log Y$ or $\sqrt{Y}$. Such a transformation results in a greater amount of shrinkage of the larger responses, leading to a reduction in heteroscedasticity. The right-hand panel of Figure 3.11 displays the residual plot after transforming the response

heterosceda-
sticity

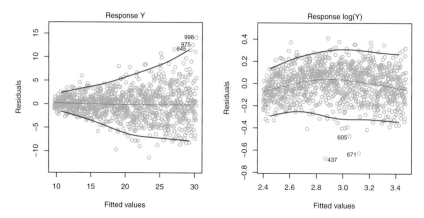

FIGURE 3.11. *Residual plots. In each plot, the red line is a smooth fit to the residuals, intended to make it easier to identify a trend. The blue lines track the outer quantiles of the residuals, and emphasize patterns. Left: The funnel shape indicates heteroscedasticity. Right: The response has been log transformed, and there is now no evidence of heteroscedasticity.*

using $\log Y$. The residuals now appear to have constant variance, though there is some evidence of a slight non-linear relationship in the data.

Sometimes we have a good idea of the variance of each response. For example, the ith response could be an average of n_i raw observations. If each of these raw observations is uncorrelated with variance σ^2, then their average has variance $\sigma_i^2 = \sigma^2/n_i$. In this case a simple remedy is to fit our model by *weighted least squares*, with weights proportional to the inverse variances—i.e. $w_i = n_i$ in this case. Most linear regression software allows for observation weights.

weighted least squares

4. Outliers

An *outlier* is a point for which y_i is far from the value predicted by the model. Outliers can arise for a variety of reasons, such as incorrect recording of an observation during data collection.

outlier

The red point (observation 20) in the left-hand panel of Figure 3.12 illustrates a typical outlier. The red solid line is the least squares regression fit, while the blue dashed line is the least squares fit after removal of the outlier. In this case, removing the outlier has little effect on the least squares line: it leads to almost no change in the slope, and a miniscule reduction in the intercept. It is typical for an outlier that does not have an unusual predictor value to have little effect on the least squares fit. However, even if an outlier does not have much effect on the least squares fit, it can cause other problems. For instance, in this example, the RSE is 1.09 when the outlier is included in the regression, but it is only 0.77 when the outlier is removed. Since the RSE is used to compute all confidence intervals and

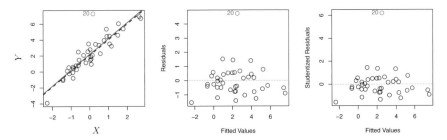

FIGURE 3.12. Left: *The least squares regression line is shown in red, and the regression line after removing the outlier is shown in blue.* Center: *The residual plot clearly identifies the outlier.* Right: *The outlier has a studentized residual of 6; typically we expect values between −3 and 3.*

p-values, such a dramatic increase caused by a single data point can have implications for the interpretation of the fit. Similarly, inclusion of the outlier causes the R^2 to decline from 0.892 to 0.805.

Residual plots can be used to identify outliers. In this example, the outlier is clearly visible in the residual plot illustrated in the center panel of Figure 3.12. But in practice, it can be difficult to decide how large a residual needs to be before we consider the point to be an outlier. To address this problem, instead of plotting the residuals, we can plot the *studentized residuals*, computed by dividing each residual e_i by its estimated standard error. Observations whose studentized residuals are greater than 3 in absolute value are possible outliers. In the right-hand panel of Figure 3.12, the outlier's studentized residual exceeds 6, while all other observations have studentized residuals between −2 and 2.

studentized residual

If we believe that an outlier has occurred due to an error in data collection or recording, then one solution is to simply remove the observation. However, care should be taken, since an outlier may instead indicate a deficiency with the model, such as a missing predictor.

5. High Leverage Points

We just saw that outliers are observations for which the response y_i is unusual given the predictor x_i. In contrast, observations with *high leverage* have an unusual value for x_i. For example, observation 41 in the left-hand panel of Figure 3.13 has high leverage, in that the predictor value for this observation is large relative to the other observations. (Note that the data displayed in Figure 3.13 are the same as the data displayed in Figure 3.12, but with the addition of a single high leverage observation.) The red solid line is the least squares fit to the data, while the blue dashed line is the fit produced when observation 41 is removed. Comparing the left-hand panels of Figures 3.12 and 3.13, we observe that removing the high leverage observation has a much more substantial impact on the least squares line

high leverage

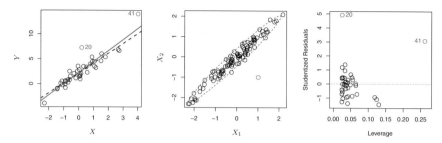

FIGURE 3.13. *Left: Observation 41 is a high leverage point, while 20 is not. The red line is the fit to all the data, and the blue line is the fit with observation 41 removed. Center: The red observation is not unusual in terms of its X_1 value or its X_2 value, but still falls outside the bulk of the data, and hence has high leverage. Right: Observation 41 has a high leverage and a high residual.*

than removing the outlier. In fact, high leverage observations tend to have a sizable impact on the estimated regression line. It is cause for concern if the least squares line is heavily affected by just a couple of observations, because any problems with these points may invalidate the entire fit. For this reason, it is important to identify high leverage observations.

In a simple linear regression, high leverage observations are fairly easy to identify, since we can simply look for observations for which the predictor value is outside of the normal range of the observations. But in a multiple linear regression with many predictors, it is possible to have an observation that is well within the range of each individual predictor's values, but that is unusual in terms of the full set of predictors. An example is shown in the center panel of Figure 3.13, for a data set with two predictors, X_1 and X_2. Most of the observations' predictor values fall within the blue dashed ellipse, but the red observation is well outside of this range. But neither its value for X_1 nor its value for X_2 is unusual. So if we examine just X_1 or just X_2, we will fail to notice this high leverage point. This problem is more pronounced in multiple regression settings with more than two predictors, because then there is no simple way to plot all dimensions of the data simultaneously.

In order to quantify an observation's leverage, we compute the *leverage statistic*. A large value of this statistic indicates an observation with high leverage. For a simple linear regression,

$$h_i = \frac{1}{n} + \frac{(x_i - \bar{x})^2}{\sum_{i'=1}^{n}(x_{i'} - \bar{x})^2}. \tag{3.37}$$

It is clear from this equation that h_i increases with the distance of x_i from $\bar{x}$. There is a simple extension of h_i to the case of multiple predictors, though we do not provide the formula here. The leverage statistic h_i is always between $1/n$ and 1, and the average leverage for all the observations is always equal to $(p+1)/n$. So if a given observation has a leverage statistic

leverage
statistic

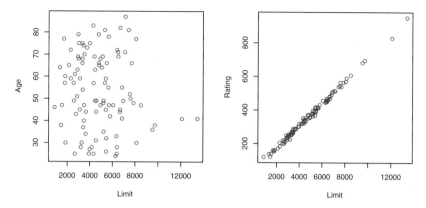

FIGURE 3.14. *Scatterplots of the observations from the* Credit *data set. Left: A plot of* age *versus* limit. *These two variables are not collinear. Right: A plot of* rating *versus* limit. *There is high collinearity.*

that greatly exceeds $(p+1)/n$, then we may suspect that the corresponding point has high leverage.

The right-hand panel of Figure 3.13 provides a plot of the studentized residuals versus h_i for the data in the left-hand panel of Figure 3.13. Observation 41 stands out as having a very high leverage statistic as well as a high studentized residual. In other words, it is an outlier as well as a high leverage observation. This is a particularly dangerous combination! This plot also reveals the reason that observation 20 had relatively little effect on the least squares fit in Figure 3.12: it has low leverage.

6. Collinearity

Collinearity refers to the situation in which two or more predictor variables   collinearity
are closely related to one another. The concept of collinearity is illustrated
in Figure 3.14 using the Credit data set. In the left-hand panel of Figure 3.14, the two predictors limit and age appear to have no obvious relationship. In contrast, in the right-hand panel of Figure 3.14, the predictors limit and rating are very highly correlated with each other, and we say that they are *collinear*. The presence of collinearity can pose problems in the regression context, since it can be difficult to separate out the individual effects of collinear variables on the response. In other words, since limit and rating tend to increase or decrease together, it can be difficult to determine how each one separately is associated with the response, balance.

Figure 3.15 illustrates some of the difficulties that can result from collinearity. The left-hand panel of Figure 3.15 is a contour plot of the RSS (3.22) associated with different possible coefficient estimates for the regression of balance on limit and age. Each ellipse represents a set of coefficients that correspond to the same RSS, with ellipses nearest to the center taking on the lowest values of RSS. The black dots and associated dashed

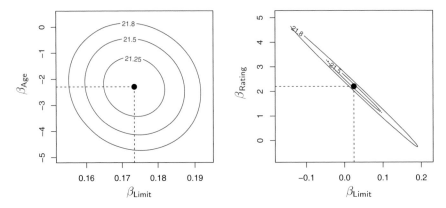

FIGURE 3.15. *Contour plots for the RSS values as a function of the parameters* β *for various regressions involving the* Credit *data set. In each plot, the black dots represent the coefficient values corresponding to the minimum RSS. Left: A contour plot of RSS for the regression of* balance *onto* age *and* limit*. The minimum value is well defined. Right: A contour plot of RSS for the regression of* balance *onto* rating *and* limit*. Because of the collinearity, there are many pairs* $(\beta_{\text{Limit}}, \beta_{\text{Rating}})$ *with a similar value for RSS.*

lines represent the coefficient estimates that result in the smallest possible RSS—in other words, these are the least squares estimates. The axes for limit and age have been scaled so that the plot includes possible coefficient estimates that are up to four standard errors on either side of the least squares estimates. Thus the plot includes all plausible values for the coefficients. For example, we see that the true limit coefficient is almost certainly somewhere between 0.15 and 0.20.

In contrast, the right-hand panel of Figure 3.15 displays contour plots of the RSS associated with possible coefficient estimates for the regression of balance onto limit and rating, which we know to be highly collinear. Now the contours run along a narrow valley; there is a broad range of values for the coefficient estimates that result in equal values for RSS. Hence a small change in the data could cause the pair of coefficient values that yield the smallest RSS—that is, the least squares estimates—to move anywhere along this valley. This results in a great deal of uncertainty in the coefficient estimates. Notice that the scale for the limit coefficient now runs from roughly −0.2 to 0.2; this is an eight-fold increase over the plausible range of the limit coefficient in the regression with age. Interestingly, even though the limit and rating coefficients now have much more individual uncertainty, they will almost certainly lie somewhere in this contour valley. For example, we would not expect the true value of the limit and rating coefficients to be −0.1 and 1 respectively, even though such a value is plausible for each coefficient individually.

		Coefficient	Std. error	t-statistic	p-value
	Intercept	−173.411	43.828	−3.957	< 0.0001
Model 1	age	−2.292	0.672	−3.407	0.0007
	limit	0.173	0.005	34.496	< 0.0001
	Intercept	−377.537	45.254	−8.343	< 0.0001
Model 2	rating	2.202	0.952	2.312	0.0213
	limit	0.025	0.064	0.384	0.7012

TABLE 3.11. *The results for two multiple regression models involving the* Credit *data set are shown. Model 1 is a regression of* balance *on* age *and* limit, *and Model 2 a regression of* balance *on* rating *and* limit. *The standard error of* $\hat{\beta}_{\text{limit}}$ *increases 12-fold in the second regression, due to collinearity.*

Since collinearity reduces the accuracy of the estimates of the regression coefficients, it causes the standard error for $\hat{\beta}_j$ to grow. Recall that the t-statistic for each predictor is calculated by dividing $\hat{\beta}_j$ by its standard error. Consequently, collinearity results in a decline in the t-statistic. As a result, in the presence of collinearity, we may fail to reject $H_0 : \beta_j = 0$. This means that the *power* of the hypothesis test—the probability of correctly detecting a *non-zero* coefficient—is reduced by collinearity.

power

Table 3.11 compares the coefficient estimates obtained from two separate multiple regression models. The first is a regression of balance on age and limit, and the second is a regression of balance on rating and limit. In the first regression, both age and limit are highly significant with very small p-values. In the second, the collinearity between limit and rating has caused the standard error for the limit coefficient estimate to increase by a factor of 12 and the p-value to increase to 0.701. In other words, the importance of the limit variable has been masked due to the presence of collinearity. To avoid such a situation, it is desirable to identify and address potential collinearity problems while fitting the model.

A simple way to detect collinearity is to look at the correlation matrix of the predictors. An element of this matrix that is large in absolute value indicates a pair of highly correlated variables, and therefore a collinearity problem in the data. Unfortunately, not all collinearity problems can be detected by inspection of the correlation matrix: it is possible for collinearity to exist between three or more variables even if no pair of variables has a particularly high correlation. We call this situation *multicollinearity*. Instead of inspecting the correlation matrix, a better way to assess multicollinearity is to compute the *variance inflation factor* (VIF). The VIF is the ratio of the variance of $\hat{\beta}_j$ when fitting the full model divided by the variance of $\hat{\beta}_j$ if fit on its own. The smallest possible value for VIF is 1, which indicates the complete absence of collinearity. Typically in practice there is a small amount of collinearity among the predictors. As a rule of thumb, a VIF value that exceeds 5 or 10 indicates a problematic amount of

multi-
collinearity

variance
inflation
factor

collinearity. The VIF for each variable can be computed using the formula

$$\text{VIF}(\hat{\beta}_j) = \frac{1}{1 - R^2_{X_j|X_{-j}}},$$

where $R^2_{X_j|X_{-j}}$ is the R^2 from a regression of X_j onto all of the other predictors. If $R^2_{X_j|X_{-j}}$ is close to one, then collinearity is present, and so the VIF will be large.

In the Credit data, a regression of balance on age, rating, and limit indicates that the predictors have VIF values of 1.01, 160.67, and 160.59. As we suspected, there is considerable collinearity in the data!

When faced with the problem of collinearity, there are two simple solutions. The first is to drop one of the problematic variables from the regression. This can usually be done without much compromise to the regression fit, since the presence of collinearity implies that the information that this variable provides about the response is redundant in the presence of the other variables. For instance, if we regress balance onto age and limit, without the rating predictor, then the resulting VIF values are close to the minimum possible value of 1, and the R^2 drops from 0.754 to 0.75. So dropping rating from the set of predictors has effectively solved the collinearity problem without compromising the fit. The second solution is to combine the collinear variables together into a single predictor. For instance, we might take the average of standardized versions of limit and rating in order to create a new variable that measures *credit worthiness*.

3.4 The Marketing Plan

We now briefly return to the seven questions about the Advertising data that we set out to answer at the beginning of this chapter.

1. *Is there a relationship between advertising sales and budget?*
   This question can be answered by fitting a multiple regression model of sales onto TV, radio, and newspaper, as in (3.20), and testing the hypothesis $H_0 : \beta_{\text{TV}} = \beta_{\text{radio}} = \beta_{\text{newspaper}} = 0$. In Section 3.2.2, we showed that the F-statistic can be used to determine whether or not we should reject this null hypothesis. In this case the p-value corresponding to the F-statistic in Table 3.6 is very low, indicating clear evidence of a relationship between advertising and sales.

2. *How strong is the relationship?*
   We discussed two measures of model accuracy in Section 3.1.3. First, the RSE estimates the standard deviation of the response from the population regression line. For the Advertising data, the RSE is 1,681

units while the mean value for the response is 14,022, indicating a percentage error of roughly 12 %. Second, the R^2 statistic records the percentage of variability in the response that is explained by the predictors. The predictors explain almost 90 % of the variance in sales. The RSE and R^2 statistics are displayed in Table 3.6.

3. *Which media contribute to sales?*
   To answer this question, we can examine the p-values associated with each predictor's t-statistic (Section 3.1.2). In the multiple linear regression displayed in Table 3.4, the p-values for TV and radio are low, but the p-value for newspaper is not. This suggests that only TV and radio are related to sales. In Chapter 6 we explore this question in greater detail.

4. *How large is the effect of each medium on sales?*
   We saw in Section 3.1.2 that the standard error of $\hat{\beta}_j$ can be used to construct confidence intervals for β_j. For the Advertising data, the 95 % confidence intervals are as follows: $(0.043, 0.049)$ for TV, $(0.172, 0.206)$ for radio, and $(-0.013, 0.011)$ for newspaper. The confidence intervals for TV and radio are narrow and far from zero, providing evidence that these media are related to sales. But the interval for newspaper includes zero, indicating that the variable is not statistically significant given the values of TV and radio.

   We saw in Section 3.3.3 that collinearity can result in very wide standard errors. Could collinearity be the reason that the confidence interval associated with newspaper is so wide? The VIF scores are 1.005, 1.145, and 1.145 for TV, radio, and newspaper, suggesting no evidence of collinearity.

   In order to assess the association of each medium individually on sales, we can perform three separate simple linear regressions. Results are shown in Tables 3.1 and 3.3. There is evidence of an extremely strong association between TV and sales and between radio and sales. There is evidence of a mild association between newspaper and sales, when the values of TV and radio are ignored.

5. *How accurately can we predict future sales?*
   The response can be predicted using (3.21). The accuracy associated with this estimate depends on whether we wish to predict an individual response, $Y = f(X) + \epsilon$, or the average response, $f(X)$ (Section 3.2.2). If the former, we use a prediction interval, and if the latter, we use a confidence interval. Prediction intervals will always be wider than confidence intervals because they account for the uncertainty associated with ϵ, the irreducible error.

6. *Is the relationship linear?*

In Section 3.3.3, we saw that residual plots can be used in order to identify non-linearity. If the relationships are linear, then the residual plots should display no pattern. In the case of the `Advertising` data, we observe a non-linear effect in Figure 3.5, though this effect could also be observed in a residual plot. In Section 3.3.2, we discussed the inclusion of transformations of the predictors in the linear regression model in order to accommodate non-linear relationships.

7. *Is there synergy among the advertising media?*

The standard linear regression model assumes an additive relationship between the predictors and the response. An additive model is easy to interpret because the effect of each predictor on the response is unrelated to the values of the other predictors. However, the additive assumption may be unrealistic for certain data sets. In Section 3.3.2, we showed how to include an interaction term in the regression model in order to accommodate non-additive relationships. A small p-value associated with the interaction term indicates the presence of such relationships. Figure 3.5 suggested that the `Advertising` data may not be additive. Including an interaction term in the model results in a substantial increase in R^2, from around 90 % to almost 97 %.

3.5 Comparison of Linear Regression with K-Nearest Neighbors

As discussed in Chapter 2, linear regression is an example of a *parametric* approach because it assumes a linear functional form for $f(X)$. Parametric methods have several advantages. They are often easy to fit, because one need estimate only a small number of coefficients. In the case of linear regression, the coefficients have simple interpretations, and tests of statistical significance can be easily performed. But parametric methods do have a disadvantage: by construction, they make strong assumptions about the form of $f(X)$. If the specified functional form is far from the truth, and prediction accuracy is our goal, then the parametric method will perform poorly. For instance, if we assume a linear relationship between X and Y but the true relationship is far from linear, then the resulting model will provide a poor fit to the data, and any conclusions drawn from it will be suspect.

In contrast, *non-parametric* methods do not explicitly assume a parametric form for $f(X)$, and thereby provide an alternative and more flexible approach for performing regression. We discuss various non-parametric methods in this book. Here we consider one of the simplest and best-known non-parametric methods, *K-nearest neighbors regression* (KNN regression).

K-nearest neighbors regression

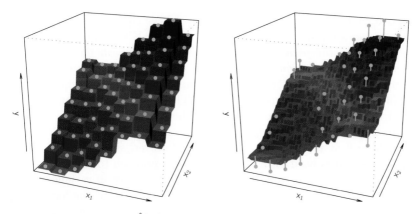

FIGURE 3.16. *Plots of $\hat{f}(X)$ using KNN regression on a two-dimensional data set with* 64 *observations (orange dots).* Left: $K = 1$ *results in a rough step function fit.* Right: $K = 9$ *produces a much smoother fit.*

The KNN regression method is closely related to the KNN classifier discussed in Chapter 2. Given a value for K and a prediction point x_0, KNN regression first identifies the K training observations that are closest to x_0, represented by $\mathcal{N}_0$. It then estimates $f(x_0)$ using the average of all the training responses in $\mathcal{N}_0$. In other words,

$$\hat{f}(x_0) = \frac{1}{K} \sum_{x_i \in \mathcal{N}_0} y_i.$$

Figure 3.16 illustrates two KNN fits on a data set with $p = 2$ predictors. The fit with $K = 1$ is shown in the left-hand panel, while the right-hand panel corresponds to $K = 9$. We see that when $K = 1$, the KNN fit perfectly interpolates the training observations, and consequently takes the form of a step function. When $K = 9$, the KNN fit still is a step function, but averaging over nine observations results in much smaller regions of constant prediction, and consequently a smoother fit. In general, the optimal value for K will depend on the *bias-variance tradeoff*, which we introduced in Chapter 2. A small value for K provides the most flexible fit, which will have low bias but high variance. This variance is due to the fact that the prediction in a given region is entirely dependent on just one observation. In contrast, larger values of K provide a smoother and less variable fit; the prediction in a region is an average of several points, and so changing one observation has a smaller effect. However, the smoothing may cause bias by masking some of the structure in $f(X)$. In Chapter 5, we introduce several approaches for estimating test error rates. These methods can be used to identify the optimal value of K in KNN regression.

In what setting will a parametric approach such as least squares linear regression outperform a non-parametric approach such as KNN regression? The answer is simple: *the parametric approach will outperform the non-parametric approach if the parametric form that has been selected is close to the true form of f.* Figure 3.17 provides an example with data generated from a one-dimensional linear regression model. The black solid lines represent $f(X)$, while the blue curves correspond to the KNN fits using $K = 1$ and $K = 9$. In this case, the $K = 1$ predictions are far too variable, while the smoother $K = 9$ fit is much closer to $f(X)$. However, since the true relationship is linear, it is hard for a non-parametric approach to compete with linear regression: a non-parametric approach incurs a cost in variance that is not offset by a reduction in bias. The blue dashed line in the left-hand panel of Figure 3.18 represents the linear regression fit to the same data. It is almost perfect. The right-hand panel of Figure 3.18 reveals that linear regression outperforms KNN for this data. The green solid line, plotted as a function of $1/K$, represents the test set mean squared error (MSE) for KNN. The KNN errors are well above the black dashed line, which is the test MSE for linear regression. When the value of K is large, then KNN performs only a little worse than least squares regression in terms of MSE. It performs far worse when K is small.

In practice, the true relationship between X and Y is rarely exactly linear. Figure 3.19 examines the relative performances of least squares regression and KNN under increasing levels of non-linearity in the relationship between X and Y. In the top row, the true relationship is nearly linear. In this case we see that the test MSE for linear regression is still superior to that of KNN for low values of K. However, for $K \geq 4$, KNN outperforms linear regression. The second row illustrates a more substantial deviation from linearity. In this situation, KNN substantially outperforms linear regression for all values of K. Note that as the extent of non-linearity increases, there is little change in the test set MSE for the non-parametric KNN method, but there is a large increase in the test set MSE of linear regression.

Figures 3.18 and 3.19 display situations in which KNN performs slightly worse than linear regression when the relationship is linear, but much better than linear regression for non-linear situations. In a real life situation in which the true relationship is unknown, one might draw the conclusion that KNN should be favored over linear regression because it will at worst be slightly inferior than linear regression if the true relationship is linear, and may give substantially better results if the true relationship is non-linear. But in reality, even when the true relationship is highly non-linear, KNN may still provide inferior results to linear regression. In particular, both Figures 3.18 and 3.19 illustrate settings with $p = 1$ predictor. But in higher dimensions, KNN often performs worse than linear regression.

Figure 3.20 considers the same strongly non-linear situation as in the second row of Figure 3.19, except that we have added additional *noise*

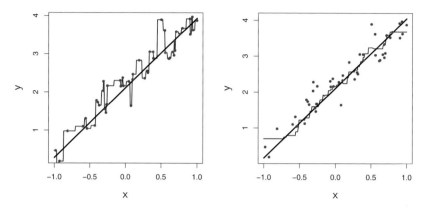

FIGURE 3.17. *Plots of $\hat{f}(X)$ using KNN regression on a one-dimensional data set with* 100 *observations. The true relationship is given by the black solid line. Left: The blue curve corresponds to $K = 1$ and interpolates (i.e. passes directly through) the training data. Right: The blue curve corresponds to $K = 9$, and represents a smoother fit.*

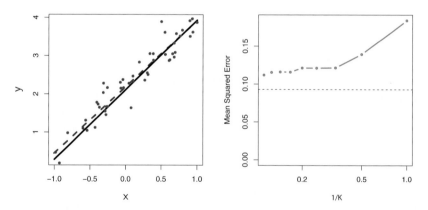

FIGURE 3.18. *The same data set shown in Figure 3.17 is investigated further. Left: The blue dashed line is the least squares fit to the data. Since $f(X)$ is in fact linear (displayed as the black line), the least squares regression line provides a very good estimate of $f(X)$. Right: The dashed horizontal line represents the least squares test set MSE, while the green solid line corresponds to the MSE for KNN as a function of $1/K$ (on the log scale). Linear regression achieves a lower test MSE than does KNN regression, since $f(X)$ is in fact linear. For KNN regression, the best results occur with a very large value of K, corresponding to a small value of $1/K$.*

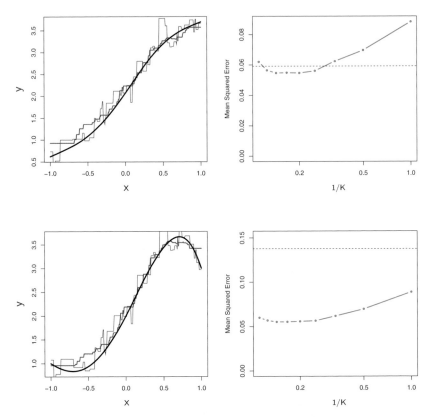

FIGURE 3.19. Top Left: *In a setting with a slightly non-linear relationship between X and Y (solid black line), the KNN fits with $K = 1$ (blue) and $K = 9$ (red) are displayed.* Top Right: *For the slightly non-linear data, the test set MSE for least squares regression (horizontal black) and KNN with various values of $1/K$ (green) are displayed.* Bottom Left and Bottom Right: *As in the top panel, but with a strongly non-linear relationship between X and Y.*

predictors that are not associated with the response. When $p = 1$ or $p = 2$, KNN outperforms linear regression. But for $p = 3$ the results are mixed, and for $p \geq 4$ linear regression is superior to KNN. In fact, the increase in dimension has only caused a small deterioration in the linear regression test set MSE, but it has caused more than a ten-fold increase in the MSE for KNN. This decrease in performance as the dimension increases is a common problem for KNN, and results from the fact that in higher dimensions there is effectively a reduction in sample size. In this data set there are 100 training observations; when $p = 1$, this provides enough information to accurately estimate $f(X)$. However, spreading 100 observations over $p = 20$ dimensions results in a phenomenon in which a given observation has no *nearby neighbors*—this is the so-called *curse of dimensionality*. That is, the K observations that are nearest to a given test observation x_0 may be very far away from x_0 in p-dimensional space when p is large, leading to a

curse of di-
mensionality

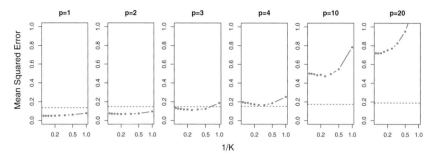

FIGURE 3.20. *Test MSE for linear regression (black dashed lines) and KNN (green curves) as the number of variables p increases. The true function is non–linear in the first variable, as in the lower panel in Figure 3.19, and does not depend on the additional variables. The performance of linear regression deteriorates slowly in the presence of these additional noise variables, whereas KNN's performance degrades much more quickly as p increases.*

very poor prediction of $f(x_0)$ and hence a poor KNN fit. As a general rule, parametric methods will tend to outperform non-parametric approaches when there is a small number of observations per predictor.

Even in problems in which the dimension is small, we might prefer linear regression to KNN from an interpretability standpoint. If the test MSE of KNN is only slightly lower than that of linear regression, we might be willing to forego a little bit of prediction accuracy for the sake of a simple model that can be described in terms of just a few coefficients, and for which p-values are available.

3.6   Lab: Linear Regression

3.6.1   *Libraries*

The library() function is used to load *libraries*, or groups of functions and data sets that are not included in the base R distribution. Basic functions that perform least squares linear regression and other simple analyses come standard with the base distribution, but more exotic functions require additional libraries. Here we load the MASS package, which is a very large collection of data sets and functions. We also load the ISLR package, which includes the data sets associated with this book.

`library()`

```
> library(MASS)
> library(ISLR)
```

If you receive an error message when loading any of these libraries, it likely indicates that the corresponding library has not yet been installed on your system. Some libraries, such as MASS, come with R and do not need to be separately installed on your computer. However, other packages, such as

ISLR, must be downloaded the first time they are used. This can be done directly from within R. For example, on a Windows system, select the Install package option under the Packages tab. After you select any mirror site, a list of available packages will appear. Simply select the package you wish to install and R will automatically download the package. Alternatively, this can be done at the R command line via install.packages("ISLR"). This installation only needs to be done the first time you use a package. However, the library() function must be called each time you wish to use a given package.

3.6.2 Simple Linear Regression

The MASS library contains the Boston data set, which records medv (median house value) for 506 neighborhoods around Boston. We will seek to predict medv using 13 predictors such as rm (average number of rooms per house), age (average age of houses), and lstat (percent of households with low socioeconomic status).

```
> fix(Boston)
> names(Boston)
 [1] "crim"    "zn"     "indus"    "chas"     "nox"     "rm"      "age"
 [8] "dis"     "rad"    "tax"      "ptratio"  "black"   "lstat"   "medv"
```

To find out more about the data set, we can type ?Boston.

We will start by using the lm() function to fit a simple linear regression model, with medv as the response and lstat as the predictor. The basic syntax is lm(y~x,data), where y is the response, x is the predictor, and data is the data set in which these two variables are kept.

lm()

```
> lm.fit=lm(medv~lstat)
Error in eval(expr, envir, enclos) : Object "medv" not found
```

The command causes an error because R does not know where to find the variables medv and lstat. The next line tells R that the variables are in Boston. If we attach Boston, the first line works fine because R now recognizes the variables.

```
> lm.fit=lm(medv~lstat,data=Boston)
> attach(Boston)
> lm.fit=lm(medv~lstat)
```

If we type lm.fit, some basic information about the model is output. For more detailed information, we use summary(lm.fit). This gives us p-values and standard errors for the coefficients, as well as the R^2 statistic and F-statistic for the model.

```
> lm.fit

Call:
lm(formula = medv ~ lstat)
```

```
Coefficients:
(Intercept)            lstat
       34.55          -0.95

> summary(lm.fit)

Call:
lm(formula = medv ~ lstat)

Residuals:
   Min      1Q Median      3Q     Max
-15.17   -3.99  -1.32    2.03   24.50

Coefficients:
             Estimate Std. Error t value Pr(>|t|)
(Intercept)   34.5538     0.5626    61.4   <2e-16 ***
lstat         -0.9500     0.0387   -24.5   <2e-16 ***
---
Signif. codes:  0 *** 0.001 ** 0.01 * 0.05 . 0.1    1

Residual standard error: 6.22 on 504 degrees of freedom
Multiple R-squared: 0.544,        Adjusted R-squared: 0.543
F-statistic:  602 on 1 and 504 DF,  p-value: <2e-16
```

We can use the names() function in order to find out what other pieces of information are stored in lm.fit. Although we can extract these quantities by name—e.g. lm.fit$coefficients—it is safer to use the extractor functions like coef() to access them.

names()

coef()

```
> names(lm.fit)
 [1] "coefficients"  "residuals"       "effects"
 [4] "rank"          "fitted.values"   "assign"
 [7] "qr"            "df.residual"     "xlevels"
[10] "call"          "terms"           "model"
> coef(lm.fit)
(Intercept)            lstat
       34.55          -0.95
```

In order to obtain a confidence interval for the coefficient estimates, we can use the confint() command.

confint()

```
> confint(lm.fit)
               2.5 % 97.5 %
(Intercept) 33.45 35.659
lstat       -1.03 -0.874
```

The predict() function can be used to produce confidence intervals and prediction intervals for the prediction of medv for a given value of lstat.

predict()

```
> predict(lm.fit,data.frame(lstat=c(5,10,15)),
       interval="confidence")
    fit   lwr   upr
1 29.80 29.01 30.60
2 25.05 24.47 25.63
3 20.30 19.73 20.87
```

```
> predict (lm.fit ,data.frame (lstat =c (5 ,10 ,15)) ,
        interval ="prediction ")
     fit    lwr    upr
1 29.80 17.566 42.04
2 25.05 12.828 37.28
3 20.30  8.078 32.53
```

For instance, the 95 % confidence interval associated with a lstat value of
10 is $(24.47, 25.63)$, and the 95 % prediction interval is $(12.828, 37.28)$. As
expected, the confidence and prediction intervals are centered around the
same point (a predicted value of 25.05 for medv when lstat equals 10), but
the latter are substantially wider.

We will now plot medv and lstat along with the least squares regression
line using the plot() and abline() functions.

abline()

```
> plot (lstat ,medv)
> abline (lm.fit )
```

There is some evidence for non-linearity in the relationship between lstat
and medv. We will explore this issue later in this lab.

The abline() function can be used to draw any line, not just the least
squares regression line. To draw a line with intercept a and slope b, we
type abline(a,b). Below we experiment with some additional settings for
plotting lines and points. The lwd=3 command causes the width of the
regression line to be increased by a factor of 3; this works for the plot()
and lines() functions also. We can also use the pch option to create different
plotting symbols.

```
> abline (lm.fit ,lwd =3)
> abline (lm.fit ,lwd =3, col ="red ")
> plot (lstat ,medv , col ="red ")
> plot (lstat ,medv , pch =20)
> plot (lstat ,medv , pch ="+")
> plot (1:20 ,1:20 , pch =1:20)
```

Next we examine some diagnostic plots, several of which were discussed
in Section 3.3.3. Four diagnostic plots are automatically produced by applying the plot() function directly to the output from lm(). In general, this
command will produce one plot at a time, and hitting *Enter* will generate
the next plot. However, it is often convenient to view all four plots together.
We can achieve this by using the par() function, which tells R to split the
display screen into separate panels so that multiple plots can be viewed simultaneously. For example, par(mfrow=c(2,2)) divides the plotting region
into a 2×2 grid of panels.

par()

```
> par (mfrow =c (2 ,2))
> plot (lm.fit )
```

Alternatively, we can compute the residuals from a linear regression fit
using the residuals() function. The function rstudent() will return the
studentized residuals, and we can use this function to plot the residuals
against the fitted values.

residuals()
rstudent()

```
> plot(predict(lm.fit), residuals(lm.fit))
> plot(predict(lm.fit), rstudent(lm.fit))
```

On the basis of the residual plots, there is some evidence of non-linearity. Leverage statistics can be computed for any number of predictors using the hatvalues() function.

hatvalues()

```
> plot(hatvalues(lm.fit))
> which.max(hatvalues(lm.fit))
375
```

The which.max() function identifies the index of the largest element of a vector. In this case, it tells us which observation has the largest leverage statistic.

which.max()

3.6.3 Multiple Linear Regression

In order to fit a multiple linear regression model using least squares, we again use the lm() function. The syntax lm(y~x1+x2+x3) is used to fit a model with three predictors, x1, x2, and x3. The summary() function now outputs the regression coefficients for all the predictors.

```
> lm.fit=lm(medv~lstat+age,data=Boston)
> summary(lm.fit)

Call:
lm(formula = medv ~ lstat + age, data = Boston)

Residuals:
    Min     1Q Median     3Q    Max
 -15.98  -3.98  -1.28   1.97  23.16

Coefficients:
              Estimate Std. Error t value Pr(>|t|)
(Intercept)   33.2228     0.7308   45.46   <2e-16 ***
lstat         -1.0321     0.0482  -21.42   <2e-16 ***
age            0.0345     0.0122    2.83   0.0049 **
---
Signif. codes:  0 *** 0.001 ** 0.01 * 0.05 . 0.1  1

Residual standard error: 6.17 on 503 degrees of freedom
Multiple R-squared: 0.551,        Adjusted R-squared: 0.549
F-statistic:  309 on 2 and 503 DF,  p-value: <2e-16
```

The Boston data set contains 13 variables, and so it would be cumbersome to have to type all of these in order to perform a regression using all of the predictors. Instead, we can use the following short-hand:

```
> lm.fit=lm(medv~.,data=Boston)
> summary(lm.fit)

Call:
lm(formula = medv ~ ., data = Boston)
```

```
Residuals:
     Min      1Q  Median      3Q     Max
 -15.594  -2.730  -0.518   1.777  26.199

Coefficients:
              Estimate Std. Error t value Pr(>|t|)
(Intercept)  3.646e+01  5.103e+00   7.144 3.28e-12 ***
crim        -1.080e-01  3.286e-02  -3.287 0.001087 **
zn           4.642e-02  1.373e-02   3.382 0.000778 ***
indus        2.056e-02  6.150e-02   0.334 0.738288
chas         2.687e+00  8.616e-01   3.118 0.001925 **
nox         -1.777e+01  3.820e+00  -4.651 4.25e-06 ***
rm           3.810e+00  4.179e-01   9.116  < 2e-16 ***
age          6.922e-04  1.321e-02   0.052 0.958229
dis         -1.476e+00  1.995e-01  -7.398 6.01e-13 ***
rad          3.060e-01  6.635e-02   4.613 5.07e-06 ***
tax         -1.233e-02  3.761e-03  -3.280 0.001112 **
ptratio     -9.527e-01  1.308e-01  -7.283 1.31e-12 ***
black        9.312e-03  2.686e-03   3.467 0.000573 ***
lstat       -5.248e-01  5.072e-02 -10.347  < 2e-16 ***
---
Signif. codes:  0 '***' 0.001 '**' 0.01 '*' 0.05 '.' 0.1 ' ' 1

Residual standard error: 4.745 on 492 degrees of freedom
Multiple R-Squared: 0.7406,    Adjusted R-squared: 0.7338
F-statistic: 108.1 on 13 and 492 DF,  p-value: < 2.2e-16
```

We can access the individual components of a summary object by name
(type ?summary.lm to see what is available). Hence summary(lm.fit)$r.sq
gives us the R^2, and summary(lm.fit)$sigma gives us the RSE. The vif()
function, part of the car package, can be used to compute variance inflation
factors. Most VIF's are low to moderate for this data. The car package is
not part of the base R installation so it must be downloaded the first time
you use it via the install.packages option in R.

vif()

```
> library(car)
> vif(lm.fit)
   crim      zn   indus    chas     nox      rm     age
   1.79    2.30    3.99    1.07    4.39    1.93    3.10
    dis     rad     tax ptratio   black   lstat
   3.96    7.48    9.01    1.80    1.35    2.94
```

What if we would like to perform a regression using all of the variables but
one? For example, in the above regression output, age has a high p-value.
So we may wish to run a regression excluding this predictor. The following
syntax results in a regression using all predictors except age.

```
> lm.fit1=lm(medv~.-age,data=Boston)
> summary(lm.fit1)
...
```

Alternatively, the update() function can be used.

update()

```
> lm.fit1=update(lm.fit, ~.-age)
```

3.6.4 Interaction Terms

It is easy to include interaction terms in a linear model using the lm() function. The syntax lstat:black tells R to include an interaction term between
lstat and black. The syntax lstat*age simultaneously includes lstat, age,
and the interaction term lstat×age as predictors; it is a shorthand for
lstat+age+lstat:age.

```
> summary(lm(medv~lstat*age,data=Boston))

Call:
lm(formula = medv ~ lstat * age, data = Boston)

Residuals:
    Min      1Q  Median      3Q     Max
 -15.81   -4.04   -1.33    2.08   27.55

Coefficients:
              Estimate Std. Error t value Pr(>|t|)
(Intercept)  36.088536   1.469835   24.55  < 2e-16 ***
lstat        -1.392117   0.167456   -8.31  8.8e-16 ***
age          -0.000721   0.019879   -0.04    0.971
lstat:age     0.004156   0.001852    2.24    0.025 *
---
Signif. codes:  0 '***' 0.001 '**' 0.01 '*' 0.05 '.' 0.1 ' ' 1

Residual standard error: 6.15 on 502 degrees of freedom
Multiple R-squared: 0.556,   Adjusted R-squared: 0.553
F-statistic:  209 on 3 and 502 DF,  p-value: <2e-16
```

3.6.5 Non-linear Transformations of the Predictors

The lm() function can also accommodate non-linear transformations of the
predictors. For instance, given a predictor X, we can create a predictor X^2
using I(X^2). The function I() is needed since the ^ has a special meaning
in a formula; wrapping as we do allows the standard usage in R, which is   I()
to raise X to the power 2. We now perform a regression of medv onto lstat
and lstat^2.

```
> lm.fit2=lm(medv~lstat+I(lstat^2))
> summary(lm.fit2)

Call:
lm(formula = medv ~ lstat + I(lstat^2))

Residuals:
    Min      1Q  Median      3Q     Max
 -15.28   -3.83   -0.53    2.31   25.41
```

```
Coefficients:
             Estimate Std. Error t value Pr(>|t|)
(Intercept) 42.86201    0.87208    49.1   <2e-16 ***
lstat       -2.33282    0.12380   -18.8   <2e-16 ***
I(lstat^2)   0.04355    0.00375    11.6   <2e-16 ***
---
Signif. codes:  0 '***' 0.001 '**' 0.01 '*' 0.05 '.' 0.1 ' ' 1

Residual standard error: 5.52 on 503 degrees of freedom
Multiple R-squared: 0.641,  Adjusted R-squared: 0.639
F-statistic:  449 on 2 and 503 DF,  p-value: <2e-16
```

The near-zero p-value associated with the quadratic term suggests that it leads to an improved model. We use the `anova()` function to further quantify the extent to which the quadratic fit is superior to the linear fit.

`anova()`

```
> lm.fit=lm(medv~lstat)
> anova(lm.fit,lm.fit2)
Analysis of Variance Table

Model 1: medv ~ lstat
Model 2: medv ~ lstat + I(lstat^2)
  Res.Df    RSS Df Sum of Sq    F Pr(>F)
1    504  19472
2    503  15347  1     4125  135 <2e-16 ***
---
Signif. codes:  0 '***' 0.001 '**' 0.01 '*' 0.05 '.' 0.1 ' ' 1
```

Here Model 1 represents the linear submodel containing only one predictor, `lstat`, while Model 2 corresponds to the larger quadratic model that has two predictors, `lstat` and $lstat^2$. The `anova()` function performs a hypothesis test comparing the two models. The null hypothesis is that the two models fit the data equally well, and the alternative hypothesis is that the full model is superior. Here the F-statistic is 135 and the associated p-value is virtually zero. This provides very clear evidence that the model containing the predictors `lstat` and $lstat^2$ is far superior to the model that only contains the predictor `lstat`. This is not surprising, since earlier we saw evidence for non-linearity in the relationship between `medv` and `lstat`. If we type

```
> par(mfrow=c(2,2))
> plot(lm.fit2)
```

then we see that when the $lstat^2$ term is included in the model, there is little discernible pattern in the residuals.

In order to create a cubic fit, we can include a predictor of the form `I(X^3)`. However, this approach can start to get cumbersome for higher-order polynomials. A better approach involves using the `poly()` function to create the polynomial within `lm()`. For example, the following command produces a fifth-order polynomial fit:

`poly()`

```
> lm.fit5=lm(medv~poly(lstat ,5))
> summary(lm.fit5)

Call:
lm(formula = medv ~ poly(lstat, 5))

Residuals :
    Min       1Q   Median      3Q      Max
-13.543   -3.104   -0.705   2.084   27.115

Coefficients:
                   Estimate Std. Error  t value Pr(>|t|)
(Intercept)         22.533      0.232    97.20  < 2e-16 ***
poly(lstat , 5)1  -152.460      5.215   -29.24  < 2e-16 ***
poly(lstat , 5)2    64.227      5.215    12.32  < 2e-16 ***
poly(lstat , 5)3   -27.051      5.215    -5.19  3.1e-07 ***
poly(lstat , 5)4    25.452      5.215     4.88  1.4e-06 ***
poly(lstat , 5)5   -19.252      5.215    -3.69  0.00025 ***
---
Signif. codes:  0 '***' 0.001 '**' 0.01 '*' 0.05 '.' 0.1 ' ' 1

Residual standard error: 5.21 on 500 degrees of freedom
Multiple R-squared: 0.682,   Adjusted R-squared: 0.679
F-statistic:  214 on 5 and 500 DF,  p-value: <2e-16
```

This suggests that including additional polynomial terms, up to fifth order, leads to an improvement in the model fit! However, further investigation of the data reveals that no polynomial terms beyond fifth order have significant p-values in a regression fit.

Of course, we are in no way restricted to using polynomial transformations of the predictors. Here we try a log transformation.

```
> summary(lm(medv~log(rm),data=Boston))
...
```

3.6.6 Qualitative Predictors

We will now examine the Carseats data, which is part of the ISLR library. We will attempt to predict Sales (child car seat sales) in 400 locations based on a number of predictors.

```
> fix(Carseats)
> names(Carseats)
 [1] "Sales"       "CompPrice "   "Income "      "Advertising "
 [5] "Population "  "Price "       "ShelveLoc "   "Age "
 [9] "Education "   "Urban "       "US "
```

The Carseats data includes qualitative predictors such as Shelveloc, an indicator of the quality of the shelving location—that is, the space within a store in which the car seat is displayed—at each location. The predictor Shelveloc takes on three possible values, *Bad*, *Medium*, and *Good*.

Given a qualitative variable such as Shelveloc, R generates dummy variables automatically. Below we fit a multiple regression model that includes some interaction terms.

```
> lm.fit=lm(Sales~.+Income:Advertising+Price:Age,data=Carseats)
> summary(lm.fit)

Call:
lm(formula = Sales ~ . + Income:Advertising + Price:Age, data =
    Carseats)

Residuals:
    Min      1Q  Median      3Q     Max
 -2.921  -0.750   0.018   0.675   3.341

Coefficients:
                   Estimate Std. Error t value Pr(>|t|)
(Intercept)        6.575565   1.008747    6.52 2.2e-10 ***
CompPrice          0.092937   0.004118   22.57 < 2e-16 ***
Income             0.010894   0.002604    4.18 3.6e-05 ***
Advertising        0.070246   0.022609    3.11 0.00203 **
Population         0.000159   0.000368    0.43 0.66533
Price             -0.100806   0.007440  -13.55 < 2e-16 ***
ShelveLocGood      4.848676   0.152838   31.72 < 2e-16 ***
ShelveLocMedium    1.953262   0.125768   15.53 < 2e-16 ***
Age               -0.057947   0.015951   -3.63 0.00032 ***
Education         -0.020852   0.019613   -1.06 0.28836
UrbanYes           0.140160   0.112402    1.25 0.21317
USYes             -0.157557   0.148923   -1.06 0.29073
Income:Advertising 0.000751   0.000278    2.70 0.00729 **
Price:Age          0.000107   0.000133    0.80 0.42381
---
Signif. codes:  0 '***' 0.001 '**' 0.01 '*' 0.05 '.' 0.1 ' ' 1

Residual standard error: 1.01 on 386 degrees of freedom
Multiple R-squared: 0.876,      Adjusted R-squared: 0.872
F-statistic:  210 on 13 and 386 DF,  p-value: <2e-16
```

The contrasts() function returns the coding that R uses for the dummy variables. contrasts()

```
> attach(Carseats)
> contrasts(ShelveLoc)
        Good Medium
Bad        0      0
Good       1      0
Medium     0      1
```

Use ?contrasts to learn about other contrasts, and how to set them.

R has created a ShelveLocGood dummy variable that takes on a value of 1 if the shelving location is good, and 0 otherwise. It has also created a ShelveLocMedium dummy variable that equals 1 if the shelving location is medium, and 0 otherwise. A bad shelving location corresponds to a zero for each of the two dummy variables. The fact that the coefficient for

ShelveLocGood in the regression output is positive indicates that a good shelving location is associated with high sales (relative to a bad location). And ShelveLocMedium has a smaller positive coefficient, indicating that a medium shelving location leads to higher sales than a bad shelving location but lower sales than a good shelving location.

3.6.7 Writing Functions

As we have seen, R comes with many useful functions, and still more functions are available by way of R libraries. However, we will often be interested in performing an operation for which no function is available. In this setting, we may want to write our own function. For instance, below we provide a simple function that reads in the ISLR and MASS libraries, called LoadLibraries(). Before we have created the function, R returns an error if we try to call it.

```
> LoadLibraries
Error: object 'LoadLibraries' not found
> LoadLibraries()
Error: could not find function "LoadLibraries"
```

We now create the function. Note that the + symbols are printed by R and should not be typed in. The { symbol informs R that multiple commands are about to be input. Hitting *Enter* after typing { will cause R to print the + symbol. We can then input as many commands as we wish, hitting *Enter* after each one. Finally the } symbol informs R that no further commands will be entered.

```
> LoadLibraries=function(){
+ library(ISLR)
+ library(MASS)
+ print("The libraries have been loaded.")
+ }
```

Now if we type in LoadLibraries, R will tell us what is in the function.

```
> LoadLibraries
function(){
library(ISLR)
library(MASS)
print("The libraries have been loaded.")
}
```

If we call the function, the libraries are loaded in and the print statement is output.

```
> LoadLibraries()
[1] "The libraries have been loaded."
```

3.7   Exercises

Conceptual

1. Describe the null hypotheses to which the p-values given in Table 3.4 correspond. Explain what conclusions you can draw based on these p-values. Your explanation should be phrased in terms of sales, TV, radio, and newspaper, rather than in terms of the coefficients of the linear model.

2. Carefully explain the differences between the KNN classifier and KNN regression methods.

3. Suppose we have a data set with five predictors, $X_1 = $ GPA, $X_2 = $ IQ, $X_3 = $ Gender (1 for Female and 0 for Male), $X_4 = $ Interaction between GPA and IQ, and $X_5 = $ Interaction between GPA and Gender. The response is starting salary after graduation (in thousands of dollars). Suppose we use least squares to fit the model, and get $\hat{\beta}_0 = 50, \hat{\beta}_1 = 20, \hat{\beta}_2 = 0.07, \hat{\beta}_3 = 35, \hat{\beta}_4 = 0.01, \hat{\beta}_5 = -10$.

   (a) Which answer is correct, and why?

      i. For a fixed value of IQ and GPA, males earn more on average than females.

      ii. For a fixed value of IQ and GPA, females earn more on average than males.

      iii. For a fixed value of IQ and GPA, males earn more on average than females provided that the GPA is high enough.

      iv. For a fixed value of IQ and GPA, females earn more on average than males provided that the GPA is high enough.

   (b) Predict the salary of a female with IQ of 110 and a GPA of 4.0.

   (c) True or false: Since the coefficient for the GPA/IQ interaction term is very small, there is very little evidence of an interaction effect. Justify your answer.

4. I collect a set of data ($n = 100$ observations) containing a single predictor and a quantitative response. I then fit a linear regression model to the data, as well as a separate cubic regression, i.e. $Y = \beta_0 + \beta_1 X + \beta_2 X^2 + \beta_3 X^3 + \epsilon$.

   (a) Suppose that the true relationship between X and Y is linear, i.e. $Y = \beta_0 + \beta_1 X + \epsilon$. Consider the training residual sum of squares (RSS) for the linear regression, and also the training RSS for the cubic regression. Would we expect one to be lower than the other, would we expect them to be the same, or is there not enough information to tell? Justify your answer.

(b) Answer (a) using test rather than training RSS.

(c) Suppose that the true relationship between X and Y is not linear, but we don't know how far it is from linear. Consider the training RSS for the linear regression, and also the training RSS for the cubic regression. Would we expect one to be lower than the other, would we expect them to be the same, or is there not enough information to tell? Justify your answer.

(d) Answer (c) using test rather than training RSS.

5. Consider the fitted values that result from performing linear regression without an intercept. In this setting, the ith fitted value takes the form

$$\hat{y}_i = x_i \hat{\beta},$$

where

$$\hat{\beta} = \left(\sum_{i=1}^{n} x_i y_i \right) / \left(\sum_{i'=1}^{n} x_{i'}^2 \right). \tag{3.38}$$

Show that we can write

$$\hat{y}_i = \sum_{i'=1}^{n} a_{i'} y_{i'}.$$

What is $a_{i'}$?

Note: We interpret this result by saying that the fitted values from linear regression are linear combinations *of the response values.*

6. Using (3.4), argue that in the case of simple linear regression, the least squares line always passes through the point $(\bar{x}, \bar{y})$.

7. It is claimed in the text that in the case of simple linear regression of Y onto X, the R^2 statistic (3.17) is equal to the square of the correlation between X and Y (3.18). Prove that this is the case. For simplicity, you may assume that $\bar{x} = \bar{y} = 0$.

Applied

8. This question involves the use of simple linear regression on the Auto data set.

(a) Use the lm() function to perform a simple linear regression with mpg as the response and horsepower as the predictor. Use the summary() function to print the results. Comment on the output. For example:

      i. Is there a relationship between the predictor and the response?

      ii. How strong is the relationship between the predictor and the response?

      iii. Is the relationship between the predictor and the response positive or negative?

      iv. What is the predicted `mpg` associated with a `horsepower` of 98? What are the associated 95 % confidence and prediction intervals?

(b) Plot the response and the predictor. Use the `abline()` function to display the least squares regression line.

(c) Use the `plot()` function to produce diagnostic plots of the least squares regression fit. Comment on any problems you see with the fit.

9. This question involves the use of multiple linear regression on the `Auto` data set.

(a) Produce a scatterplot matrix which includes all of the variables in the data set.

(b) Compute the matrix of correlations between the variables using the function `cor()`. You will need to exclude the `name` variable, which is qualitative.   `cor()`

(c) Use the `lm()` function to perform a multiple linear regression with `mpg` as the response and all other variables except `name` as the predictors. Use the `summary()` function to print the results. Comment on the output. For instance:

      i. Is there a relationship between the predictors and the response?

      ii. Which predictors appear to have a statistically significant relationship to the response?

      iii. What does the coefficient for the `year` variable suggest?

(d) Use the `plot()` function to produce diagnostic plots of the linear regression fit. Comment on any problems you see with the fit. Do the residual plots suggest any unusually large outliers? Does the leverage plot identify any observations with unusually high leverage?

(e) Use the `*` and `:` symbols to fit linear regression models with interaction effects. Do any interactions appear to be statistically significant?

(f) Try a few different transformations of the variables, such as $\log(X)$, $\sqrt{X}$, X^2. Comment on your findings.

10. This question should be answered using the `Carseats` data set.

    (a) Fit a multiple regression model to predict `Sales` using `Price`, `Urban`, and `US`.

    (b) Provide an interpretation of each coefficient in the model. Be careful—some of the variables in the model are qualitative!

    (c) Write out the model in equation form, being careful to handle the qualitative variables properly.

    (d) For which of the predictors can you reject the null hypothesis $H_0 : \beta_j = 0$?

    (e) On the basis of your response to the previous question, fit a smaller model that only uses the predictors for which there is evidence of association with the outcome.

    (f) How well do the models in (a) and (e) fit the data?

    (g) Using the model from (e), obtain 95 % confidence intervals for the coefficient(s).

    (h) Is there evidence of outliers or high leverage observations in the model from (e)?

11. In this problem we will investigate the t-statistic for the null hypothesis $H_0 : \beta = 0$ in simple linear regression without an intercept. To begin, we generate a predictor x and a response y as follows.

```
> set.seed(1)
> x=rnorm(100)
> y=2*x+rnorm(100)
```

    (a) Perform a simple linear regression of y onto x, *without* an intercept. Report the coefficient estimate $\hat{\beta}$, the standard error of this coefficient estimate, and the t-statistic and p-value associated with the null hypothesis $H_0 : \beta = 0$. Comment on these results. (You can perform regression without an intercept using the command `lm(y~x+0)`.)

    (b) Now perform a simple linear regression of x onto y without an intercept, and report the coefficient estimate, its standard error, and the corresponding t-statistic and p-values associated with the null hypothesis $H_0 : \beta = 0$. Comment on these results.

    (c) What is the relationship between the results obtained in (a) and (b)?

    (d) For the regression of Y onto X without an intercept, the t-statistic for $H_0 : \beta = 0$ takes the form $\hat{\beta}/\mathrm{SE}(\hat{\beta})$, where $\hat{\beta}$ is given by (3.38), and where

$$\mathrm{SE}(\hat{\beta}) = \sqrt{\frac{\sum_{i=1}^{n}(y_i - x_i\hat{\beta})^2}{(n-1)\sum_{i'=1}^{n}x_{i'}^2}}.$$

(These formulas are slightly different from those given in Sections 3.1.1 and 3.1.2, since here we are performing regression without an intercept.) Show algebraically, and confirm numerically in R, that the t-statistic can be written as

$$\frac{(\sqrt{n-1})\sum_{i=1}^{n} x_i y_i}{\sqrt{(\sum_{i=1}^{n} x_i^2)(\sum_{i'=1}^{n} y_{i'}^2) - (\sum_{i'=1}^{n} x_{i'} y_{i'})^2}}.$$

(e) Using the results from (d), argue that the t-statistic for the regression of y onto x is the same as the t-statistic for the regression of x onto y.

(f) In R, show that when regression is performed *with* an intercept, the t-statistic for $H_0 : \beta_1 = 0$ is the same for the regression of y onto x as it is for the regression of x onto y.

12. This problem involves simple linear regression without an intercept.

(a) Recall that the coefficient estimate $\hat{\beta}$ for the linear regression of Y onto X without an intercept is given by (3.38). Under what circumstance is the coefficient estimate for the regression of X onto Y the same as the coefficient estimate for the regression of Y onto X?

(b) Generate an example in R with $n = 100$ observations in which the coefficient estimate for the regression of X onto Y is *different from* the coefficient estimate for the regression of Y onto X.

(c) Generate an example in R with $n = 100$ observations in which the coefficient estimate for the regression of X onto Y is *the same as* the coefficient estimate for the regression of Y onto X.

13. In this exercise you will create some simulated data and will fit simple linear regression models to it. Make sure to use set.seed(1) prior to starting part (a) to ensure consistent results.

(a) Using the rnorm() function, create a vector, x, containing 100 observations drawn from a $N(0, 1)$ distribution. This represents a feature, X.

(b) Using the rnorm() function, create a vector, eps, containing 100 observations drawn from a $N(0, 0.25)$ distribution i.e. a normal distribution with mean zero and variance 0.25.

(c) Using x and eps, generate a vector y according to the model

$$Y = -1 + 0.5X + \epsilon. \tag{3.39}$$

What is the length of the vector y? What are the values of β_0 and β_1 in this linear model?

(d) Create a scatterplot displaying the relationship between x and y. Comment on what you observe.

(e) Fit a least squares linear model to predict y using x. Comment on the model obtained. How do $\hat{\beta}_0$ and $\hat{\beta}_1$ compare to β_0 and β_1?

(f) Display the least squares line on the scatterplot obtained in (d). Draw the population regression line on the plot, in a different color. Use the legend() command to create an appropriate legend.

(g) Now fit a polynomial regression model that predicts y using x and x^2. Is there evidence that the quadratic term improves the model fit? Explain your answer.

(h) Repeat (a)–(f) after modifying the data generation process in such a way that there is *less* noise in the data. The model (3.39) should remain the same. You can do this by decreasing the variance of the normal distribution used to generate the error term ϵ in (b). Describe your results.

(i) Repeat (a)–(f) after modifying the data generation process in such a way that there is *more* noise in the data. The model (3.39) should remain the same. You can do this by increasing the variance of the normal distribution used to generate the error term ϵ in (b). Describe your results.

(j) What are the confidence intervals for β_0 and β_1 based on the original data set, the noisier data set, and the less noisy data set? Comment on your results.

14. This problem focuses on the *collinearity* problem.

(a) Perform the following commands in R:

```
> set.seed(1)
> x1=runif(100)
> x2=0.5*x1+rnorm(100)/10
> y=2+2*x1+0.3*x2+rnorm(100)
```

The last line corresponds to creating a linear model in which y is a function of x1 and x2. Write out the form of the linear model. What are the regression coefficients?

(b) What is the correlation between x1 and x2? Create a scatterplot displaying the relationship between the variables.

(c) Using this data, fit a least squares regression to predict y using x1 and x2. Describe the results obtained. What are $\hat{\beta}_0$, $\hat{\beta}_1$, and $\hat{\beta}_2$? How do these relate to the true β_0, β_1, and β_2? Can you reject the null hypothesis $H_0 : \beta_1 = 0$? How about the null hypothesis $H_0 : \beta_2 = 0$?

(d) Now fit a least squares regression to predict y using only x1. Comment on your results. Can you reject the null hypothesis $H_0 : \beta_1 = 0$?

(e) Now fit a least squares regression to predict y using only x2. Comment on your results. Can you reject the null hypothesis $H_0 : \beta_1 = 0$?

(f) Do the results obtained in (c)–(e) contradict each other? Explain your answer.

(g) Now suppose we obtain one additional observation, which was unfortunately mismeasured.

```
> x1=c(x1, 0.1)
> x2=c(x2, 0.8)
> y=c(y,6)
```

Re-fit the linear models from (c) to (e) using this new data. What effect does this new observation have on the each of the models? In each model, is this observation an outlier? A high-leverage point? Both? Explain your answers.

15. This problem involves the Boston data set, which we saw in the lab for this chapter. We will now try to predict per capita crime rate using the other variables in this data set. In other words, per capita crime rate is the response, and the other variables are the predictors.

(a) For each predictor, fit a simple linear regression model to predict the response. Describe your results. In which of the models is there a statistically significant association between the predictor and the response? Create some plots to back up your assertions.

(b) Fit a multiple regression model to predict the response using all of the predictors. Describe your results. For which predictors can we reject the null hypothesis $H_0 : \beta_j = 0$?

(c) How do your results from (a) compare to your results from (b)? Create a plot displaying the univariate regression coefficients from (a) on the x-axis, and the multiple regression coefficients from (b) on the y-axis. That is, each predictor is displayed as a single point in the plot. Its coefficient in a simple linear regression model is shown on the x-axis, and its coefficient estimate in the multiple linear regression model is shown on the y-axis.

(d) Is there evidence of non-linear association between any of the predictors and the response? To answer this question, for each predictor X, fit a model of the form

$$Y = \beta_0 + \beta_1 X + \beta_2 X^2 + \beta_3 X^3 + \epsilon.$$

4

Classification

The linear regression model discussed in Chapter 3 assumes that the response variable Y is quantitative. But in many situations, the response variable is instead *qualitative*. For example, eye color is qualitative, taking on values blue, brown, or green. Often qualitative variables are referred to as *categorical*; we will use these terms interchangeably. In this chapter, we study approaches for predicting qualitative responses, a process that is known as *classification*. Predicting a qualitative response for an observation can be referred to as *classifying* that observation, since it involves assigning the observation to a category, or class. On the other hand, often the methods used for classification first predict the probability of each of the categories of a qualitative variable, as the basis for making the classification. In this sense they also behave like regression methods.

qualitative

classification

There are many possible classification techniques, or *classifiers*, that one might use to predict a qualitative response. We touched on some of these in Sections 2.1.5 and 2.2.3. In this chapter we discuss three of the most widely-used classifiers: *logistic regression, linear discriminant analysis,* and *K-nearest neighbors*. We discuss more computer-intensive methods in later chapters, such as generalized additive models (Chapter 7), trees, random forests, and boosting (Chapter 8), and support vector machines (Chapter 9).

classifier

logistic regression

linear discriminant analysis

K-nearest neighbors

G. James et al., *An Introduction to Statistical Learning: with Applications in R*,   127
Springer Texts in Statistics, DOI 10.1007/978-1-4614-7138-7_4,
© Springer Science+Business Media New York 2013

4.1 An Overview of Classification

Classification problems occur often, perhaps even more so than regression problems. Some examples include:

1. A person arrives at the emergency room with a set of symptoms that could possibly be attributed to one of three medical conditions. Which of the three conditions does the individual have?

2. An online banking service must be able to determine whether or not a transaction being performed on the site is fraudulent, on the basis of the user's IP address, past transaction history, and so forth.

3. On the basis of DNA sequence data for a number of patients with and without a given disease, a biologist would like to figure out which DNA mutations are deleterious (disease-causing) and which are not.

Just as in the regression setting, in the classification setting we have a set of training observations $(x_1, y_1), \ldots, (x_n, y_n)$ that we can use to build a classifier. We want our classifier to perform well not only on the training data, but also on test observations that were not used to train the classifier.

In this chapter, we will illustrate the concept of classification using the simulated Default data set. We are interested in predicting whether an individual will default on his or her credit card payment, on the basis of annual income and monthly credit card balance. The data set is displayed in Figure 4.1. We have plotted annual income and monthly credit card balance for a subset of 10,000 individuals. The left-hand panel of Figure 4.1 displays individuals who defaulted in a given month in orange, and those who did not in blue. (The overall default rate is about 3%, so we have plotted only a fraction of the individuals who did not default.) It appears that individuals who defaulted tended to have higher credit card balances than those who did not. In the right-hand panel of Figure 4.1, two pairs of boxplots are shown. The first shows the distribution of balance split by the binary default variable; the second is a similar plot for income. In this chapter, we learn how to build a model to predict default (Y) for any given value of balance (X_1) and income (X_2). Since Y is not quantitative, the simple linear regression model of Chapter 3 is not appropriate.

It is worth noting that Figure 4.1 displays a very pronounced relationship between the predictor balance and the response default. In most real applications, the relationship between the predictor and the response will not be nearly so strong. However, for the sake of illustrating the classification procedures discussed in this chapter, we use an example in which the relationship between the predictor and the response is somewhat exaggerated.

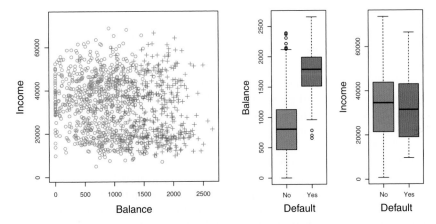

FIGURE 4.1. *The* Default *data set. Left: The annual incomes and monthly credit card balances of a number of individuals. The individuals who defaulted on their credit card payments are shown in orange, and those who did not are shown in blue. Center: Boxplots of* balance *as a function of* default *status. Right: Boxplots of* income *as a function of* default *status.*

4.2 Why Not Linear Regression?

We have stated that linear regression is not appropriate in the case of a qualitative response. Why not?

Suppose that we are trying to predict the medical condition of a patient in the emergency room on the basis of her symptoms. In this simplified example, there are three possible diagnoses: stroke, drug overdose, and epileptic seizure. We could consider encoding these values as a quantitative response variable, Y, as follows:

$$Y = \begin{cases} 1 & \text{if stroke;} \\ 2 & \text{if drug overdose;} \\ 3 & \text{if epileptic seizure.} \end{cases}$$

Using this coding, least squares could be used to fit a linear regression model to predict Y on the basis of a set of predictors $X_1, \ldots, X_p$. Unfortunately, this coding implies an ordering on the outcomes, putting drug overdose in between stroke and epileptic seizure, and insisting that the difference between stroke and drug overdose is the same as the difference between drug overdose and epileptic seizure. In practice there is no particular reason that this needs to be the case. For instance, one could choose an equally reasonable coding,

$$Y = \begin{cases} 1 & \text{if epileptic seizure;} \\ 2 & \text{if stroke;} \\ 3 & \text{if drug overdose.} \end{cases}$$

which would imply a totally different relationship among the three conditions. Each of these codings would produce fundamentally different linear models that would ultimately lead to different sets of predictions on test observations.

If the response variable's values did take on a natural ordering, such as *mild, moderate,* and *severe,* and we felt the gap between mild and moderate was similar to the gap between moderate and severe, then a 1, 2, 3 coding would be reasonable. Unfortunately, in general there is no natural way to convert a qualitative response variable with more than two levels into a quantitative response that is ready for linear regression.

For a *binary* (two level) qualitative response, the situation is better. For instance, perhaps there are only two possibilities for the patient's medical condition: stroke and drug overdose. We could then potentially use the *dummy variable* approach from Section 3.3.1 to code the response as follows:

$$Y = \begin{cases} 0 & \text{if stroke;} \\ 1 & \text{if drug overdose.} \end{cases}$$

We could then fit a linear regression to this binary response, and predict drug overdose if $\hat{Y} > 0.5$ and stroke otherwise. In the binary case it is not hard to show that even if we flip the above coding, linear regression will produce the same final predictions.

For a binary response with a 0/1 coding as above, regression by least squares does make sense; it can be shown that the $X\hat{\beta}$ obtained using linear regression is in fact an estimate of $\Pr(\text{drug overdose}|X)$ in this special case. However, if we use linear regression, some of our estimates might be outside the $[0, 1]$ interval (see Figure 4.2), making them hard to interpret as probabilities! Nevertheless, the predictions provide an ordering and can be interpreted as crude probability estimates. Curiously, it turns out that the classifications that we get if we use linear regression to predict a binary response will be the same as for the linear discriminant analysis (LDA) procedure we discuss in Section 4.4.

However, the dummy variable approach cannot be easily extended to accommodate qualitative responses with more than two levels. For these reasons, it is preferable to use a classification method that is truly suited for qualitative response values, such as the ones presented next.

4.3 Logistic Regression

Consider again the Default data set, where the response default falls into one of two categories, Yes or No. Rather than modeling this response Y directly, logistic regression models the *probability* that Y belongs to a particular category.

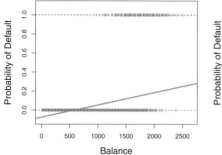 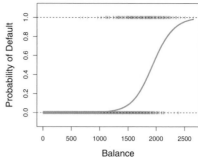

FIGURE 4.2. *Classification using the* Default *data. Left: Estimated probability of* default *using linear regression. Some estimated probabilities are negative! The orange ticks indicate the 0/1 values coded for* default *(No or Yes). Right: Predicted probabilities of* default *using logistic regression. All probabilities lie between 0 and 1.*

For the Default data, logistic regression models the probability of default. For example, the probability of default given balance can be written as

$$\Pr(\texttt{default} = \texttt{Yes}|\texttt{balance}).$$

The values of $\Pr(\texttt{default} = \texttt{Yes}|\texttt{balance})$, which we abbreviate $p(\texttt{balance})$, will range between 0 and 1. Then for any given value of balance, a prediction can be made for default. For example, one might predict default = Yes for any individual for whom $p(\texttt{balance}) > 0.5$. Alternatively, if a company wishes to be conservative in predicting individuals who are at risk for default, then they may choose to use a lower threshold, such as $p(\texttt{balance}) > 0.1$.

4.3.1 The Logistic Model

How should we model the relationship between $p(X) = \Pr(Y = 1|X)$ and X? (For convenience we are using the generic 0/1 coding for the response). In Section 4.2 we talked of using a linear regression model to represent these probabilities:

$$p(X) = \beta_0 + \beta_1 X. \tag{4.1}$$

If we use this approach to predict default=Yes using balance, then we obtain the model shown in the left-hand panel of Figure 4.2. Here we see the problem with this approach: for balances close to zero we predict a negative probability of default; if we were to predict for very large balances, we would get values bigger than 1. These predictions are not sensible, since of course the true probability of default, regardless of credit card balance, must fall between 0 and 1. This problem is not unique to the credit default data. Any time a straight line is fit to a binary response that is coded as

0 or 1, in principle we can always predict $p(X) < 0$ for some values of X and $p(X) > 1$ for others (unless the range of X is limited).

To avoid this problem, we must model $p(X)$ using a function that gives outputs between 0 and 1 for all values of X. Many functions meet this description. In logistic regression, we use the *logistic function*,

logistic function

$$p(X) = \frac{e^{\beta_0 + \beta_1 X}}{1 + e^{\beta_0 + \beta_1 X}}. \tag{4.2}$$

To fit the model (4.2), we use a method called *maximum likelihood*, which we discuss in the next section. The right-hand panel of Figure 4.2 illustrates the fit of the logistic regression model to the `Default` data. Notice that for low balances we now predict the probability of default as close to, but never below, zero. Likewise, for high balances we predict a default probability close to, but never above, one. The logistic function will always produce an *S-shaped* curve of this form, and so regardless of the value of X, we will obtain a sensible prediction. We also see that the logistic model is better able to capture the range of probabilities than is the linear regression model in the left-hand plot. The average fitted probability in both cases is 0.0333 (averaged over the training data), which is the same as the overall proportion of defaulters in the data set.

maximum likelihood

After a bit of manipulation of (4.2), we find that

$$\frac{p(X)}{1 - p(X)} = e^{\beta_0 + \beta_1 X}. \tag{4.3}$$

The quantity $p(X)/[1-p(X)]$ is called the *odds*, and can take on any value between 0 and ∞. Values of the odds close to 0 and ∞ indicate very low and very high probabilities of default, respectively. For example, on average 1 in 5 people with an odds of $1/4$ will default, since $p(X) = 0.2$ implies an odds of $\frac{0.2}{1-0.2} = 1/4$. Likewise on average nine out of every ten people with an odds of 9 will default, since $p(X) = 0.9$ implies an odds of $\frac{0.9}{1-0.9} = 9$. Odds are traditionally used instead of probabilities in horse-racing, since they relate more naturally to the correct betting strategy.

odds

By taking the logarithm of both sides of (4.3), we arrive at

$$\log\left(\frac{p(X)}{1 - p(X)}\right) = \beta_0 + \beta_1 X. \tag{4.4}$$

The left-hand side is called the *log-odds* or *logit*. We see that the logistic regression model (4.2) has a logit that is linear in X.

log-odds logit

Recall from Chapter 3 that in a linear regression model, β_1 gives the average change in Y associated with a one-unit increase in X. In contrast, in a logistic regression model, increasing X by one unit changes the log odds by β_1 (4.4), or equivalently it multiplies the odds by e^{β_1} (4.3). However, because the relationship between $p(X)$ and X in (4.2) is not a straight line,

β_1 does *not* correspond to the change in $p(X)$ associated with a one-unit increase in X. The amount that $p(X)$ changes due to a one-unit change in X will depend on the current value of X. But regardless of the value of X, if β_1 is positive then increasing X will be associated with increasing $p(X)$, and if β_1 is negative then increasing X will be associated with decreasing $p(X)$. The fact that there is not a straight-line relationship between $p(X)$ and X, and the fact that the rate of change in $p(X)$ per unit change in X depends on the current value of X, can also be seen by inspection of the right-hand panel of Figure 4.2.

4.3.2 Estimating the Regression Coefficients

The coefficients β_0 and β_1 in (4.2) are unknown, and must be estimated based on the available training data. In Chapter 3, we used the least squares approach to estimate the unknown linear regression coefficients. Although we could use (non-linear) least squares to fit the model (4.4), the more general method of *maximum likelihood* is preferred, since it has better statistical properties. The basic intuition behind using maximum likelihood to fit a logistic regression model is as follows: we seek estimates for β_0 and β_1 such that the predicted probability $\hat{p}(x_i)$ of default for each individual, using (4.2), corresponds as closely as possible to the individual's observed default status. In other words, we try to find $\hat{\beta}_0$ and $\hat{\beta}_1$ such that plugging these estimates into the model for $p(X)$, given in (4.2), yields a number close to one for all individuals who defaulted, and a number close to zero for all individuals who did not. This intuition can be formalized using a mathematical equation called a *likelihood function*:

$$\ell(\beta_0, \beta_1) = \prod_{i:y_i=1} p(x_i) \prod_{i':y_{i'}=0} (1 - p(x_{i'})). \qquad (4.5)$$

likelihood function

The estimates $\hat{\beta}_0$ and $\hat{\beta}_1$ are chosen to *maximize* this likelihood function.

Maximum likelihood is a very general approach that is used to fit many of the non-linear models that we examine throughout this book. In the linear regression setting, the least squares approach is in fact a special case of maximum likelihood. The mathematical details of maximum likelihood are beyond the scope of this book. However, in general, logistic regression and other models can be easily fit using a statistical software package such as R, and so we do not need to concern ourselves with the details of the maximum likelihood fitting procedure.

Table 4.1 shows the coefficient estimates and related information that result from fitting a logistic regression model on the Default data in order to predict the probability of default=Yes using balance. We see that $\hat{\beta}_1 = 0.0055$; this indicates that an increase in balance is associated with an increase in the probability of default. To be precise, a one-unit increase in balance is associated with an increase in the log odds of default by 0.0055 units.

	Coefficient	Std. error	Z-statistic	P-value
Intercept	−10.6513	0.3612	−29.5	<0.0001
balance	0.0055	0.0002	24.9	<0.0001

TABLE 4.1. *For the* `Default` *data, estimated coefficients of the logistic regression model that predicts the probability of* `default` *using* `balance`*. A one-unit increase in* `balance` *is associated with an increase in the log odds of* `default` *by* 0.0055 *units.*

Many aspects of the logistic regression output shown in Table 4.1 are similar to the linear regression output of Chapter 3. For example, we can measure the accuracy of the coefficient estimates by computing their standard errors. The z-statistic in Table 4.1 plays the same role as the t-statistic in the linear regression output, for example in Table 3.1 on page 68. For instance, the z-statistic associated with β_1 is equal to $\hat{\beta}_1/SE(\hat{\beta}_1)$, and so a large (absolute) value of the z-statistic indicates evidence against the null hypothesis $H_0 : \beta_1 = 0$. This null hypothesis implies that $p(X) = \frac{e^{\beta_0}}{1+e^{\beta_0}}$— in other words, that the probability of `default` does not depend on `balance`. Since the p-value associated with `balance` in Table 4.1 is tiny, we can reject H_0. In other words, we conclude that there is indeed an association between `balance` and probability of `default`. The estimated intercept in Table 4.1 is typically not of interest; its main purpose is to adjust the average fitted probabilities to the proportion of ones in the data.

4.3.3 Making Predictions

Once the coefficients have been estimated, it is a simple matter to compute the probability of `default` for any given credit card balance. For example, using the coefficient estimates given in Table 4.1, we predict that the default probability for an individual with a `balance` of $\$1,000$ is

$$\hat{p}(X) = \frac{e^{\hat{\beta}_0+\hat{\beta}_1 X}}{1 + e^{\hat{\beta}_0+\hat{\beta}_1 X}} = \frac{e^{-10.6513+0.0055\times1,000}}{1 + e^{-10.6513+0.0055\times1,000}} = 0.00576,$$

which is below 1 %. In contrast, the predicted probability of default for an individual with a balance of $\$2,000$ is much higher, and equals 0.586 or 58.6 %.

One can use qualitative predictors with the logistic regression model using the dummy variable approach from Section 3.3.1. As an example, the `Default` data set contains the qualitative variable `student`. To fit the model we simply create a dummy variable that takes on a value of 1 for students and 0 for non-students. The logistic regression model that results from predicting probability of default from student status can be seen in Table 4.2. The coefficient associated with the dummy variable is positive,

	Coefficient	Std. error	Z-statistic	P-value
Intercept	−3.5041	0.0707	−49.55	<0.0001
student[Yes]	0.4049	0.1150	3.52	0.0004

TABLE 4.2. *For the* Default *data, estimated coefficients of the logistic regression model that predicts the probability of* default *using student status. Student status is encoded as a dummy variable, with a value of 1 for a student and a value of 0 for a non-student, and represented by the variable* student[Yes] *in the table.*

and the associated p-value is statistically significant. This indicates that students tend to have higher default probabilities than non-students:

$$\widehat{\Pr}(\texttt{default=Yes}|\texttt{student=Yes}) = \frac{e^{-3.5041+0.4049\times 1}}{1+e^{-3.5041+0.4049\times 1}} = 0.0431,$$

$$\widehat{\Pr}(\texttt{default=Yes}|\texttt{student=No}) = \frac{e^{-3.5041+0.4049\times 0}}{1+e^{-3.5041+0.4049\times 0}} = 0.0292.$$

4.3.4 Multiple Logistic Regression

We now consider the problem of predicting a binary response using multiple predictors. By analogy with the extension from simple to multiple linear regression in Chapter 3, we can generalize (4.4) as follows:

$$\log\left(\frac{p(X)}{1-p(X)}\right) = \beta_0 + \beta_1 X_1 + \cdots + \beta_p X_p, \tag{4.6}$$

where $X = (X_1, \ldots, X_p)$ are p predictors. Equation 4.6 can be rewritten as

$$p(X) = \frac{e^{\beta_0+\beta_1 X_1+\cdots+\beta_p X_p}}{1+e^{\beta_0+\beta_1 X_1+\cdots+\beta_p X_p}}. \tag{4.7}$$

Just as in Section 4.3.2, we use the maximum likelihood method to estimate $\beta_0, \beta_1, \ldots, \beta_p$.

Table 4.3 shows the coefficient estimates for a logistic regression model that uses balance, income (in thousands of dollars), and student status to predict probability of default. There is a surprising result here. The p-values associated with balance and the dummy variable for student status are very small, indicating that each of these variables is associated with the probability of default. However, the coefficient for the dummy variable is negative, indicating that students are less likely to default than non-students. In contrast, the coefficient for the dummy variable is positive in Table 4.2. How is it possible for student status to be associated with an *increase* in probability of default in Table 4.2 and a *decrease* in probability of default in Table 4.3? The left-hand panel of Figure 4.3 provides a graphical illustration of this apparent paradox. The orange and blue solid lines show the average default rates for students and non-students, respectively,

	Coefficient	Std. error	Z-statistic	P-value
Intercept	−10.8690	0.4923	−22.08	<0.0001
balance	0.0057	0.0002	24.74	<0.0001
income	0.0030	0.0082	0.37	0.7115
student[Yes]	−0.6468	0.2362	−2.74	0.0062

TABLE 4.3. *For the* Default *data, estimated coefficients of the logistic regression model that predicts the probability of* default *using* balance, income, *and* student *status. Student status is encoded as a dummy variable* student[Yes], *with a value of* 1 *for a student and a value of* 0 *for a non-student. In fitting this model,* income *was measured in thousands of dollars.*

as a function of credit card balance. The negative coefficient for student in the multiple logistic regression indicates that *for a fixed value of* balance *and* income, a student is less likely to default than a non-student. Indeed, we observe from the left-hand panel of Figure 4.3 that the student default rate is at or below that of the non-student default rate for every value of balance. But the horizontal broken lines near the base of the plot, which show the default rates for students and non-students averaged over all values of balance and income, suggest the opposite effect: the overall student default rate is higher than the non-student default rate. Consequently, there is a positive coefficient for student in the single variable logistic regression output shown in Table 4.2.

The right-hand panel of Figure 4.3 provides an explanation for this discrepancy. The variables student and balance are correlated. Students tend to hold higher levels of debt, which is in turn associated with higher probability of default. In other words, students are more likely to have large credit card balances, which, as we know from the left-hand panel of Figure 4.3, tend to be associated with high default rates. Thus, even though an individual student with a given credit card balance will tend to have a lower probability of default than a non-student with the same credit card balance, the fact that students on the whole tend to have higher credit card balances means that overall, students tend to default at a higher rate than non-students. This is an important distinction for a credit card company that is trying to determine to whom they should offer credit. A student is riskier than a non-student if no information about the student's credit card balance is available. However, that student is less risky than a non-student *with the same credit card balance!*

This simple example illustrates the dangers and subtleties associated with performing regressions involving only a single predictor when other predictors may also be relevant. As in the linear regression setting, the results obtained using one predictor may be quite different from those obtained using multiple predictors, especially when there is correlation among the predictors. In general, the phenomenon seen in Figure 4.3 is known as *confounding.*

confounding

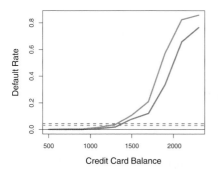 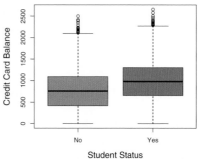

FIGURE 4.3. *Confounding in the* Default *data. Left: Default rates are shown for students (orange) and non-students (blue). The solid lines display default rate as a function of* balance, *while the horizontal broken lines display the overall default rates. Right: Boxplots of* balance *for students (orange) and non-students (blue) are shown.*

By substituting estimates for the regression coefficients from Table 4.3 into (4.7), we can make predictions. For example, a student with a credit card balance of $1,500 and an income of $40,000 has an estimated probability of default of

$$\hat{p}(X) = \frac{e^{-10.869+0.00574\times1,500+0.003\times40-0.6468\times1}}{1+e^{-10.869+0.00574\times1,500+0.003\times40-0.6468\times1}} = 0.058. \qquad (4.8)$$

A non-student with the same balance and income has an estimated probability of default of

$$\hat{p}(X) = \frac{e^{-10.869+0.00574\times1,500+0.003\times40-0.6468\times0}}{1+e^{-10.869+0.00574\times1,500+0.003\times40-0.6468\times0}} = 0.105. \qquad (4.9)$$

(Here we multiply the income coefficient estimate from Table 4.3 by 40, rather than by 40,000, because in that table the model was fit with income measured in units of $1,000.)

4.3.5 *Logistic Regression for >2 Response Classes*

We sometimes wish to classify a response variable that has more than two classes. For example, in Section 4.2 we had three categories of medical condition in the emergency room: stroke, drug overdose, epileptic seizure. In this setting, we wish to model both $\Pr(Y = \text{stroke}|X)$ and $\Pr(Y = \text{drug overdose}|X)$, with the remaining $\Pr(Y = \text{epileptic seizure}|X) = 1 - \Pr(Y = \text{stroke}|X) - \Pr(Y = \text{drug overdose}|X)$. The two-class logistic regression models discussed in the previous sections have multiple-class extensions, but in practice they tend not to be used all that often. One of the reasons is that the method we discuss in the next section, *discriminant analysis*, is popular for multiple-class classification. So we do not go into the details of multiple-class logistic regression here, but simply note that such an approach is possible, and that software for it is available in R.

4.4   Linear Discriminant Analysis

Logistic regression involves directly modeling $\Pr(Y = k|X = x)$ using the logistic function, given by (4.7) for the case of two response classes. In statistical jargon, we model the conditional distribution of the response Y, given the predictor(s) X. We now consider an alternative and less direct approach to estimating these probabilities. In this alternative approach, we model the distribution of the predictors X separately in each of the response classes (i.e. given Y), and then use Bayes' theorem to flip these around into estimates for $\Pr(Y = k|X = x)$. When these distributions are assumed to be normal, it turns out that the model is very similar in form to logistic regression.

Why do we need another method, when we have logistic regression? There are several reasons:

- When the classes are well-separated, the parameter estimates for the logistic regression model are surprisingly unstable. Linear discriminant analysis does not suffer from this problem.

- If n is small and the distribution of the predictors X is approximately normal in each of the classes, the linear discriminant model is again more stable than the logistic regression model.

- As mentioned in Section 4.3.5, linear discriminant analysis is popular when we have more than two response classes.

4.4.1   Using Bayes' Theorem for Classification

Suppose that we wish to classify an observation into one of K classes, where $K \geq 2$. In other words, the qualitative response variable Y can take on K possible distinct and unordered values. Let π_k represent the overall or *prior* probability that a randomly chosen observation comes from the kth class; this is the probability that a given observation is associated with the kth category of the response variable Y. Let $f_k(x) \equiv \Pr(X = x|Y = k)$[1] denote the *density function* of X for an observation that comes from the kth class. In other words, $f_k(x)$ is relatively large if there is a high probability that an observation in the kth class has $X \approx x$, and $f_k(x)$ is small if it is very

prior

density function

---

[1]Technically this definition is only correct if X is a discrete random variable. If X is continuous then $f_k(x)dx$ would correspond to the probability of X falling in in a small region dx around x.

unlikely that an observation in the kth class has $X \approx x$. Then *Bayes' theorem* states that

$$\Pr(Y = k | X = x) = \frac{\pi_k f_k(x)}{\sum_{l=1}^{K} \pi_l f_l(x)}. \qquad (4.10)$$

In accordance with our earlier notation, we will use the abbreviation $p_k(X)$ $= \Pr(Y = k | X)$. This suggests that instead of directly computing $p_k(X)$ as in Section 4.3.1, we can simply plug in estimates of π_k and $f_k(X)$ into (4.10). In general, estimating π_k is easy if we have a random sample of Ys from the population: we simply compute the fraction of the training observations that belong to the kth class. However, estimating $f_k(X)$ tends to be more challenging, unless we assume some simple forms for these densities. We refer to $p_k(x)$ as the *posterior* probability that an observation $X = x$ belongs to the kth class. That is, it is the probability that the observation belongs to the kth class, *given* the predictor value for that observation.

We know from Chapter 2 that the Bayes classifier, which classifies an observation to the class for which $p_k(X)$ is largest, has the lowest possible error rate out of all classifiers. (This is of course only true if the terms in (4.10) are all correctly specified.) Therefore, if we can find a way to estimate $f_k(X)$, then we can develop a classifier that approximates the Bayes classifier. Such an approach is the topic of the following sections.

4.4.2 Linear Discriminant Analysis for $p = 1$

For now, assume that $p = 1$—that is, we have only one predictor. We would like to obtain an estimate for $f_k(x)$ that we can plug into (4.10) in order to estimate $p_k(x)$. We will then classify an observation to the class for which $p_k(x)$ is greatest. In order to estimate $f_k(x)$, we will first make some assumptions about its form.

Suppose we assume that $f_k(x)$ is *normal* or *Gaussian*. In the one-dimensional setting, the normal density takes the form

$$f_k(x) = \frac{1}{\sqrt{2\pi}\sigma_k} \exp\left(-\frac{1}{2\sigma_k^2}(x - \mu_k)^2\right), \qquad (4.11)$$

where μ_k and σ_k^2 are the mean and variance parameters for the kth class. For now, let us further assume that $\sigma_1^2 = \ldots = \sigma_K^2$: that is, there is a shared variance term across all K classes, which for simplicity we can denote by σ^2. Plugging (4.11) into (4.10), we find that

$$p_k(x) = \frac{\pi_k \frac{1}{\sqrt{2\pi}\sigma} \exp\left(-\frac{1}{2\sigma^2}(x - \mu_k)^2\right)}{\sum_{l=1}^{K} \pi_l \frac{1}{\sqrt{2\pi}\sigma} \exp\left(-\frac{1}{2\sigma^2}(x - \mu_l)^2\right)}. \qquad (4.12)$$

(Note that in (4.12), π_k denotes the prior probability that an observation belongs to the kth class, not to be confused with $\pi \approx 3.14159$, the mathematical constant.) The Bayes classifier involves assigning an observation

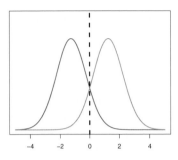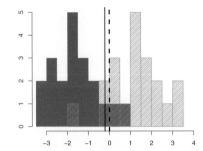

FIGURE 4.4. Left: *Two one-dimensional normal density functions are shown. The dashed vertical line represents the Bayes decision boundary.* Right: *20 observations were drawn from each of the two classes, and are shown as histograms. The Bayes decision boundary is again shown as a dashed vertical line. The solid vertical line represents the LDA decision boundary estimated from the training data.*

$X = x$ to the class for which (4.12) is largest. Taking the log of (4.12) and rearranging the terms, it is not hard to show that this is equivalent to assigning the observation to the class for which

$$\delta_k(x) = x \cdot \frac{\mu_k}{\sigma^2} - \frac{\mu_k^2}{2\sigma^2} + \log(\pi_k) \qquad (4.13)$$

is largest. For instance, if $K = 2$ and $\pi_1 = \pi_2$, then the Bayes classifier assigns an observation to class 1 if $2x(\mu_1 - \mu_2) > \mu_1^2 - \mu_2^2$, and to class 2 otherwise. In this case, the Bayes decision boundary corresponds to the point where

$$x = \frac{\mu_1^2 - \mu_2^2}{2(\mu_1 - \mu_2)} = \frac{\mu_1 + \mu_2}{2}. \qquad (4.14)$$

An example is shown in the left-hand panel of Figure 4.4. The two normal density functions that are displayed, $f_1(x)$ and $f_2(x)$, represent two distinct classes. The mean and variance parameters for the two density functions are $\mu_1 = -1.25$, $\mu_2 = 1.25$, and $\sigma_1^2 = \sigma_2^2 = 1$. The two densities overlap, and so given that $X = x$, there is some uncertainty about the class to which the observation belongs. If we assume that an observation is equally likely to come from either class—that is, $\pi_1 = \pi_2 = 0.5$—then by inspection of (4.14), we see that the Bayes classifier assigns the observation to class 1 if $x < 0$ and class 2 otherwise. Note that in this case, we can compute the Bayes classifier because we know that X is drawn from a Gaussian distribution within each class, and we know all of the parameters involved. In a real-life situation, we are not able to calculate the Bayes classifier.

In practice, even if we are quite certain of our assumption that X is drawn from a Gaussian distribution within each class, we still have to estimate the parameters $\mu_1, \ldots, \mu_K$, $\pi_1, \ldots, \pi_K$, and σ^2. The *linear discriminant*

analysis (LDA) method approximates the Bayes classifier by plugging estimates for π_k, μ_k, and σ^2 into (4.13). In particular, the following estimates
are used:

linear
discriminant
analysis

$$\hat{\mu}_k = \frac{1}{n_k} \sum_{i:y_i=k} x_i$$

$$\hat{\sigma}^2 = \frac{1}{n-K} \sum_{k=1}^{K} \sum_{i:y_i=k} (x_i - \hat{\mu}_k)^2 \qquad (4.15)$$

where n is the total number of training observations, and n_k is the number
of training observations in the kth class. The estimate for μ_k is simply the
average of all the training observations from the kth class, while $\hat{\sigma}^2$ can
be seen as a weighted average of the sample variances for each of the K
classes. Sometimes we have knowledge of the class membership probabilities $\pi_1, \ldots, \pi_K$, which can be used directly. In the absence of any additional
information, LDA estimates π_k using the proportion of the training observations that belong to the kth class. In other words,

$$\hat{\pi}_k = n_k/n. \qquad (4.16)$$

The LDA classifier plugs the estimates given in (4.15) and (4.16) into (4.13),
and assigns an observation $X = x$ to the class for which

$$\hat{\delta}_k(x) = x \cdot \frac{\hat{\mu}_k}{\hat{\sigma}^2} - \frac{\hat{\mu}_k^2}{2\hat{\sigma}^2} + \log(\hat{\pi}_k) \qquad (4.17)$$

is largest. The word *linear* in the classifier's name stems from the fact
that the *discriminant functions* $\hat{\delta}_k(x)$ in (4.17) are linear functions of x (as
opposed to a more complex function of x).

discriminant
function

The right-hand panel of Figure 4.4 displays a histogram of a random
sample of 20 observations from each class. To implement LDA, we began
by estimating π_k, μ_k, and σ^2 using (4.15) and (4.16). We then computed the
decision boundary, shown as a black solid line, that results from assigning
an observation to the class for which (4.17) is largest. All points to the left
of this line will be assigned to the green class, while points to the right of
this line are assigned to the purple class. In this case, since $n_1 = n_2 = 20$,
we have $\hat{\pi}_1 = \hat{\pi}_2$. As a result, the decision boundary corresponds to the
midpoint between the sample means for the two classes, $(\hat{\mu}_1 + \hat{\mu}_2)/2$. The
figure indicates that the LDA decision boundary is slightly to the left of
the optimal Bayes decision boundary, which instead equals $(\mu_1 + \mu_2)/2 =
0$. How well does the LDA classifier perform on this data? Since this is
simulated data, we can generate a large number of test observations in order
to compute the Bayes error rate and the LDA test error rate. These are
10.6 % and 11.1 %, respectively. In other words, the LDA classifier's error
rate is only 0.5 % above the smallest possible error rate! This indicates that
LDA is performing pretty well on this data set.

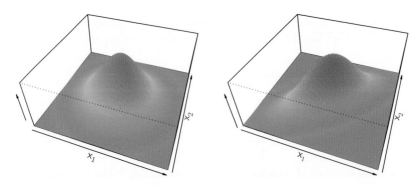

FIGURE 4.5. *Two multivariate Gaussian density functions are shown, with*
$p = 2$. Left: *The two predictors are uncorrelated.* Right: *The two variables have
a correlation of* 0.7.

To reiterate, the LDA classifier results from assuming that the observations within each class come from a normal distribution with a class-specific
mean vector and a common variance σ^2, and plugging estimates for these
parameters into the Bayes classifier. In Section 4.4.4, we will consider a less
stringent set of assumptions, by allowing the observations in the kth class
to have a class-specific variance, σ_k^2.

4.4.3   Linear Discriminant Analysis for $p > 1$

We now extend the LDA classifier to the case of multiple predictors. To
do this, we will assume that $X = (X_1, X_2, \ldots, X_p)$ is drawn from a *multivariate Gaussian* (or multivariate normal) distribution, with a class-specific
mean vector and a common covariance matrix. We begin with a brief review
of such a distribution.

    The multivariate Gaussian distribution assumes that each individual pre-
dictor follows a one-dimensional normal distribution, as in (4.11), with some
correlation between each pair of predictors. Two examples of multivariate
Gaussian distributions with $p = 2$ are shown in Figure 4.5. The height of
the surface at any particular point represents the probability that both X_1
and X_2 fall in a small region around that point. In either panel, if the surface is cut along the X_1 axis or along the X_2 axis, the resulting cross-section
will have the shape of a one-dimensional normal distribution. The left-hand
panel of Figure 4.5 illustrates an example in which $\text{Var}(X_1) = \text{Var}(X_2)$ and
$\text{Cor}(X_1, X_2) = 0$; this surface has a characteristic *bell shape*. However, the
bell shape will be distorted if the predictors are correlated or have unequal
variances, as is illustrated in the right-hand panel of Figure 4.5. In this
situation, the base of the bell will have an elliptical, rather than circular,

multivariate
Gaussian

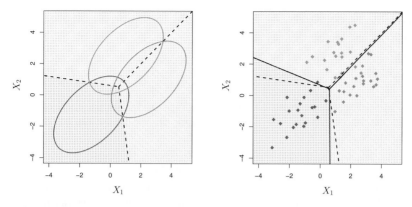

FIGURE 4.6. *An example with three classes. The observations from each class are drawn from a multivariate Gaussian distribution with $p = 2$, with a class-specific mean vector and a common covariance matrix. Left: Ellipses that contain 95 % of the probability for each of the three classes are shown. The dashed lines are the Bayes decision boundaries. Right: 20 observations were generated from each class, and the corresponding LDA decision boundaries are indicated using solid black lines. The Bayes decision boundaries are once again shown as dashed lines.*

shape. To indicate that a p-dimensional random variable X has a multivariate Gaussian distribution, we write $X \sim N(\mu, \boldsymbol{\Sigma})$. Here $E(X) = \mu$ is the mean of X (a vector with p components), and $\mathrm{Cov}(X) = \boldsymbol{\Sigma}$ is the $p \times p$ covariance matrix of X. Formally, the multivariate Gaussian density is defined as

$$f(x) = \frac{1}{(2\pi)^{p/2}|\boldsymbol{\Sigma}|^{1/2}} \exp\left(-\frac{1}{2}(x - \mu)^T \boldsymbol{\Sigma}^{-1}(x - \mu)\right). \qquad (4.18)$$

In the case of $p > 1$ predictors, the LDA classifier assumes that the observations in the kth class are drawn from a multivariate Gaussian distribution $N(\mu_k, \boldsymbol{\Sigma})$, where μ_k is a class-specific mean vector, and $\boldsymbol{\Sigma}$ is a covariance matrix that is common to all K classes. Plugging the density function for the kth class, $f_k(X = x)$, into (4.10) and performing a little bit of algebra reveals that the Bayes classifier assigns an observation $X = x$ to the class for which

$$\delta_k(x) = x^T \boldsymbol{\Sigma}^{-1} \mu_k - \frac{1}{2}\mu_k^T \boldsymbol{\Sigma}^{-1}\mu_k + \log \pi_k \qquad (4.19)$$

is largest. This is the vector/matrix version of (4.13).

An example is shown in the left-hand panel of Figure 4.6. Three equally-sized Gaussian classes are shown with class-specific mean vectors and a common covariance matrix. The three ellipses represent regions that contain 95 % of the probability for each of the three classes. The dashed lines

are the Bayes decision boundaries. In other words, they represent the set
of values x for which $\delta_k(x) = \delta_\ell(x)$; i.e.

$$x^T\mathbf{\Sigma}^{-1}\mu_k - \frac{1}{2}\mu_k^T\mathbf{\Sigma}^{-1}\mu_k = x^T\mathbf{\Sigma}^{-1}\mu_l - \frac{1}{2}\mu_l^T\mathbf{\Sigma}^{-1}\mu_l \qquad (4.20)$$

for $k \neq l$. (The $\log \pi_k$ term from (4.19) has disappeared because each of
the three classes has the same number of training observations; i.e. π_k is
the same for each class.) Note that there are three lines representing the
Bayes decision boundaries because there are three *pairs of classes* among
the three classes. That is, one Bayes decision boundary separates class 1
from class 2, one separates class 1 from class 3, and one separates class 2
from class 3. These three Bayes decision boundaries divide the predictor
space into three regions. The Bayes classifier will classify an observation
according to the region in which it is located.

Once again, we need to estimate the unknown parameters $\mu_1, \ldots, \mu_K$,
$\pi_1, \ldots, \pi_K$, and $\mathbf{\Sigma}$; the formulas are similar to those used in the one-
dimensional case, given in (4.15). To assign a new observation $X = x$,
LDA plugs these estimates into (4.19) and classifies to the class for which
$\hat{\delta}_k(x)$ is largest. Note that in (4.19) $\delta_k(x)$ is a linear function of x; that is,
the LDA decision rule depends on x only through a linear combination of
its elements. Once again, this is the reason for the word *linear* in LDA.

In the right-hand panel of Figure 4.6, 20 observations drawn from each of
the three classes are displayed, and the resulting LDA decision boundaries
are shown as solid black lines. Overall, the LDA decision boundaries are
pretty close to the Bayes decision boundaries, shown again as dashed lines.
The test error rates for the Bayes and LDA classifiers are 0.0746 and 0.0770,
respectively. This indicates that LDA is performing well on this data.

We can perform LDA on the `Default` data in order to predict whether
or not an individual will default on the basis of credit card balance and
student status. The LDA model fit to the 10,000 training samples results
in a *training* error rate of 2.75 %. This sounds like a low error rate, but two
caveats must be noted.

- First of all, training error rates will usually be lower than test error
  rates, which are the real quantity of interest. In other words, we
  might expect this classifier to perform worse if we use it to predict
  whether or not a new set of individuals will default. The reason is
  that we specifically adjust the parameters of our model to do well on
  the training data. The higher the ratio of parameters p to number
  of samples n, the more we expect this *overfitting* to play a role. For   *overfitting*
  these data we don't expect this to be a problem, since $p = 2$ and
  $n = 10,000$.

- Second, since only 3.33 % of the individuals in the training sample
  defaulted, a simple but useless classifier that always predicts that

		True default status		
		No	Yes	Total
Predicted	No	9,644	252	9,896
default status	Yes	23	81	104
	Total	9,667	333	10,000

TABLE 4.4. *A confusion matrix compares the LDA predictions to the true default statuses for the* 10,000 *training observations in the* Default *data set. Elements on the diagonal of the matrix represent individuals whose default statuses were correctly predicted, while off-diagonal elements represent individuals that were misclassified. LDA made incorrect predictions for 23 individuals who did not default and for 252 individuals who did default.*

each individual will not default, regardless of his or her credit card balance and student status, will result in an error rate of 3.33%. In other words, the trivial *null* classifier will achieve an error rate that is only a bit higher than the LDA training set error rate.

null

In practice, a binary classifier such as this one can make two types of errors: it can incorrectly assign an individual who defaults to the *no default* category, or it can incorrectly assign an individual who does not default to the *default* category. It is often of interest to determine which of these two types of errors are being made. A *confusion matrix*, shown for the Default data in Table 4.4, is a convenient way to display this information. The table reveals that LDA predicted that a total of 104 people would default. Of these people, 81 actually defaulted and 23 did not. Hence only 23 out of 9,667 of the individuals who did not default were incorrectly labeled. This looks like a pretty low error rate! However, of the 333 individuals who defaulted, 252 (or 75.7%) were missed by LDA. So while the overall error rate is low, the error rate among individuals who defaulted is very high. From the perspective of a credit card company that is trying to identify high-risk individuals, an error rate of $252/333 = 75.7\%$ among individuals who default may well be unacceptable.

confusion matrix

Class-specific performance is also important in medicine and biology, where the terms *sensitivity* and *specificity* characterize the performance of a classifier or screening test. In this case the sensitivity is the percentage of true defaulters that are identified, a low 24.3% in this case. The specificity is the percentage of non-defaulters that are correctly identified, here $(1 - 23/9,667) \times 100 = 99.8\%$.

sensitivity
specificity

Why does LDA do such a poor job of classifying the customers who default? In other words, why does it have such a low sensitivity? As we have seen, LDA is trying to approximate the Bayes classifier, which has the lowest *total* error rate out of all classifiers (if the Gaussian model is correct). That is, the Bayes classifier will yield the smallest possible total number of misclassified observations, irrespective of which class the errors come from. That is, some misclassifications will result from incorrectly assigning

| | | True default status | | |
		No	Yes	Total
Predicted	No	9, 432	138	9, 570
default status	Yes	235	195	430
	Total	9, 667	333	10, 000

TABLE 4.5. *A confusion matrix compares the LDA predictions to the true default statuses for the* 10, 000 *training observations in the* Default *data set, using a modified threshold value that predicts default for any individuals whose posterior default probability exceeds* 20 %.

a customer who does not default to the default class, and others will result from incorrectly assigning a customer who defaults to the non-default class. In contrast, a credit card company might particularly wish to avoid incorrectly classifying an individual who will default, whereas incorrectly classifying an individual who will not default, though still to be avoided, is less problematic. We will now see that it is possible to modify LDA in order to develop a classifier that better meets the credit card company's needs.

The Bayes classifier works by assigning an observation to the class for which the posterior probability $p_k(X)$ is greatest. In the two-class case, this amounts to assigning an observation to the *default* class if

$$\Pr(\texttt{default} = \texttt{Yes}|X = x) > 0.5. \qquad (4.21)$$

Thus, the Bayes classifier, and by extension LDA, uses a threshold of 50 % for the posterior probability of default in order to assign an observation to the *default* class. However, if we are concerned about incorrectly predicting the default status for individuals who default, then we can consider lowering this threshold. For instance, we might label any customer with a posterior probability of default above 20 % to the *default* class. In other words, instead of assigning an observation to the *default* class if (4.21) holds, we could instead assign an observation to this class if

$$\Pr(\texttt{default} = \texttt{Yes}|X = x) > 0.2. \qquad (4.22)$$

The error rates that result from taking this approach are shown in Table 4.5. Now LDA predicts that 430 individuals will default. Of the 333 individuals who default, LDA correctly predicts all but 138, or 41.4 %. This is a vast improvement over the error rate of 75.7 % that resulted from using the threshold of 50 %. However, this improvement comes at a cost: now 235 individuals who do not default are incorrectly classified. As a result, the overall error rate has increased slightly to 3.73 %. But a credit card company may consider this slight increase in the total error rate to be a small price to pay for more accurate identification of individuals who do indeed default.

Figure 4.7 illustrates the trade-off that results from modifying the threshold value for the posterior probability of default. Various error rates are

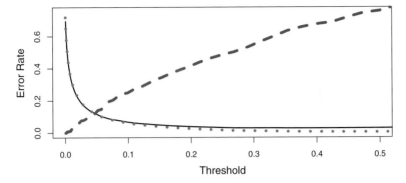

FIGURE 4.7. *For the* Default *data set, error rates are shown as a function of the threshold value for the posterior probability that is used to perform the assignment. The black solid line displays the overall error rate. The blue dashed line represents the fraction of defaulting customers that are incorrectly classified, and the orange dotted line indicates the fraction of errors among the non-defaulting customers.*

shown as a function of the threshold value. Using a threshold of 0.5, as in (4.21), minimizes the overall error rate, shown as a black solid line. This is to be expected, since the Bayes classifier uses a threshold of 0.5 and is known to have the lowest overall error rate. But when a threshold of 0.5 is used, the error rate among the individuals who default is quite high (blue dashed line). As the threshold is reduced, the error rate among individuals who default decreases steadily, but the error rate among the individuals who do not default increases. How can we decide which threshold value is best? Such a decision must be based on *domain knowledge*, such as detailed information about the costs associated with default.

The *ROC curve* is a popular graphic for simultaneously displaying the ROC curve
two types of errors for all possible thresholds. The name "ROC" is historic, and comes from communications theory. It is an acronym for *receiver operating characteristics*. Figure 4.8 displays the ROC curve for the LDA classifier on the training data. The overall performance of a classifier, summarized over all possible thresholds, is given by the *area under the (ROC) curve* (AUC). An ideal ROC curve will hug the top left corner, so the larger area under the AUC the better the classifier. For this data the AUC is 0.95, which is the (ROC) close to the maximum of one so would be considered very good. We expect curve a classifier that performs no better than chance to have an AUC of 0.5 (when evaluated on an independent test set not used in model training). ROC curves are useful for comparing different classifiers, since they take into account all possible thresholds. It turns out that the ROC curve for the logistic regression model of Section 4.3.4 fit to these data is virtually indistinguishable from this one for the LDA model, so we do not display it here.

As we have seen above, varying the classifier threshold changes its true positive and false positive rate. These are also called the *sensitivity* and one sensitivity

ROC Curve

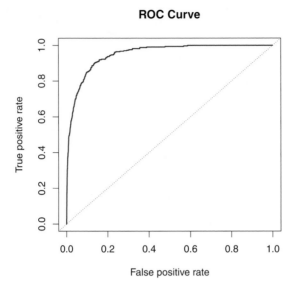

FIGURE 4.8. *A ROC curve for the LDA classifier on the* Default *data. It traces out two types of error as we vary the threshold value for the posterior probability of default. The actual thresholds are not shown. The true positive rate is the sensitivity: the fraction of defaulters that are correctly identified, using a given threshold value. The false positive rate is 1-specificity: the fraction of non-defaulters that we classify incorrectly as defaulters, using that same threshold value. The ideal ROC curve hugs the top left corner, indicating a high true positive rate and a low false positive rate. The dotted line represents the "no information" classifier; this is what we would expect if student status and credit card balance are not associated with probability of default.*

		Predicted class		Total
		− or Null	+ or Non-null	
True	− or Null	True Neg. (TN)	False Pos. (FP)	N
class	+ or Non-null	False Neg. (FN)	True Pos. (TP)	P
	Total	N*	P*	

TABLE 4.6. *Possible results when applying a classifier or diagnostic test to a population.*

minus the *specificity* of our classifier. Since there is an almost bewildering array of terms used in this context, we now give a summary. Table 4.6 shows the possible results when applying a classifier (or diagnostic test) to a population. To make the connection with the epidemiology literature, we think of "+" as the "disease" that we are trying to detect, and "−" as the "non-disease" state. To make the connection to the classical hypothesis testing literature, we think of "−" as the null hypothesis and "+" as the alternative (non-null) hypothesis. In the context of the Default data, "+" indicates an individual who defaults, and "−" indicates one who does not.

specificity

Name	Definition	Synonyms
False Pos. rate	FP/N	Type I error, 1−Specificity
True Pos. rate	TP/P	1−Type II error, power, sensitivity, recall
Pos. Pred. value	TP/P*	Precision, 1−false discovery proportion
Neg. Pred. value	TN/N*	

TABLE 4.7. *Important measures for classification and diagnostic testing, derived from quantities in Table 4.6.*

Table 4.7 lists many of the popular performance measures that are used in this context. The denominators for the false positive and true positive rates are the actual population counts in each class. In contrast, the denominators for the positive predictive value and the negative predictive value are the total predicted counts for each class.

4.4.4 Quadratic Discriminant Analysis

As we have discussed, LDA assumes that the observations within each class are drawn from a multivariate Gaussian distribution with a class-specific mean vector and a covariance matrix that is common to all K classes. *Quadratic discriminant analysis* (QDA) provides an alternative approach. Like LDA, the QDA classifier results from assuming that the observations from each class are drawn from a Gaussian distribution, and plugging estimates for the parameters into Bayes' theorem in order to perform prediction. However, unlike LDA, QDA assumes that each class has its own covariance matrix. That is, it assumes that an observation from the kth class is of the form $X \sim N(\mu_k, \boldsymbol{\Sigma}_k)$, where $\boldsymbol{\Sigma}_k$ is a covariance matrix for the kth class. Under this assumption, the Bayes classifier assigns an observation $X = x$ to the class for which

$$
\begin{aligned}
\delta_k(x) &= -\frac{1}{2}(x - \mu_k)^T \boldsymbol{\Sigma}_k^{-1}(x - \mu_k) - \frac{1}{2}\log|\boldsymbol{\Sigma}_k| + \log \pi_k \\
&= -\frac{1}{2}x^T \boldsymbol{\Sigma}_k^{-1} x + x^T \boldsymbol{\Sigma}_k^{-1} \mu_k - \frac{1}{2}\mu_k^T \boldsymbol{\Sigma}_k^{-1} \mu_k - \frac{1}{2}\log|\boldsymbol{\Sigma}_k| + \log \pi_k
\end{aligned}
\tag{4.23}
$$

is largest. So the QDA classifier involves plugging estimates for $\boldsymbol{\Sigma}_k$, μ_k, and π_k into (4.23), and then assigning an observation $X = x$ to the class for which this quantity is largest. Unlike in (4.19), the quantity x appears as a *quadratic* function in (4.23). This is where QDA gets its name.

Why does it matter whether or not we assume that the K classes share a common covariance matrix? In other words, why would one prefer LDA to QDA, or vice-versa? The answer lies in the bias-variance trade-off. When there are p predictors, then estimating a covariance matrix requires estimating $p(p+1)/2$ parameters. QDA estimates a separate covariance matrix for each class, for a total of $Kp(p+1)/2$ parameters. With 50 predictors this

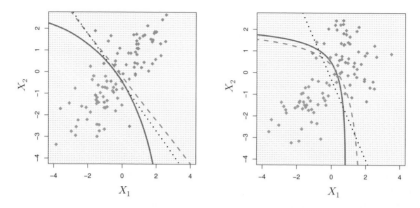

FIGURE 4.9. Left: *The Bayes (purple dashed), LDA (black dotted), and QDA (green solid) decision boundaries for a two-class problem with* $\Sigma_1 = \Sigma_2$. *The shading indicates the QDA decision rule. Since the Bayes decision boundary is linear, it is more accurately approximated by LDA than by QDA. Right: Details are as given in the left-hand panel, except that* $\Sigma_1 \neq \Sigma_2$. *Since the Bayes decision boundary is non-linear, it is more accurately approximated by QDA than by LDA.*

is some multiple of 1,275, which is a lot of parameters. By instead assuming that the K classes share a common covariance matrix, the LDA model becomes linear in x, which means there are Kp linear coefficients to estimate. Consequently, LDA is a much less flexible classifier than QDA, and so has substantially lower variance. This can potentially lead to improved prediction performance. But there is a trade-off: if LDA's assumption that the K classes share a common covariance matrix is badly off, then LDA can suffer from high bias. Roughly speaking, LDA tends to be a better bet than QDA if there are relatively few training observations and so reducing variance is crucial. In contrast, QDA is recommended if the training set is very large, so that the variance of the classifier is not a major concern, or if the assumption of a common covariance matrix for the K classes is clearly untenable.

Figure 4.9 illustrates the performances of LDA and QDA in two scenarios. In the left-hand panel, the two Gaussian classes have a common correlation of 0.7 between X_1 and X_2. As a result, the Bayes decision boundary is linear and is accurately approximated by the LDA decision boundary. The QDA decision boundary is inferior, because it suffers from higher variance without a corresponding decrease in bias. In contrast, the right-hand panel displays a situation in which the orange class has a correlation of 0.7 between the variables and the blue class has a correlation of -0.7. Now the Bayes decision boundary is quadratic, and so QDA more accurately approximates this boundary than does LDA.

4.5   A Comparison of Classification Methods

In this chapter, we have considered three different classification approaches: logistic regression, LDA, and QDA. In Chapter 2, we also discussed the K-nearest neighbors (KNN) method. We now consider the types of scenarios in which one approach might dominate the others.

Though their motivations differ, the logistic regression and LDA methods are closely connected. Consider the two-class setting with $p = 1$ predictor, and let $p_1(x)$ and $p_2(x) = 1 - p_1(x)$ be the probabilities that the observation $X = x$ belongs to class 1 and class 2, respectively. In the LDA framework, we can see from (4.12) to (4.13) (and a bit of simple algebra) that the log odds is given by

$$\log\left(\frac{p_1(x)}{1 - p_1(x)}\right) = \log\left(\frac{p_1(x)}{p_2(x)}\right) = c_0 + c_1 x, \qquad (4.24)$$

where c_0 and c_1 are functions of μ_1, μ_2, and σ^2. From (4.4), we know that in logistic regression,

$$\log\left(\frac{p_1}{1 - p_1}\right) = \beta_0 + \beta_1 x. \qquad (4.25)$$

Both (4.24) and (4.25) are linear functions of x. Hence, both logistic regression and LDA produce linear decision boundaries. The only difference between the two approaches lies in the fact that β_0 and β_1 are estimated using maximum likelihood, whereas c_0 and c_1 are computed using the estimated mean and variance from a normal distribution. This same connection between LDA and logistic regression also holds for multidimensional data with $p > 1$.

Since logistic regression and LDA differ only in their fitting procedures, one might expect the two approaches to give similar results. This is often, but not always, the case. LDA assumes that the observations are drawn from a Gaussian distribution with a common covariance matrix in each class, and so can provide some improvements over logistic regression when this assumption approximately holds. Conversely, logistic regression can outperform LDA if these Gaussian assumptions are not met.

Recall from Chapter 2 that KNN takes a completely different approach from the classifiers seen in this chapter. In order to make a prediction for an observation $X = x$, the K training observations that are closest to x are identified. Then X is assigned to the class to which the plurality of these observations belong. Hence KNN is a completely non-parametric approach: no assumptions are made about the shape of the decision boundary. There- fore, we can expect this approach to dominate LDA and logistic regression when the decision boundary is highly non-linear. On the other hand, KNN does not tell us which predictors are important; we don't get a table of coefficients as in Table 4.3.

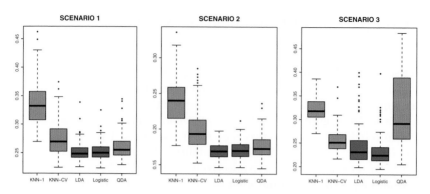

FIGURE 4.10. *Boxplots of the test error rates for each of the linear scenarios described in the main text.*

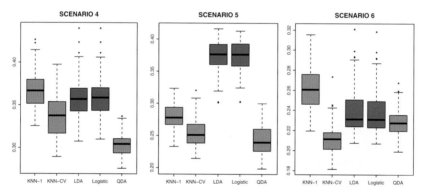

FIGURE 4.11. *Boxplots of the test error rates for each of the non-linear scenarios described in the main text.*

Finally, QDA serves as a compromise between the non-parametric KNN method and the linear LDA and logistic regression approaches. Since QDA assumes a quadratic decision boundary, it can accurately model a wider range of problems than can the linear methods. Though not as flexible as KNN, QDA can perform better in the presence of a limited number of training observations because it does make some assumptions about the form of the decision boundary.

To illustrate the performances of these four classification approaches, we generated data from six different scenarios. In three of the scenarios, the Bayes decision boundary is linear, and in the remaining scenarios it is non-linear. For each scenario, we produced 100 random training data sets. On each of these training sets, we fit each method to the data and computed the resulting test error rate on a large test set. Results for the linear scenarios are shown in Figure 4.10, and the results for the non-linear scenarios are in Figure 4.11. The KNN method requires selection of K, the number of neighbors. We performed KNN with two values of K: $K = 1$, and a value of K that was chosen automatically using an approach called *cross-validation*, which we discuss further in Chapter 5.

In each of the six scenarios, there were $p = 2$ predictors. The scenarios were as follows:

Scenario 1: There were 20 training observations in each of two classes. The observations within each class were uncorrelated random normal variables with a different mean in each class. The left-hand panel of Figure 4.10 shows that LDA performed well in this setting, as one would expect since this is the model assumed by LDA. KNN performed poorly because it paid a price in terms of variance that was not offset by a reduction in bias. QDA also performed worse than LDA, since it fit a more flexible classifier than necessary. Since logistic regression assumes a linear decision boundary, its results were only slightly inferior to those of LDA.

Scenario 2: Details are as in Scenario 1, except that within each class, the two predictors had a correlation of -0.5. The center panel of Figure 4.10 indicates little change in the relative performances of the methods as compared to the previous scenario.

Scenario 3: We generated X_1 and X_2 from the *t-distribution*, with 50 observations per class. The t-distribution has a similar shape to the normal distribution, but it has a tendency to yield more extreme points—that is, more points that are far from the mean. In this setting, the decision boundary was still linear, and so fit into the logistic regression framework. The set-up violated the assumptions of LDA, since the observations were not drawn from a normal distribution. The right-hand panel of Figure 4.10 shows that logistic regression outperformed LDA, though both methods were superior to the other approaches. In particular, the QDA results deteriorated considerably as a consequence of non-normality. *t*-distribution

Scenario 4: The data were generated from a normal distribution, with a correlation of 0.5 between the predictors in the first class, and correlation of -0.5 between the predictors in the second class. This setup corresponded to the QDA assumption, and resulted in quadratic decision boundaries. The left-hand panel of Figure 4.11 shows that QDA outperformed all of the other approaches.

Scenario 5: Within each class, the observations were generated from a normal distribution with uncorrelated predictors. However, the responses were sampled from the logistic function using X_1^2, X_2^2, and $X_1 \times X_2$ as predictors. Consequently, there is a quadratic decision boundary. The center panel of Figure 4.11 indicates that QDA once again performed best, followed closely by KNN-CV. The linear methods had poor performance.

Scenario 6: Details are as in the previous scenario, but the responses were sampled from a more complicated non-linear function. As a result, even the quadratic decision boundaries of QDA could not adequately model the data. The right-hand panel of Figure 4.11 shows that QDA gave slightly better results than the linear methods, while the much more flexible KNN-CV method gave the best results. But KNN with $K = 1$ gave the worst results out of all methods. This highlights the fact that even when the data exhibits a complex non-linear relationship, a non-parametric method such as KNN can still give poor results if the level of smoothness is not chosen correctly.

These six examples illustrate that no one method will dominate the others in every situation. When the true decision boundaries are linear, then the LDA and logistic regression approaches will tend to perform well. When the boundaries are moderately non-linear, QDA may give better results. Finally, for much more complicated decision boundaries, a non-parametric approach such as KNN can be superior. But the level of smoothness for a non-parametric approach must be chosen carefully. In the next chapter we examine a number of approaches for choosing the correct level of smoothness and, in general, for selecting the best overall method.

Finally, recall from Chapter 3 that in the regression setting we can accommodate a non-linear relationship between the predictors and the response by performing regression using transformations of the predictors. A similar approach could be taken in the classification setting. For instance, we could create a more flexible version of logistic regression by including X^2, X^3, and even X^4 as predictors. This may or may not improve logistic regression's performance, depending on whether the increase in variance due to the added flexibility is offset by a sufficiently large reduction in bias. We could do the same for LDA. If we added all possible quadratic terms and cross-products to LDA, the form of the model would be the same as the QDA model, although the parameter estimates would be different. This device allows us to move somewhere between an LDA and a QDA model.

4.6   Lab: Logistic Regression, LDA, QDA, and KNN

4.6.1   The Stock Market Data

We will begin by examining some numerical and graphical summaries of the Smarket data, which is part of the ISLR library. This data set consists of percentage returns for the S&P 500 stock index over $1,250$ days, from the beginning of 2001 until the end of 2005. For each date, we have recorded the percentage returns for each of the five previous trading days, Lag1 through Lag5. We have also recorded Volume (the number of shares traded

on the previous day, in billions), Today (the percentage return on the date
in question) and Direction (whether the market was Up or Down on this
date).

```
> library(ISLR)
> names(Smarket)
[1] "Year"      "Lag1"      "Lag2"      "Lag3"      "Lag4"
[6] "Lag5"      "Volume"    "Today"     "Direction"
> dim(Smarket)
[1] 1250      9
> summary(Smarket)
      Year                Lag1                 Lag2
 Min.   :2001    Min.   :-4.92200    Min.   :-4.92200
 1st Qu.:2002    1st Qu.:-0.63950    1st Qu.:-0.63950
 Median :2003    Median : 0.03900    Median : 0.03900
 Mean   :2003    Mean   : 0.00383    Mean   : 0.00392
 3rd Qu.:2004    3rd Qu.: 0.59675    3rd Qu.: 0.59675
 Max.   :2005    Max.   : 5.73300    Max.   : 5.73300
      Lag3                 Lag4                 Lag5
 Min.   :-4.92200    Min.   :-4.92200    Min.   :-4.92200
 1st Qu.:-0.64000    1st Qu.:-0.64000    1st Qu.:-0.64000
 Median : 0.03850    Median : 0.03850    Median : 0.03850
 Mean   : 0.00172    Mean   : 0.00164    Mean   : 0.00561
 3rd Qu.: 0.59675    3rd Qu.: 0.59675    3rd Qu.: 0.59700
 Max.   : 5.73300    Max.   : 5.73300    Max.   : 5.73300
      Volume              Today              Direction
 Min.   :0.356    Min.   :-4.92200    Down:602
 1st Qu.:1.257    1st Qu.:-0.63950    Up  :648
 Median :1.423    Median : 0.03850
 Mean   :1.478    Mean   : 0.00314
 3rd Qu.:1.642    3rd Qu.: 0.59675
 Max.   :3.152    Max.   : 5.73300
> pairs(Smarket)
```

The cor() function produces a matrix that contains all of the pairwise
correlations among the predictors in a data set. The first command below
gives an error message because the Direction variable is qualitative.

```
> cor(Smarket)
Error in cor(Smarket) : 'x' must be numeric
> cor(Smarket[,-9])
           Year     Lag1     Lag2     Lag3     Lag4     Lag5
Year    1.0000  0.02970  0.03060  0.03319  0.03569  0.02979
Lag1    0.0297  1.00000 -0.02629 -0.01080 -0.00299 -0.00567
Lag2    0.0306 -0.02629  1.00000 -0.02590 -0.01085 -0.00356
Lag3    0.0332 -0.01080 -0.02590  1.00000 -0.02405 -0.01881
Lag4    0.0357 -0.00299 -0.01085 -0.02405  1.00000 -0.02708
Lag5    0.0298 -0.00567 -0.00356 -0.01881 -0.02708  1.00000
Volume  0.5390  0.04091 -0.04338 -0.04182 -0.04841 -0.02200
Today   0.0301 -0.02616 -0.01025 -0.00245 -0.00690 -0.03486
          Volume    Today
Year      0.5390  0.03010
```

```
Lag1      0.0409  -0.02616
Lag2     -0.0434  -0.01025
Lag3     -0.0418  -0.00245
Lag4     -0.0484  -0.00690
Lag5     -0.0220  -0.03486
Volume    1.0000   0.01459
Today     0.0146   1.00000
```

As one would expect, the correlations between the lag variables and to-day's returns are close to zero. In other words, there appears to be little correlation between today's returns and previous days' returns. The only substantial correlation is between Year and Volume. By plotting the data we see that Volume is increasing over time. In other words, the average number of shares traded daily increased from 2001 to 2005.

```
> attach(Smarket)
> plot(Volume)
```

4.6.2 Logistic Regression

Next, we will fit a logistic regression model in order to predict Direction using Lag1 through Lag5 and Volume. The glm() function fits *generalized* *linear models*, a class of models that includes logistic regression. The syntax of the glm() function is similar to that of lm(), except that we must pass in the argument family=binomial in order to tell R to run a logistic regression rather than some other type of generalized linear model.

glm()

generalized
linear model

```
> glm.fits=glm(Direction~Lag1+Lag2+Lag3+Lag4+Lag5+Volume ,
     data=Smarket ,family=binomial )
> summary(glm.fits)

Call:
glm(formula = Direction ~ Lag1 + Lag2 + Lag3 + Lag4 + Lag5
    + Volume , family = binomial , data = Smarket )

Deviance Residuals :
   Min     1Q   Median     3Q      Max
 -1.45   -1.20    1.07    1.15     1.33

Coefficients:
            Estimate Std. Error  z value  Pr(>|z|)
(Intercept) -0.12600    0.24074    -0.52     0.60
Lag1        -0.07307    0.05017    -1.46     0.15
Lag2        -0.04230    0.05009    -0.84     0.40
Lag3         0.01109    0.04994     0.22     0.82
Lag4         0.00936    0.04997     0.19     0.85
Lag5         0.01031    0.04951     0.21     0.83
Volume       0.13544    0.15836     0.86     0.39
```

```
(Dispersion parameter for binomial family taken to be 1)

    Null deviance: 1731.2  on 1249  degrees of freedom
Residual deviance: 1727.6  on 1243  degrees of freedom
AIC: 1742

Number of Fisher Scoring iterations: 3
```

The smallest p-value here is associated with Lag1. The negative coefficient for this predictor suggests that if the market had a positive return yesterday, then it is less likely to go up today. However, at a value of 0.15, the p-value is still relatively large, and so there is no clear evidence of a real association between Lag1 and Direction.

We use the coef() function in order to access just the coefficients for this fitted model. We can also use the summary() function to access particular aspects of the fitted model, such as the p-values for the coefficients.

```
> coef(glm.fits)
(Intercept)          Lag1          Lag2          Lag3          Lag4
   -0.12600       -0.07307       -0.04230        0.01109        0.00936
       Lag5        Volume
    0.01031       0.13544
> summary(glm.fits)$coef
             Estimate Std. Error z value Pr(>|z|)
(Intercept)  -0.12600     0.2407  -0.523    0.601
Lag1         -0.07307     0.0502  -1.457    0.145
Lag2         -0.04230     0.0501  -0.845    0.398
Lag3          0.01109     0.0499   0.222    0.824
Lag4          0.00936     0.0500   0.187    0.851
Lag5          0.01031     0.0495   0.208    0.835
Volume        0.13544     0.1584   0.855    0.392
> summary(glm.fits)$coef[,4]
(Intercept)          Lag1          Lag2          Lag3          Lag4
      0.601         0.145         0.398         0.824         0.851
       Lag5        Volume
      0.835         0.392
```

The predict() function can be used to predict the probability that the market will go up, given values of the predictors. The type="response" option tells R to output probabilities of the form $P(Y = 1|X)$, as opposed to other information such as the logit. If no data set is supplied to the predict() function, then the probabilities are computed for the training data that was used to fit the logistic regression model. Here we have printed only the first ten probabilities. We know that these values correspond to the probability of the market going up, rather than down, because the contrasts() function indicates that R has created a dummy variable with a 1 for Up.

```
> glm.probs=predict(glm.fits,type="response")
> glm.probs[1:10]
    1     2     3     4     5     6     7     8     9    10
0.507 0.481 0.481 0.515 0.511 0.507 0.493 0.509 0.518 0.489
```

```
> contrasts(Direction)
       Up
Down   0
Up     1
```

In order to make a prediction as to whether the market will go up or down on a particular day, we must convert these predicted probabilities into class labels, Up or Down. The following two commands create a vector of class predictions based on whether the predicted probability of a market increase is greater than or less than 0.5.

```
> glm.pred=rep("Down",1250)
> glm.pred[glm.probs>.5]="Up"
```

The first command creates a vector of 1,250 Down elements. The second line transforms to Up all of the elements for which the predicted probability of a market increase exceeds 0.5. Given these predictions, the table() function can be used to produce a confusion matrix in order to determine how many observations were correctly or incorrectly classified.   `table()`

```
> table(glm.pred,Direction)
          Direction
glm.pred Down   Up
    Down   145 141
     Up    457 507
> (507+145)/1250
[1]  0.5216
> mean(glm.pred==Direction)
[1]  0.5216
```

The diagonal elements of the confusion matrix indicate correct predictions, while the off-diagonals represent incorrect predictions. Hence our model correctly predicted that the market would go up on 507 days and that it would go down on 145 days, for a total of $507 + 145 = 652$ correct predictions. The mean() function can be used to compute the fraction of days for which the prediction was correct. In this case, logistic regression correctly predicted the movement of the market 52.2 % of the time.

At first glance, it appears that the logistic regression model is working a little better than random guessing. However, this result is misleading because we trained and tested the model on the same set of 1, 250 observations. In other words, $100 - 52.2 = 47.8$ % is the *training* error rate. As we have seen previously, the training error rate is often overly optimistic—it tends to underestimate the test error rate. In order to better assess the accuracy of the logistic regression model in this setting, we can fit the model using part of the data, and then examine how well it predicts the *held out* data. This will yield a more realistic error rate, in the sense that in practice we will be interested in our model's performance not on the data that we used to fit the model, but rather on days in the future for which the market's movements are unknown.

To implement this strategy, we will first create a vector corresponding to the observations from 2001 through 2004. We will then use this vector to create a held out data set of observations from 2005.

```
> train=(Year<2005)
> Smarket.2005=Smarket[!train,]
> dim(Smarket.2005)
[1] 252    9
> Direction.2005=Direction[!train]
```

The object train is a vector of 1,250 elements, corresponding to the observations in our data set. The elements of the vector that correspond to observations that occurred before 2005 are set to TRUE, whereas those that correspond to observations in 2005 are set to FALSE. The object train is a *Boolean* vector, since its elements are TRUE and FALSE. Boolean vectors can be used to obtain a subset of the rows or columns of a matrix. For instance, the command Smarket[train,] would pick out a submatrix of the stock market data set, corresponding only to the dates before 2005, since those are the ones for which the elements of train are TRUE. The ! symbol can be used to reverse all of the elements of a Boolean vector. That is, !train is a vector similar to train, except that the elements that are TRUE in train get swapped to FALSE in !train, and the elements that are FALSE in train get swapped to TRUE in !train. Therefore, Smarket[!train,] yields a submatrix of the stock market data containing only the observations for which train is FALSE—that is, the observations with dates in 2005. The output above indicates that there are 252 such observations.

boolean

We now fit a logistic regression model using only the subset of the observations that correspond to dates before 2005, using the subset argument. We then obtain predicted probabilities of the stock market going up for each of the days in our test set—that is, for the days in 2005.

```
> glm.fits=glm(Direction~Lag1+Lag2+Lag3+Lag4+Lag5+Volume ,
    data=Smarket ,family=binomial ,subset=train)
> glm.probs=predict(glm.fits ,Smarket.2005 ,type="response")
```

Notice that we have trained and tested our model on two completely separate data sets: training was performed using only the dates before 2005, and testing was performed using only the dates in 2005. Finally, we compute the predictions for 2005 and compare them to the actual movements of the market over that time period.

```
> glm.pred=rep("Down",252)
> glm.pred[glm.probs>.5]="Up"
> table(glm.pred,Direction.2005)
         Direction.2005
glm.pred Down Up
    Down   77 97
    Up     34 44
> mean(glm.pred==Direction.2005)
```

```
[1] 0.48
> mean(glm.pred!=Direction.2005)
[1] 0.52
```

The != notation means *not equal to*, and so the last command computes the test set error rate. The results are rather disappointing: the test error rate is 52%, which is worse than random guessing! Of course this result is not all that surprising, given that one would not generally expect to be able to use previous days' returns to predict future market performance. (After all, if it were possible to do so, then the authors of this book would be out striking it rich rather than writing a statistics textbook.)

We recall that the logistic regression model had very underwhelming p-values associated with all of the predictors, and that the smallest p-value, though not very small, corresponded to Lag1. Perhaps by removing the variables that appear not to be helpful in predicting Direction, we can obtain a more effective model. After all, using predictors that have no relationship with the response tends to cause a deterioration in the test error rate (since such predictors cause an increase in variance without a corresponding decrease in bias), and so removing such predictors may in turn yield an improvement. Below we have refit the logistic regression using just Lag1 and Lag2, which seemed to have the highest predictive power in the original logistic regression model.

```
> glm.fits=glm(Direction~Lag1+Lag2,data=Smarket,family=binomial,
    subset=train)
> glm.probs=predict(glm.fits,Smarket.2005,type="response")
> glm.pred=rep("Down",252)
> glm.pred[glm.probs>.5]="Up"
> table(glm.pred,Direction.2005)
         Direction.2005
glm.pred Down   Up
    Down   35   35
    Up     76  106
> mean(glm.pred==Direction.2005)
[1] 0.56
> 106/(106+76)
[1] 0.582
```

Now the results appear to be a little better: 56% of the daily movements have been correctly predicted. It is worth noting that in this case, a much simpler strategy of predicting that the market will increase every day will also be correct 56% of the time! Hence, in terms of overall error rate, the logistic regression method is no better than the naïve approach. However, the confusion matrix shows that on days when logistic regression predicts an increase in the market, it has a 58% accuracy rate. This suggests a possible trading strategy of buying on days when the model predicts an increasing market, and avoiding trades on days when a decrease is predicted. Of course one would need to investigate more carefully whether this small improvement was real or just due to random chance.

Suppose that we want to predict the returns associated with particular values of Lag1 and Lag2. In particular, we want to predict Direction on a day when Lag1 and Lag2 equal 1.2 and 1.1, respectively, and on a day when they equal 1.5 and −0.8. We do this using the predict() function.

```
> predict(glm.fits,newdata=data.frame(Lag1=c(1.2,1.5),
    Lag2=c(1.1,-0.8)),type="response")
       1         2
  0.4791    0.4961
```

4.6.3 Linear Discriminant Analysis

Now we will perform LDA on the Smarket data. In R, we fit an LDA model using the lda() function, which is part of the MASS library. Notice that the syntax for the lda() function is identical to that of lm(), and to that of glm() except for the absence of the family option. We fit the model using only the observations before 2005.

lda()

```
> library(MASS)
> lda.fit=lda(Direction~Lag1+Lag2,data=Smarket,subset=train)
> lda.fit
Call:
lda(Direction ~ Lag1 + Lag2, data = Smarket, subset = train)

Prior probabilities of groups:
 Down      Up
0.492   0.508

Group means:
          Lag1      Lag2
Down    0.0428    0.0339
Up     -0.0395   -0.0313

Coefficients of linear discriminants:
         LD1
Lag1  -0.642
Lag2  -0.514
> plot(lda.fit)
```

The LDA output indicates that $\hat{\pi}_1 = 0.492$ and $\hat{\pi}_2 = 0.508$; in other words, 49.2 % of the training observations correspond to days during which the market went down. It also provides the group means; these are the average of each predictor within each class, and are used by LDA as estimates of μ_k. These suggest that there is a tendency for the previous 2 days' returns to be negative on days when the market increases, and a tendency for the previous days' returns to be positive on days when the market declines. The *coefficients of linear discriminants* output provides the linear combination of Lag1 and Lag2 that are used to form the LDA decision rule. In other words, these are the multipliers of the elements of $X = x$ in (4.19). If $-0.642 \times Lag1 - 0.514 \times Lag2$ is large, then the LDA classifier will predict a market increase, and if it is small, then the LDA classifier will predict a market decline. The plot() function produces plots of the *linear discriminants*, obtained by computing $-0.642 \times$ Lag1 $- 0.514 \times$ Lag2 for each of the training observations.

The predict() function returns a list with three elements. The first element, class, contains LDA's predictions about the movement of the market. The second element, posterior, is a matrix whose kth column contains the posterior probability that the corresponding observation belongs to the kth class, computed from (4.10). Finally, x contains the linear discriminants, described earlier.

```
> lda.pred=predict(lda.fit, Smarket.2005)
> names(lda.pred)
[1] "class"      "posterior" "x"
```

As we observed in Section 4.5, the LDA and logistic regression predictions are almost identical.

```
> lda.class=lda.pred$class
> table(lda.class,Direction.2005)
         Direction.2005
lda.pred Down  Up
    Down   35  35
    Up     76 106
> mean(lda.class==Direction.2005)
[1] 0.56
```

Applying a 50 % threshold to the posterior probabilities allows us to recreate the predictions contained in lda.pred$class.

```
> sum(lda.pred$posterior[,1]>=.5)
[1] 70
> sum(lda.pred$posterior[,1]<.5)
[1] 182
```

Notice that the posterior probability output by the model corresponds to the probability that the market will *decrease*:

```
> lda.pred$posterior[1:20,1]
> lda.class[1:20]
```

If we wanted to use a posterior probability threshold other than 50 % in order to make predictions, then we could easily do so. For instance, suppose that we wish to predict a market decrease only if we are very certain that the market will indeed decrease on that day—say, if the posterior probability is at least 90 %.

```
> sum(lda.pred$posterior[,1]>.9)
[1] 0
```

No days in 2005 meet that threshold! In fact, the greatest posterior probability of decrease in all of 2005 was 52.02 %.

4.6.4 Quadratic Discriminant Analysis

We will now fit a QDA model to the Smarket data. QDA is implemented
in R using the qda() function, which is also part of the MASS library. The
syntax is identical to that of lda().

qda()

```
> qda.fit=qda(Direction~Lag1+Lag2,data=Smarket,subset=train)
> qda.fit
Call:
qda(Direction ~ Lag1 + Lag2, data = Smarket, subset = train)

Prior probabilities of groups:
 Down    Up
0.492 0.508

Group means:
        Lag1     Lag2
Down   0.0428   0.0339
Up    -0.0395  -0.0313
```

The output contains the group means. But it does not contain the coefficients of the linear discriminants, because the QDA classifier involves a
quadratic, rather than a linear, function of the predictors. The predict()
function works in exactly the same fashion as for LDA.

```
> qda.class=predict(qda.fit,Smarket.2005)$class
> table(qda.class,Direction.2005)
          Direction.2005
qda.class Down  Up
    Down   30   20
      Up   81  121
> mean(qda.class==Direction.2005)
[1] 0.599
```

Interestingly, the QDA predictions are accurate almost 60% of the time,
even though the 2005 data was not used to fit the model. This level of accuracy is quite impressive for stock market data, which is known to be quite
hard to model accurately. This suggests that the quadratic form assumed
by QDA may capture the true relationship more accurately than the linear
forms assumed by LDA and logistic regression. However, we recommend
evaluating this method's performance on a larger test set before betting
that this approach will consistently beat the market!

4.6.5 K-Nearest Neighbors

We will now perform KNN using the knn() function, which is part of the
class library. This function works rather differently from the other model-
fitting functions that we have encountered thus far. Rather than a two-step
approach in which we first fit the model and then we use the model to make
predictions, knn() forms predictions using a single command. The function
requires four inputs.

knn()

1. A matrix containing the predictors associated with the training data, labeled train.X below.

2. A matrix containing the predictors associated with the data for which we wish to make predictions, labeled test.X below.

3. A vector containing the class labels for the training observations, labeled train.Direction below.

4. A value for K, the number of nearest neighbors to be used by the classifier.

We use the cbind() function, short for *column bind*, to bind the Lag1 and Lag2 variables together into two matrices, one for the training set and the other for the test set.

cbind()

```
> library(class)
> train.X=cbind(Lag1,Lag2)[train,]
> test.X=cbind(Lag1,Lag2)[!train,]
> train.Direction=Direction[train]
```

Now the knn() function can be used to predict the market's movement for the dates in 2005. We set a random seed before we apply knn() because if several observations are tied as nearest neighbors, then R will randomly break the tie. Therefore, a seed must be set in order to ensure reproducibility of results.

```
> set.seed(1)
> knn.pred=knn(train.X,test.X,train.Direction,k=1)
> table(knn.pred,Direction.2005)
        Direction.2005
knn.pred Down Up
    Down   43 58
      Up   68 83
> (83+43)/252
[1] 0.5
```

The results using $K = 1$ are not very good, since only 50% of the observations are correctly predicted. Of course, it may be that $K = 1$ results in an overly flexible fit to the data. Below, we repeat the analysis using $K = 3$.

```
> knn.pred=knn(train.X,test.X,train.Direction,k=3)
> table(knn.pred,Direction.2005)
        Direction.2005
knn.pred Down Up
    Down   48 54
      Up   63 87
> mean(knn.pred==Direction.2005)
[1] 0.536
```

The results have improved slightly. But increasing K further turns out to provide no further improvements. It appears that for this data, QDA provides the best results of the methods that we have examined so far.

4.6.6 An Application to Caravan Insurance Data

Finally, we will apply the KNN approach to the Caravan data set, which is part of the ISLR library. This data set includes 85 predictors that measure demographic characteristics for 5,822 individuals. The response variable is Purchase, which indicates whether or not a given individual purchases a caravan insurance policy. In this data set, only 6 % of people purchased caravan insurance.

```
> dim(Caravan)
[1] 5822    86
> attach(Caravan)
> summary(Purchase)
  No   Yes
5474   348
> 348/5822
[1] 0.0598
```

Because the KNN classifier predicts the class of a given test observation by identifying the observations that are nearest to it, the scale of the variables matters. Any variables that are on a large scale will have a much larger effect on the *distance* between the observations, and hence on the KNN classifier, than variables that are on a small scale. For instance, imagine a data set that contains two variables, salary and age (measured in dollars and years, respectively). As far as KNN is concerned, a difference of $1,000 in salary is enormous compared to a difference of 50 years in age. Consequently, salary will drive the KNN classification results, and age will have almost no effect. This is contrary to our intuition that a salary difference of $1,000 is quite small compared to an age difference of 50 years. Furthermore, the importance of scale to the KNN classifier leads to another issue: if we measured salary in Japanese yen, or if we measured age in minutes, then we'd get quite different classification results from what we get if these two variables are measured in dollars and years.

A good way to handle this problem is to *standardize* the data so that all variables are given a mean of zero and a standard deviation of one. Then all variables will be on a comparable scale. The scale() function does just this. In standardizing the data, we exclude column 86, because that is the qualitative Purchase variable.

standardize

scale()

```
> standardized.X=scale(Caravan[,-86])
> var(Caravan[,1])
[1] 165
> var(Caravan[,2])
[1] 0.165
> var(standardized.X[,1])
[1] 1
> var(standardized.X[,2])
[1] 1
```

Now every column of standardized.X has a standard deviation of one and a mean of zero.

We now split the observations into a test set, containing the first 1,000 observations, and a training set, containing the remaining observations. We fit a KNN model on the training data using $K = 1$, and evaluate its performance on the test data.

```
> test=1:1000
> train.X=standardized.X[-test,]
> test.X=standardized.X[test,]
> train.Y=Purchase[-test]
> test.Y=Purchase[test]
> set.seed(1)
> knn.pred=knn(train.X,test.X,train.Y,k=1)
> mean(test.Y!=knn.pred)
[1] 0.118
> mean(test.Y!="No")
[1] 0.059
```

The vector test is numeric, with values from 1 through 1,000. Typing standardized.X[test,] yields the submatrix of the data containing the observations whose indices range from 1 to 1,000, whereas typing standardized.X[-test,] yields the submatrix containing the observations whose indices do *not* range from 1 to 1,000. The KNN error rate on the 1,000 test observations is just under 12 %. At first glance, this may appear to be fairly good. However, since only 6 % of customers purchased insurance, we could get the error rate down to 6 % by always predicting No regardless of the values of the predictors!

Suppose that there is some non-trivial cost to trying to sell insurance to a given individual. For instance, perhaps a salesperson must visit each potential customer. If the company tries to sell insurance to a random selection of customers, then the success rate will be only 6 %, which may be far too low given the costs involved. Instead, the company would like to try to sell insurance only to customers who are likely to buy it. So the overall error rate is not of interest. Instead, the fraction of individuals that are correctly predicted to buy insurance is of interest.

It turns out that KNN with $K = 1$ does far better than random guessing among the customers that are predicted to buy insurance. Among 77 such customers, 9, or 11.7 %, actually do purchase insurance. This is double the rate that one would obtain from random guessing.

```
> table(knn.pred,test.Y)
         test.Y
knn.pred  No  Yes
     No  873   50
    Yes   68    9
> 9/(68+9)
[1] 0.117
```

Using $K = 3$, the success rate increases to 19 %, and with $K = 5$ the rate is 26.7 %. This is over four times the rate that results from random guessing. It appears that KNN is finding some real patterns in a difficult data set!

```
> knn.pred=knn(train.X,test.X,train.Y,k=3)
> table(knn.pred,test.Y)
          test.Y
knn.pred   No Yes
      No  920  54
      Yes  21   5
> 5/26
[1] 0.192
> knn.pred=knn(train.X,test.X,train.Y,k=5)
> table(knn.pred,test.Y)
          test.Y
knn.pred   No Yes
      No  930  55
      Yes  11   4
> 4/15
[1] 0.267
```

As a comparison, we can also fit a logistic regression model to the data. If we use 0.5 as the predicted probability cut-off for the classifier, then we have a problem: only seven of the test observations are predicted to purchase insurance. Even worse, we are wrong about all of these! However, we are not required to use a cut-off of 0.5. If we instead predict a purchase any time the predicted probability of purchase exceeds 0.25, we get much better results: we predict that 33 people will purchase insurance, and we are correct for about 33 % of these people. This is over five times better than random guessing!

```
> glm.fits=glm(Purchase~.,data=Caravan ,family=binomial ,
    subset =-test)
Warning message:
glm.fits: fitted probabilities numerically 0 or 1 occurred
> glm.probs=predict(glm.fits,Caravan[test ,],type="response")
> glm.pred=rep("No" ,1000)
> glm.pred[glm.probs >.5]="Yes"
> table(glm.pred,test.Y)
          test.Y
glm.pred   No Yes
      No  934  59
      Yes   7   0
> glm.pred=rep("No" ,1000)
> glm.pred[glm.probs >.25]="Yes"
> table(glm.pred,test.Y)
          test.Y
glm.pred   No Yes
      No  919  48
      Yes  22  11
> 11/(22+11)
[1] 0.333
```

4.7 Exercises

Conceptual

1. Using a little bit of algebra, prove that (4.2) is equivalent to (4.3). In other words, the logistic function representation and logit representation for the logistic regression model are equivalent.

2. It was stated in the text that classifying an observation to the class for which (4.12) is largest is equivalent to classifying an observation to the class for which (4.13) is largest. Prove that this is the case. In other words, under the assumption that the observations in the kth class are drawn from a $N(\mu_k, \sigma^2)$ distribution, the Bayes' classifier assigns an observation to the class for which the discriminant function is maximized.

3. This problem relates to the QDA model, in which the observations within each class are drawn from a normal distribution with a class-specific mean vector and a class specific covariance matrix. We consider the simple case where $p = 1$; i.e. there is only one feature.

   Suppose that we have K classes, and that if an observation belongs to the kth class then X comes from a one-dimensional normal distribution, $X \sim N(\mu_k, \sigma_k^2)$. Recall that the density function for the one-dimensional normal distribution is given in (4.11). Prove that in this case, the Bayes' classifier is *not* linear. Argue that it is in fact quadratic.

   Hint: For this problem, you should follow the arguments laid out in Section 4.4.2, but without making the assumption that $\sigma_1^2 = \ldots = \sigma_K^2$.

4. When the number of features p is large, there tends to be a deterioration in the performance of KNN and other *local* approaches that perform prediction using only observations that are *near* the test observation for which a prediction must be made. This phenomenon is known as the *curse of dimensionality*, and it ties into the fact that non-parametric approaches often perform poorly when p is large. We will now investigate this curse.

   curse of dimensionality

   (a) Suppose that we have a set of observations, each with measurements on $p = 1$ feature, X. We assume that X is uniformly (evenly) distributed on $[0, 1]$. Associated with each observation is a response value. Suppose that we wish to predict a test observation's response using only observations that are within 10 % of the range of X closest to that test observation. For instance, in order to predict the response for a test observation with $X = 0.6$, we will use observations in the range $[0.55, 0.65]$. On average, what fraction of the available observations will we use to make the prediction?

(b) Now suppose that we have a set of observations, each with measurements on $p = 2$ features, X_1 and X_2. We assume that (X_1, X_2) are uniformly distributed on $[0, 1] \times [0, 1]$. We wish to predict a test observation's response using only observations that are within 10 % of the range of X_1 and within 10 % of the range of X_2 closest to that test observation. For instance, in order to predict the response for a test observation with $X_1 = 0.6$ and $X_2 = 0.35$, we will use observations in the range $[0.55, 0.65]$ for X_1 and in the range $[0.3, 0.4]$ for X_2. On average, what fraction of the available observations will we use to make the prediction?

(c) Now suppose that we have a set of observations on $p = 100$ features. Again the observations are uniformly distributed on each feature, and again each feature ranges in value from 0 to 1. We wish to predict a test observation's response using observations within the 10 % of each feature's range that is closest to that test observation. What fraction of the available observations will we use to make the prediction?

(d) Using your answers to parts (a)–(c), argue that a drawback of KNN when p is large is that there are very few training observations "near" any given test observation.

(e) Now suppose that we wish to make a prediction for a test observation by creating a p-dimensional hypercube centered around the test observation that contains, on average, 10 % of the training observations. For $p = 1, 2$, and 100, what is the length of each side of the hypercube? Comment on your answer.

Note: A hypercube is a generalization of a cube to an arbitrary number of dimensions. When $p = 1$, a hypercube is simply a line segment, when $p = 2$ it is a square, and when $p = 100$ it is a 100-dimensional cube.

5. We now examine the differences between LDA and QDA.

(a) If the Bayes decision boundary is linear, do we expect LDA or QDA to perform better on the training set? On the test set?

(b) If the Bayes decision boundary is non-linear, do we expect LDA or QDA to perform better on the training set? On the test set?

(c) In general, as the sample size n increases, do we expect the test prediction accuracy of QDA relative to LDA to improve, decline, or be unchanged? Why?

(d) True or False: Even if the Bayes decision boundary for a given problem is linear, we will probably achieve a superior test error rate using QDA rather than LDA because QDA is flexible enough to model a linear decision boundary. Justify your answer.

6. Suppose we collect data for a group of students in a statistics class with variables $X_1 =$ hours studied, $X_2 =$ undergrad GPA, and $Y =$ receive an A. We fit a logistic regression and produce estimated coefficient, $\hat{\beta}_0 = -6, \hat{\beta}_1 = 0.05, \hat{\beta}_2 = 1$.

   (a) Estimate the probability that a student who studies for 40 h and has an undergrad GPA of 3.5 gets an A in the class.

   (b) How many hours would the student in part (a) need to study to have a 50 % chance of getting an A in the class?

7. Suppose that we wish to predict whether a given stock will issue a dividend this year ("Yes" or "No") based on X, last year's percent profit. We examine a large number of companies and discover that the mean value of X for companies that issued a dividend was $\bar{X} = 10$, while the mean for those that didn't was $\bar{X} = 0$. In addition, the variance of X for these two sets of companies was $\hat{\sigma}^2 = 36$. Finally, 80 % of companies issued dividends. Assuming that X follows a normal distribution, predict the probability that a company will issue a dividend this year given that its percentage profit was $X = 4$ last year.

   Hint: Recall that the density function for a normal random variable is $f(x) = \frac{1}{\sqrt{2\pi\sigma^2}}e^{-(x-\mu)^2/2\sigma^2}$. You will need to use Bayes' theorem.

8. Suppose that we take a data set, divide it into equally-sized training and test sets, and then try out two different classification procedures. First we use logistic regression and get an error rate of 20 % on the training data and 30 % on the test data. Next we use 1-nearest neighbors (i.e. $K = 1$) and get an average error rate (averaged over both test and training data sets) of 18 %. Based on these results, which method should we prefer to use for classification of new observations? Why?

9. This problem has to do with *odds*.

   (a) On average, what fraction of people with an odds of 0.37 of defaulting on their credit card payment will in fact default?

   (b) Suppose that an individual has a 16 % chance of defaulting on her credit card payment. What are the odds that she will default?

Applied

10. This question should be answered using the Weekly data set, which is part of the ISLR package. This data is similar in nature to the Smarket data from this chapter's lab, except that it contains 1,089 weekly returns for 21 years, from the beginning of 1990 to the end of 2010.

    (a) Produce some numerical and graphical summaries of the Weekly data. Do there appear to be any patterns?

    (b) Use the full data set to perform a logistic regression with Direction as the response and the five lag variables plus Volume as predictors. Use the summary function to print the results. Do any of the predictors appear to be statistically significant? If so, which ones?

    (c) Compute the confusion matrix and overall fraction of correct predictions. Explain what the confusion matrix is telling you about the types of mistakes made by logistic regression.

    (d) Now fit the logistic regression model using a training data period from 1990 to 2008, with Lag2 as the only predictor. Compute the confusion matrix and the overall fraction of correct predictions for the held out data (that is, the data from 2009 and 2010).

    (e) Repeat (d) using LDA.

    (f) Repeat (d) using QDA.

    (g) Repeat (d) using KNN with $K = 1$.

    (h) Which of these methods appears to provide the best results on this data?

    (i) Experiment with different combinations of predictors, including possible transformations and interactions, for each of the methods. Report the variables, method, and associated confusion matrix that appears to provide the best results on the held out data. Note that you should also experiment with values for K in the KNN classifier.

11. In this problem, you will develop a model to predict whether a given car gets high or low gas mileage based on the Auto data set.

    (a) Create a binary variable, mpg01, that contains a 1 if mpg contains a value above its median, and a 0 if mpg contains a value below its median. You can compute the median using the median() function. Note you may find it helpful to use the data.frame() function to create a single data set containing both mpg01 and the other Auto variables.

(b) Explore the data graphically in order to investigate the association between mpg01 and the other features. Which of the other features seem most likely to be useful in predicting mpg01? Scatterplots and boxplots may be useful tools to answer this question. Describe your findings.

(c) Split the data into a training set and a test set.

(d) Perform LDA on the training data in order to predict mpg01 using the variables that seemed most associated with mpg01 in (b). What is the test error of the model obtained?

(e) Perform QDA on the training data in order to predict mpg01 using the variables that seemed most associated with mpg01 in (b). What is the test error of the model obtained?

(f) Perform logistic regression on the training data in order to predict mpg01 using the variables that seemed most associated with mpg01 in (b). What is the test error of the model obtained?

(g) Perform KNN on the training data, with several values of K, in order to predict mpg01. Use only the variables that seemed most associated with mpg01 in (b). What test errors do you obtain? Which value of K seems to perform the best on this data set?

12. This problem involves writing functions.

(a) Write a function, Power(), that prints out the result of raising 2 to the 3rd power. In other words, your function should compute 2^3 and print out the results.
   Hint: Recall that x^a raises x to the power a. Use the print() *function to output the result.*

(b) Create a new function, Power2(), that allows you to pass *any* two numbers, x and a, and prints out the value of x^a. You can do this by beginning your function with the line

```
> Power2=function(x,a){
```

You should be able to call your function by entering, for instance,

```
> Power2(3,8)
```

on the command line. This should output the value of 3^8, namely, 6, 561.

(c) Using the Power2() function that you just wrote, compute 10^3, 8^{17}, and 131^3.

(d) Now create a new function, Power3(), that actually *returns* the result x^a as an R object, rather than simply printing it to the screen. That is, if you store the value x^a in an object called result within your function, then you can simply return() this result, using the following line: return()

```
return(result)
```

The line above should be the last line in your function, before the } symbol.

(e) Now using the `Power3()` function, create a plot of $f(x) = x^2$. The x-axis should display a range of integers from 1 to 10, and the y-axis should display x^2. Label the axes appropriately, and use an appropriate title for the figure. Consider displaying either the x-axis, the y-axis, or both on the log-scale. You can do this by using `log=''x''`, `log=''y''`, or `log=''xy''` as arguments to the `plot()` function.

(f) Create a function, `PlotPower()`, that allows you to create a plot of x against x^a for a fixed a and for a range of values of x. For instance, if you call

```
> PlotPower(1:10,3)
```

then a plot should be created with an x-axis taking on values $1, 2, \ldots, 10$, and a y-axis taking on values $1^3, 2^3, \ldots, 10^3$.

13. Using the `Boston` data set, fit classification models in order to predict whether a given suburb has a crime rate above or below the median. Explore logistic regression, LDA, and KNN models using various subsets of the predictors. Describe your findings.

5

Resampling Methods

Resampling methods are an indispensable tool in modern statistics. They involve repeatedly drawing samples from a training set and refitting a model of interest on each sample in order to obtain additional information about the fitted model. For example, in order to estimate the variability of a linear regression fit, we can repeatedly draw different samples from the training data, fit a linear regression to each new sample, and then examine the extent to which the resulting fits differ. Such an approach may allow us to obtain information that would not be available from fitting the model only once using the original training sample.

Resampling approaches can be computationally expensive, because they involve fitting the same statistical method multiple times using different subsets of the training data. However, due to recent advances in computing power, the computational requirements of resampling methods generally are not prohibitive. In this chapter, we discuss two of the most commonly used resampling methods, *cross-validation* and the *bootstrap*. Both methods are important tools in the practical application of many statistical learning procedures. For example, cross-validation can be used to estimate the test error associated with a given statistical learning method in order to evaluate its performance, or to select the appropriate level of flexibility. The process of evaluating a model's performance is known as *model assessment*, whereas the process of selecting the proper level of flexibility for a model is known as *model selection*. The bootstrap is used in several contexts, most commonly to provide a measure of accuracy of a parameter estimate or of a given statistical learning method.

model assessment

model selection

G. James et al., *An Introduction to Statistical Learning: with Applications in R*, 175
Springer Texts in Statistics, DOI 10.1007/978-1-4614-7138-7_5,
© Springer Science+Business Media New York 2013

5.1   Cross-Validation

In Chapter 2 we discuss the distinction between the *test error rate* and the *training error rate*. The test error is the average error that results from using a statistical learning method to predict the response on a new observation— that is, a measurement that was not used in training the method. Given a data set, the use of a particular statistical learning method is warranted if it results in a low test error. The test error can be easily calculated if a designated test set is available. Unfortunately, this is usually not the case. In contrast, the training error can be easily calculated by applying the statistical learning method to the observations used in its training. But as we saw in Chapter 2, the training error rate often is quite different from the test error rate, and in particular the former can dramatically underestimate the latter.

In the absence of a very large designated test set that can be used to directly estimate the test error rate, a number of techniques can be used to estimate this quantity using the available training data. Some methods make a mathematical adjustment to the training error rate in order to estimate the test error rate. Such approaches are discussed in Chapter 6. In this section, we instead consider a class of methods that estimate the test error rate by *holding out* a subset of the training observations from the fitting process, and then applying the statistical learning method to those held out observations.

In Sections 5.1.1–5.1.4, for simplicity we assume that we are interested in performing regression with a quantitative response. In Section 5.1.5 we consider the case of classification with a qualitative response. As we will see, the key concepts remain the same regardless of whether the response is quantitative or qualitative.

5.1.1   *The Validation Set Approach*

Suppose that we would like to estimate the test error associated with fitting a particular statistical learning method on a set of observations. The *validation set approach*, displayed in Figure 5.1, is a very simple strategy for this task. It involves randomly dividing the available set of observations into two parts, a *training set* and a *validation set* or *hold-out set*. The model is fit on the training set, and the fitted model is used to predict the responses for the observations in the validation set. The resulting validation set error rate—typically assessed using MSE in the case of a quantitative response—provides an estimate of the test error rate.

validation set approach

validation set

hold-out set

We illustrate the validation set approach on the Auto data set. Recall from Chapter 3 that there appears to be a non-linear relationship between mpg and horsepower, and that a model that predicts mpg using horsepower and horsepower^2 gives better results than a model that uses only a linear term. It is natural to wonder whether a cubic or higher-order fit might provide

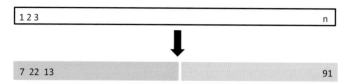

FIGURE 5.1. *A schematic display of the validation set approach. A set of n observations are randomly split into a training set (shown in blue, containing observations 7, 22, and 13, among others) and a validation set (shown in beige, and containing observation 91, among others). The statistical learning method is fit on the training set, and its performance is evaluated on the validation set.*

even better results. We answer this question in Chapter 3 by looking at the p-values associated with a cubic term and higher-order polynomial terms in a linear regression. But we could also answer this question using the validation method. We randomly split the 392 observations into two sets, a training set containing 196 of the data points, and a validation set containing the remaining 196 observations. The validation set error rates that result from fitting various regression models on the training sample and evaluating their performance on the validation sample, using MSE as a measure of validation set error, are shown in the left-hand panel of Figure 5.2. The validation set MSE for the quadratic fit is considerably smaller than for the linear fit. However, the validation set MSE for the cubic fit is actually slightly larger than for the quadratic fit. This implies that including a cubic term in the regression does not lead to better prediction than simply using a quadratic term.

Recall that in order to create the left-hand panel of Figure 5.2, we randomly divided the data set into two parts, a training set and a validation set. If we repeat the process of randomly splitting the sample set into two parts, we will get a somewhat different estimate for the test MSE. As an illustration, the right-hand panel of Figure 5.2 displays ten different validation set MSE curves from the Auto data set, produced using ten different random splits of the observations into training and validation sets. All ten curves indicate that the model with a quadratic term has a dramatically smaller validation set MSE than the model with only a linear term. Furthermore, all ten curves indicate that there is not much benefit in including cubic or higher-order polynomial terms in the model. But it is worth noting that each of the ten curves results in a different test MSE estimate for each of the ten regression models considered. And there is no consensus among the curves as to which model results in the smallest validation set MSE. Based on the variability among these curves, all that we can conclude with any confidence is that the linear fit is not adequate for this data.

The validation set approach is conceptually simple and is easy to implement. But it has two potential drawbacks:

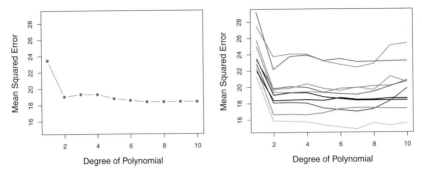

FIGURE 5.2. *The validation set approach was used on the* Auto *data set in order to estimate the test error that results from predicting* mpg *using polynomial functions of* horsepower. *Left: Validation error estimates for a single split into training and validation data sets. Right: The validation method was repeated ten times, each time using a different random split of the observations into a training set and a validation set. This illustrates the variability in the estimated test MSE that results from this approach.*

1. As is shown in the right-hand panel of Figure 5.2, the validation estimate of the test error rate can be highly variable, depending on precisely which observations are included in the training set and which observations are included in the validation set.

2. In the validation approach, only a subset of the observations—those that are included in the training set rather than in the validation set—are used to fit the model. Since statistical methods tend to perform worse when trained on fewer observations, this suggests that the validation set error rate may tend to *overestimate* the test error rate for the model fit on the entire data set.

In the coming subsections, we will present *cross-validation*, a refinement of the validation set approach that addresses these two issues.

5.1.2   *Leave-One-Out Cross-Validation*

Leave-one-out cross-validation (LOOCV) is closely related to the validation set approach of Section 5.1.1, but it attempts to address that method's drawbacks.

leave-one-out cross-validation

Like the validation set approach, LOOCV involves splitting the set of observations into two parts. However, instead of creating two subsets of comparable size, a single observation (x_1, y_1) is used for the validation set, and the remaining observations $\{(x_2, y_2), \ldots, (x_n, y_n)\}$ make up the training set. The statistical learning method is fit on the $n-1$ training observations, and a prediction $\hat{y}_1$ is made for the excluded observation, using its value x_1. Since (x_1, y_1) was not used in the fitting process, $\text{MSE}_1 =$

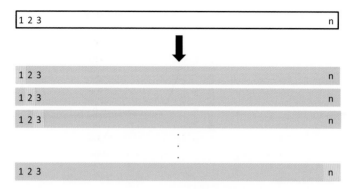

FIGURE 5.3. *A schematic display of LOOCV. A set of n data points is repeatedly split into a training set (shown in blue) containing all but one observation, and a validation set that contains only that observation (shown in beige). The test error is then estimated by averaging the n resulting MSE's. The first training set contains all but observation 1, the second training set contains all but observation 2, and so forth.*

$(y_1 - \hat{y}_1)^2$ provides an approximately unbiased estimate for the test error. But even though MSE_1 is unbiased for the test error, it is a poor estimate because it is highly variable, since it is based upon a single observation (x_1, y_1).

We can repeat the procedure by selecting (x_2, y_2) for the validation data, training the statistical learning procedure on the $n-1$ observations $\{(x_1, y_1), (x_3, y_3), \ldots, (x_n, y_n)\}$, and computing $\text{MSE}_2 = (y_2 - \hat{y}_2)^2$. Repeating this approach n times produces n squared errors, $\text{MSE}_1, \ldots, \text{MSE}_n$. The LOOCV estimate for the test MSE is the average of these n test error estimates:

$$\text{CV}_{(n)} = \frac{1}{n} \sum_{i=1}^{n} \text{MSE}_i. \tag{5.1}$$

A schematic of the LOOCV approach is illustrated in Figure 5.3.

LOOCV has a couple of major advantages over the validation set approach. First, it has far less bias. In LOOCV, we repeatedly fit the statistical learning method using training sets that contain $n-1$ observations, almost as many as are in the entire data set. This is in contrast to the validation set approach, in which the training set is typically around half the size of the original data set. Consequently, the LOOCV approach tends not to overestimate the test error rate as much as the validation set approach does. Second, in contrast to the validation approach which will yield different results when applied repeatedly due to randomness in the training/validation set splits, performing LOOCV multiple times will

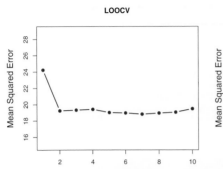
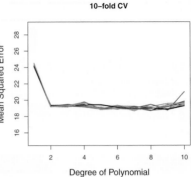

FIGURE 5.4. *Cross-validation was used on the* `Auto` *data set in order to estimate the test error that results from predicting* `mpg` *using polynomial functions of* `horsepower`. *Left: The LOOCV error curve. Right: 10-fold CV was run nine separate times, each with a different random split of the data into ten parts. The figure shows the nine slightly different CV error curves.*

always yield the same results: there is no randomness in the training/validation set splits.

We used LOOCV on the `Auto` data set in order to obtain an estimate of the test set MSE that results from fitting a linear regression model to predict `mpg` using polynomial functions of `horsepower`. The results are shown in the left-hand panel of Figure 5.4.

LOOCV has the potential to be expensive to implement, since the model has to be fit n times. This can be very time consuming if n is large, and if each individual model is slow to fit. With least squares linear or polynomial regression, an amazing shortcut makes the cost of LOOCV the same as that of a single model fit! The following formula holds:

$$\mathrm{CV}_{(n)} = \frac{1}{n}\sum_{i=1}^{n}\left(\frac{y_i - \hat{y}_i}{1 - h_i}\right)^2, \tag{5.2}$$

where $\hat{y}_i$ is the ith fitted value from the original least squares fit, and h_i is the leverage defined in (3.37) on page 98. This is like the ordinary MSE, except the ith residual is divided by $1 - h_i$. The leverage lies between $1/n$ and 1, and reflects the amount that an observation influences its own fit. Hence the residuals for high-leverage points are inflated in this formula by exactly the right amount for this equality to hold.

LOOCV is a very general method, and can be used with any kind of predictive modeling. For example we could use it with logistic regression or linear discriminant analysis, or any of the methods discussed in later

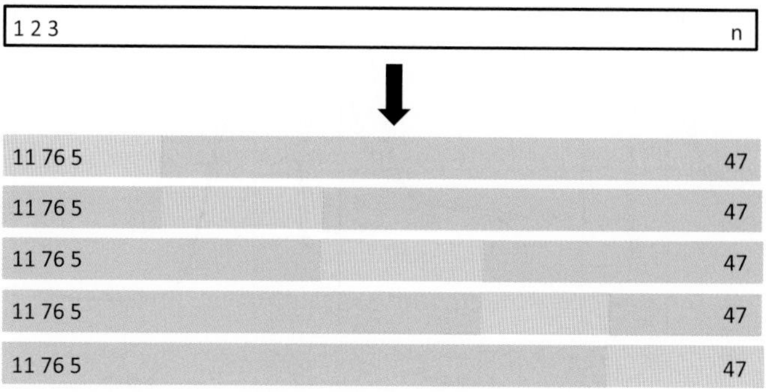

FIGURE 5.5. *A schematic display of 5-fold CV. A set of n observations is randomly split into five non-overlapping groups. Each of these fifths acts as a validation set (shown in beige), and the remainder as a training set (shown in blue). The test error is estimated by averaging the five resulting MSE estimates.*

chapters. The magic formula (5.2) does not hold in general, in which case the model has to be refit n times.

5.1.3 k-Fold Cross-Validation

An alternative to LOOCV is *k-fold CV*. This approach involves randomly dividing the set of observations into k groups, or *folds*, of approximately equal size. The first fold is treated as a validation set, and the method is fit on the remaining $k-1$ folds. The mean squared error, MSE_1, is then computed on the observations in the held-out fold. This procedure is repeated k times; each time, a different group of observations is treated as a validation set. This process results in k estimates of the test error, $\text{MSE}_1, \text{MSE}_2, \ldots, \text{MSE}_k$. The k-fold CV estimate is computed by averaging these values,

k-fold CV

$$\text{CV}_{(k)} = \frac{1}{k}\sum_{i=1}^{k}\text{MSE}_i. \tag{5.3}$$

Figure 5.5 illustrates the k-fold CV approach.

It is not hard to see that LOOCV is a special case of k-fold CV in which k is set to equal n. In practice, one typically performs k-fold CV using $k = 5$ or $k = 10$. What is the advantage of using $k = 5$ or $k = 10$ rather than $k = n$? The most obvious advantage is computational. LOOCV requires fitting the statistical learning method n times. This has the potential to be computationally expensive (except for linear models fit by least squares, in which case formula (5.2) can be used). But cross-validation is a very general approach that can be applied to almost any statistical learning method. Some statistical learning methods have computationally intensive fitting procedures, and so performing LOOCV may pose computational problems, especially if n is extremely large. In contrast, performing 10-fold

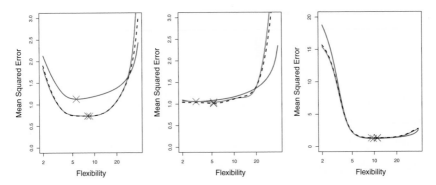

FIGURE 5.6. *True and estimated test MSE for the simulated data sets in Figures 2.9 (left), 2.10 (center), and 2.11 (right). The true test MSE is shown in blue, the LOOCV estimate is shown as a black dashed line, and the 10-fold CV estimate is shown in orange. The crosses indicate the minimum of each of the MSE curves.*

CV requires fitting the learning procedure only ten times, which may be much more feasible. As we see in Section 5.1.4, there also can be other non-computational advantages to performing 5-fold or 10-fold CV, which involve the bias-variance trade-off.

The right-hand panel of Figure 5.4 displays nine different 10-fold CV estimates for the Auto data set, each resulting from a different random split of the observations into ten folds. As we can see from the figure, there is some variability in the CV estimates as a result of the variability in how the observations are divided into ten folds. But this variability is typically much lower than the variability in the test error estimates that results from the validation set approach (right-hand panel of Figure 5.2).

When we examine real data, we do not know the *true* test MSE, and so it is difficult to determine the accuracy of the cross-validation estimate. However, if we examine simulated data, then we can compute the true test MSE, and can thereby evaluate the accuracy of our cross-validation results. In Figure 5.6, we plot the cross-validation estimates and true test error rates that result from applying smoothing splines to the simulated data sets illustrated in Figures 2.9–2.11 of Chapter 2. The true test MSE is displayed in blue. The black dashed and orange solid lines respectively show the estimated LOOCV and 10-fold CV estimates. In all three plots, the two cross-validation estimates are very similar. In the right-hand panel of Figure 5.6, the true test MSE and the cross-validation curves are almost identical. In the center panel of Figure 5.6, the two sets of curves are similar at the lower degrees of flexibility, while the CV curves overestimate the test set MSE for higher degrees of flexibility. In the left-hand panel of Figure 5.6, the CV curves have the correct general shape, but they underestimate the true test MSE.

When we perform cross-validation, our goal might be to determine how well a given statistical learning procedure can be expected to perform on independent data; in this case, the actual estimate of the test MSE is of interest. But at other times we are interested only in the location of the *minimum point in the estimated test MSE curve*. This is because we might be performing cross-validation on a number of statistical learning methods, or on a single method using different levels of flexibility, in order to identify the method that results in the lowest test error. For this purpose, the location of the minimum point in the estimated test MSE curve is important, but the actual value of the estimated test MSE is not. We find in Figure 5.6 that despite the fact that they sometimes underestimate the true test MSE, all of the CV curves come close to identifying the correct level of flexibility—that is, the flexibility level corresponding to the smallest test MSE.

5.1.4 Bias-Variance Trade-Off for k-Fold Cross-Validation

We mentioned in Section 5.1.3 that k-fold CV with $k < n$ has a computational advantage to LOOCV. But putting computational issues aside, a less obvious but potentially more important advantage of k-fold CV is that it often gives more accurate estimates of the test error rate than does LOOCV. This has to do with a bias-variance trade-off.

It was mentioned in Section 5.1.1 that the validation set approach can lead to overestimates of the test error rate, since in this approach the training set used to fit the statistical learning method contains only half the observations of the entire data set. Using this logic, it is not hard to see that LOOCV will give approximately unbiased estimates of the test error, since each training set contains $n - 1$ observations, which is almost as many as the number of observations in the full data set. And performing k-fold CV for, say, $k = 5$ or $k = 10$ will lead to an intermediate level of bias, since each training set contains $(k - 1)n/k$ observations—fewer than in the LOOCV approach, but substantially more than in the validation set approach. Therefore, from the perspective of bias reduction, it is clear that LOOCV is to be preferred to k-fold CV.

However, we know that bias is not the only source for concern in an estimating procedure; we must also consider the procedure's variance. It turns out that LOOCV has higher variance than does k-fold CV with $k < n$. Why is this the case? When we perform LOOCV, we are in effect averaging the outputs of n fitted models, each of which is trained on an almost identical set of observations; therefore, these outputs are highly (positively) correlated with each other. In contrast, when we perform k-fold CV with $k < n$, we are averaging the outputs of k fitted models that are somewhat less correlated with each other, since the overlap between the training sets in each model is smaller. Since the mean of many highly correlated quantities has higher variance than does the mean of many quantities that are not as highly correlated, the test error estimate resulting from LOOCV tends to have higher variance than does the test error estimate resulting from k-fold CV.

To summarize, there is a bias-variance trade-off associated with the choice of k in k-fold cross-validation. Typically, given these considerations, one performs k-fold cross-validation using $k = 5$ or $k = 10$, as these values have been shown empirically to yield test error rate estimates that suffer neither from excessively high bias nor from very high variance.

5.1.5 Cross-Validation on Classification Problems

In this chapter so far, we have illustrated the use of cross-validation in the regression setting where the outcome Y is quantitative, and so have used MSE to quantify test error. But cross-validation can also be a very useful approach in the classification setting when Y is qualitative. In this setting, cross-validation works just as described earlier in this chapter, except that rather than using MSE to quantify test error, we instead use the number of misclassified observations. For instance, in the classification setting, the LOOCV error rate takes the form

$$\mathrm{CV}_{(n)} = \frac{1}{n} \sum_{i=1}^{n} \mathrm{Err}_i, \qquad (5.4)$$

where $\mathrm{Err}_i = I(y_i \neq \hat{y}_i)$. The k-fold CV error rate and validation set error rates are defined analogously.

As an example, we fit various logistic regression models on the two-dimensional classification data displayed in Figure 2.13. In the top-left panel of Figure 5.7, the black solid line shows the estimated decision boundary resulting from fitting a standard logistic regression model to this data set. Since this is simulated data, we can compute the *true* test error rate, which takes a value of 0.201 and so is substantially larger than the Bayes error rate of 0.133. Clearly logistic regression does not have enough flexibility to model the Bayes decision boundary in this setting. We can easily extend logistic regression to obtain a non-linear decision boundary by using polynomial functions of the predictors, as we did in the regression setting in Section 3.3.2. For example, we can fit a *quadratic* logistic regression model, given by

$$\log \left(\frac{p}{1 - p} \right) = \beta_0 + \beta_1 X_1 + \beta_2 X_1^2 + \beta_3 X_2 + \beta_4 X_2^2. \qquad (5.5)$$

The top-right panel of Figure 5.7 displays the resulting decision boundary, which is now curved. However, the test error rate has improved only slightly, to 0.197. A much larger improvement is apparent in the bottom-left panel

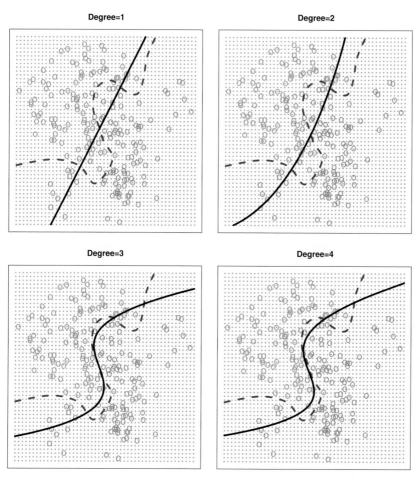

FIGURE 5.7. *Logistic regression fits on the two-dimensional classification data displayed in Figure 2.13. The Bayes decision boundary is represented using a purple dashed line. Estimated decision boundaries from linear, quadratic, cubic and quartic (degrees 1–4) logistic regressions are displayed in black. The test error rates for the four logistic regression fits are respectively 0.201, 0.197, 0.160, and 0.162, while the Bayes error rate is 0.133.*

of Figure 5.7, in which we have fit a logistic regression model involving cubic polynomials of the predictors. Now the test error rate has decreased to 0.160. Going to a quartic polynomial (bottom-right) slightly increases the test error.

In practice, for real data, the Bayes decision boundary and the test error rates are unknown. So how might we decide between the four logistic regression models displayed in Figure 5.7? We can use cross-validation in order to make this decision. The left-hand panel of Figure 5.8 displays in

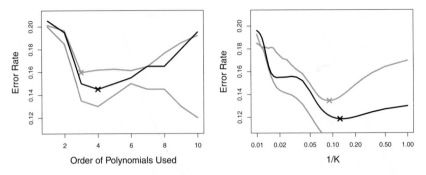

FIGURE 5.8. *Test error (brown), training error (blue), and 10-fold CV error (black) on the two-dimensional classification data displayed in Figure 5.7. Left: Logistic regression using polynomial functions of the predictors. The order of the polynomials used is displayed on the x-axis. Right: The KNN classifier with different values of K, the number of neighbors used in the KNN classifier.*

black the 10-fold CV error rates that result from fitting ten logistic regression models to the data, using polynomial functions of the predictors up to tenth order. The true test errors are shown in brown, and the training errors are shown in blue. As we have seen previously, the training error tends to decrease as the flexibility of the fit increases. (The figure indicates that though the training error rate doesn't quite decrease monotonically, it tends to decrease on the whole as the model complexity increases.) In contrast, the test error displays a characteristic U-shape. The 10-fold CV error rate provides a pretty good approximation to the test error rate. While it somewhat underestimates the error rate, it reaches a minimum when fourth-order polynomials are used, which is very close to the minimum of the test curve, which occurs when third-order polynomials are used. In fact, using fourth-order polynomials would likely lead to good test set performance, as the true test error rate is approximately the same for third, fourth, fifth, and sixth-order polynomials.

The right-hand panel of Figure 5.8 displays the same three curves using the KNN approach for classification, as a function of the value of K (which in this context indicates the number of neighbors used in the KNN classifier, rather than the number of CV folds used). Again the training error rate declines as the method becomes more flexible, and so we see that the training error rate cannot be used to select the optimal value for K. Though the cross-validation error curve slightly underestimates the test error rate, it takes on a minimum very close to the best value for K.

5.2   The Bootstrap

The *bootstrap* is a widely applicable and extremely powerful statistical tool
that can be used to quantify the uncertainty associated with a given estimator or statistical learning method. As a simple example, the bootstrap
can be used to estimate the standard errors of the coefficients from a linear
regression fit. In the specific case of linear regression, this is not particularly
useful, since we saw in Chapter 3 that standard statistical software such as
R outputs such standard errors automatically. However, the power of the
bootstrap lies in the fact that it can be easily applied to a wide range of
statistical learning methods, including some for which a measure of variability is otherwise difficult to obtain and is not automatically output by
statistical software.

bootstrap

In this section we illustrate the bootstrap on a toy example in which we
wish to determine the best investment allocation under a simple model.
In Section 5.3 we explore the use of the bootstrap to assess the variability
associated with the regression coefficients in a linear model fit.

Suppose that we wish to invest a fixed sum of money in two financial
assets that yield returns of X and Y, respectively, where X and Y are
random quantities. We will invest a fraction α of our money in X, and will
invest the remaining $1 - \alpha$ in Y. Since there is variability associated with
the returns on these two assets, we wish to choose α to minimize the total
risk, or variance, of our investment. In other words, we want to minimize
$\mathrm{Var}(\alpha X + (1 - \alpha)Y)$. One can show that the value that minimizes the risk
is given by

$$\alpha = \frac{\sigma_Y^2 - \sigma_{XY}}{\sigma_X^2 + \sigma_Y^2 - 2\sigma_{XY}}, \tag{5.6}$$

where $\sigma_X^2 = \mathrm{Var}(X), \sigma_Y^2 = \mathrm{Var}(Y)$, and $\sigma_{XY} = \mathrm{Cov}(X, Y)$.

In reality, the quantities σ_X^2, σ_Y^2, and σ_{XY} are unknown. We can compute
estimates for these quantities, $\hat{\sigma}_X^2$, $\hat{\sigma}_Y^2$, and $\hat{\sigma}_{XY}$, using a data set that
contains past measurements for X and Y. We can then estimate the value
of α that minimizes the variance of our investment using

$$\hat{\alpha} = \frac{\hat{\sigma}_Y^2 - \hat{\sigma}_{XY}}{\hat{\sigma}_X^2 + \hat{\sigma}_Y^2 - 2\hat{\sigma}_{XY}}. \tag{5.7}$$

Figure 5.9 illustrates this approach for estimating α on a simulated data
set. In each panel, we simulated 100 pairs of returns for the investments
X and Y. We used these returns to estimate σ_X^2, σ_Y^2, and σ_{XY}, which we
then substituted into (5.7) in order to obtain estimates for α. The value of
$\hat{\alpha}$ resulting from each simulated data set ranges from 0.532 to 0.657.

It is natural to wish to quantify the accuracy of our estimate of α. To
estimate the standard deviation of $\hat{\alpha}$, we repeated the process of simu-
lating 100 paired observations of X and Y, and estimating α using (5.7),

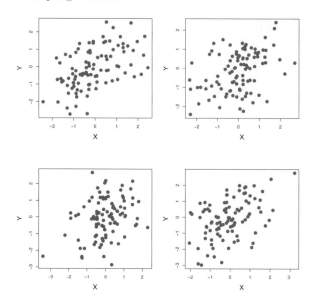

FIGURE 5.9. *Each panel displays* 100 *simulated returns for investments* X *and* Y. *From left to right and top to bottom, the resulting estimates for* α *are* 0.576, 0.532, 0.657, *and* 0.651.

1,000 times. We thereby obtained 1,000 estimates for α, which we can call $\hat{\alpha}_1, \hat{\alpha}_2, \ldots, \hat{\alpha}_{1,000}$. The left-hand panel of Figure 5.10 displays a histogram of the resulting estimates. For these simulations the parameters were set to $\sigma_X^2 = 1, \sigma_Y^2 = 1.25$, and $\sigma_{XY} = 0.5$, and so we know that the true value of α is 0.6. We indicated this value using a solid vertical line on the histogram. The mean over all 1,000 estimates for α is

$$\bar{\alpha} = \frac{1}{1,000} \sum_{r=1}^{1,000} \hat{\alpha}_r = 0.5996,$$

very close to $\alpha = 0.6$, and the standard deviation of the estimates is

$$\sqrt{\frac{1}{1,000-1} \sum_{r=1}^{1,000} (\hat{\alpha}_r - \bar{\alpha})^2} = 0.083.$$

This gives us a very good idea of the accuracy of $\hat{\alpha}$: SE($\hat{\alpha}$) ≈ 0.083. So roughly speaking, for a random sample from the population, we would expect $\hat{\alpha}$ to differ from α by approximately 0.08, on average.

In practice, however, the procedure for estimating SE($\hat{\alpha}$) outlined above cannot be applied, because for real data we cannot generate new samples from the original population. However, the bootstrap approach allows us to use a computer to emulate the process of obtaining new sample sets,

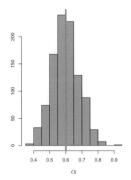 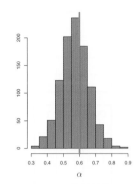 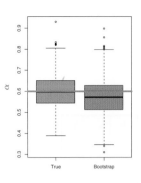

FIGURE 5.10. Left: *A histogram of the estimates of α obtained by generating 1,000 simulated data sets from the true population.* Center: *A histogram of the estimates of α obtained from 1,000 bootstrap samples from a single data set.* Right: *The estimates of α displayed in the left and center panels are shown as boxplots. In each panel, the pink line indicates the true value of α.*

so that we can estimate the variability of $\hat{\alpha}$ without generating additional samples. Rather than repeatedly obtaining independent data sets from the population, we instead obtain distinct data sets by repeatedly sampling observations *from the original data set.*

This approach is illustrated in Figure 5.11 on a simple data set, which we call Z, that contains only $n = 3$ observations. We randomly select n observations from the data set in order to produce a bootstrap data set, Z^{*1}. The sampling is performed with *replacement*, which means that the same observation can occur more than once in the bootstrap data set. In this example, Z^{*1} contains the third observation twice, the first observation once, and no instances of the second observation. Note that if an observation is contained in Z^{*1}, then both its X and Y values are included. We can use Z^{*1} to produce a new bootstrap estimate for α, which we call $\hat{\alpha}^{*1}$. This procedure is repeated B times for some large value of B, in order to produce B different bootstrap data sets, $Z^{*1}, Z^{*2}, \ldots, Z^{*B}$, and B corresponding α estimates, $\hat{\alpha}^{*1}, \hat{\alpha}^{*2}, \ldots, \hat{\alpha}^{*B}$. We can compute the standard error of these bootstrap estimates using the formula

replacement

$$\text{SE}_B(\hat{\alpha}) = \sqrt{\frac{1}{B-1}\sum_{r=1}^{B}\left(\hat{\alpha}^{*r} - \frac{1}{B}\sum_{r'=1}^{B}\hat{\alpha}^{*r'}\right)^2}. \qquad (5.8)$$

This serves as an estimate of the standard error of $\hat{\alpha}$ estimated from the original data set.

The bootstrap approach is illustrated in the center panel of Figure 5.10, which displays a histogram of 1,000 bootstrap estimates of α, each computed using a distinct bootstrap data set. This panel was constructed on the basis of a single data set, and hence could be created using real data.

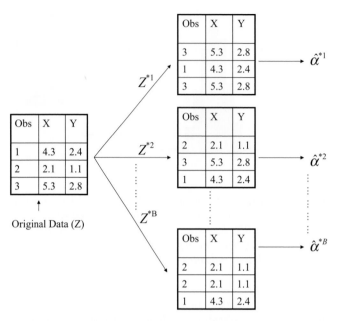

FIGURE 5.11. *A graphical illustration of the bootstrap approach on a small sample containing* $n = 3$ *observations. Each bootstrap data set contains* n *observations, sampled with replacement from the original data set. Each bootstrap data set is used to obtain an estimate of* α.

Note that the histogram looks very similar to the left-hand panel which displays the idealized histogram of the estimates of α obtained by generating 1,000 simulated data sets from the true population. In particular the bootstrap estimate $SE(\hat{\alpha})$ from (5.8) is 0.087, very close to the estimate of 0.083 obtained using 1,000 simulated data sets. The right-hand panel displays the information in the center and left panels in a different way, via boxplots of the estimates for α obtained by generating 1,000 simulated data sets from the true population and using the bootstrap approach. Again, the boxplots are quite similar to each other, indicating that the bootstrap approach can be used to effectively estimate the variability associated with $\hat{\alpha}$.

5.3 Lab: Cross-Validation and the Bootstrap

In this lab, we explore the resampling techniques covered in this chapter. Some of the commands in this lab may take a while to run on your computer.

5.3.1 The Validation Set Approach

We explore the use of the validation set approach in order to estimate the test error rates that result from fitting various linear models on the `Auto` data set.

Before we begin, we use the `set.seed()` function in order to set a *seed* for R's random number generator, so that the reader of this book will obtain precisely the same results as those shown below. It is generally a good idea to set a random seed when performing an analysis such as cross-validation that contains an element of randomness, so that the results obtained can be reproduced precisely at a later time.

seed

We begin by using the `sample()` function to split the set of observations into two halves, by selecting a random subset of 196 observations out of the original 392 observations. We refer to these observations as the training set.

`sample()`

```
> library(ISLR)
> set.seed(1)
> train=sample(392,196)
```

(Here we use a shortcut in the sample command; see ?sample for details.) We then use the subset option in `lm()` to fit a linear regression using only the observations corresponding to the training set.

```
> lm.fit=lm(mpg~horsepower,data=Auto,subset=train)
```

We now use the `predict()` function to estimate the response for all 392 observations, and we use the `mean()` function to calculate the MSE of the 196 observations in the validation set. Note that the –train index below selects only the observations that are not in the training set.

```
> attach(Auto)
> mean((mpg-predict(lm.fit,Auto))[-train]^2)
[1] 26.14
```

Therefore, the estimated test MSE for the linear regression fit is 26.14. We can use the `poly()` function to estimate the test error for the quadratic and cubic regressions.

```
> lm.fit2=lm(mpg~poly(horsepower,2),data=Auto,subset=train)
> mean((mpg-predict(lm.fit2,Auto))[-train]^2)
[1] 19.82
> lm.fit3=lm(mpg~poly(horsepower,3),data=Auto,subset=train)
> mean((mpg-predict(lm.fit3,Auto))[-train]^2)
[1] 19.78
```

These error rates are 19.82 and 19.78, respectively. If we choose a different training set instead, then we will obtain somewhat different errors on the validation set.

```
> set.seed(2)
> train=sample(392,196)
> lm.fit=lm(mpg~horsepower,subset=train)
```

```
> mean((mpg-predict(lm.fit,Auto))[-train]^2)
[1] 23.30
> lm.fit2=lm(mpg~poly(horsepower,2),data=Auto,subset=train)
> mean((mpg-predict(lm.fit2,Auto))[-train]^2)
[1] 18.90
> lm.fit3=lm(mpg~poly(horsepower,3),data=Auto,subset=train)
> mean((mpg-predict(lm.fit3,Auto))[-train]^2)
[1] 19.26
```

Using this split of the observations into a training set and a validation set, we find that the validation set error rates for the models with linear, quadratic, and cubic terms are 23.30, 18.90, and 19.26, respectively.

These results are consistent with our previous findings: a model that predicts mpg using a quadratic function of horsepower performs better than a model that involves only a linear function of horsepower, and there is little evidence in favor of a model that uses a cubic function of horsepower.

5.3.2 Leave-One-Out Cross-Validation

The LOOCV estimate can be automatically computed for any generalized linear model using the glm() and cv.glm() functions. In the lab for Chapter 4, we used the glm() function to perform logistic regression by passing in the family="binomial" argument. But if we use glm() to fit a model without passing in the family argument, then it performs linear regression, just like the lm() function. So for instance,

`cv.glm()`

```
> glm.fit=glm(mpg~horsepower,data=Auto)
> coef(glm.fit)
(Intercept)    horsepower
     39.936        -0.158
```

and

```
> lm.fit=lm(mpg~horsepower,data=Auto)
> coef(lm.fit)
(Intercept)    horsepower
     39.936        -0.158
```

yield identical linear regression models. In this lab, we will perform linear regression using the glm() function rather than the lm() function because the former can be used together with cv.glm(). The cv.glm() function is part of the boot library.

```
> library(boot)
> glm.fit=glm(mpg~horsepower,data=Auto)
> cv.err=cv.glm(Auto,glm.fit)
> cv.err$delta
       1        1
   24.23    24.23
```

The cv.glm() function produces a list with several components. The two numbers in the delta vector contain the cross-validation results. In this

case the numbers are identical (up to two decimal places) and correspond
to the LOOCV statistic given in (5.1). Below, we discuss a situation in
which the two numbers differ. Our cross-validation estimate for the test
error is approximately 24.23.

We can repeat this procedure for increasingly complex polynomial fits.
To automate the process, we use the `for()` function to initiate a *for loop*   `for()`
which iteratively fits polynomial regressions for polynomials of order $i = 1$ for loop
to $i = 5$, computes the associated cross-validation error, and stores it in
the ith element of the vector cv.error. We begin by initializing the vector.
This command will likely take a couple of minutes to run.

```
> cv.error=rep(0,5)
> for (i in 1:5){
+ glm.fit=glm(mpg~poly(horsepower,i),data=Auto)
+ cv.error[i]=cv.glm(Auto,glm.fit)$delta[1]
+ }
> cv.error
[1] 24.23 19.25 19.33 19.42 19.03
```

As in Figure 5.4, we see a sharp drop in the estimated test MSE between
the linear and quadratic fits, but then no clear improvement from using
higher-order polynomials.

5.3.3 k-Fold Cross-Validation

The `cv.glm()` function can also be used to implement k-fold CV. Below we
use $k = 10$, a common choice for k, on the Auto data set. We once again set
a random seed and initialize a vector in which we will store the CV errors
corresponding to the polynomial fits of orders one to ten.

```
> set.seed(17)
> cv.error.10=rep(0,10)
> for (i in 1:10){
+ glm.fit=glm(mpg~poly(horsepower,i),data=Auto)
+ cv.error.10[i]=cv.glm(Auto,glm.fit,K=10)$delta[1]
+ }
> cv.error.10
[1] 24.21 19.19 19.31 19.34 18.88 19.02 18.90 19.71 18.95 19.50
```

Notice that the computation time is much shorter than that of LOOCV.
(In principle, the computation time for LOOCV for a least squares linear
model should be faster than for k-fold CV, due to the availability of the
formula (5.2) for LOOCV; however, unfortunately the `cv.glm()` function
does not make use of this formula.) We still see little evidence that using
cubic or higher-order polynomial terms leads to lower test error than simply
using a quadratic fit.

We saw in Section 5.3.2 that the two numbers associated with delta are
essentially the same when LOOCV is performed. When we instead perform
k-fold CV, then the two numbers associated with delta differ slightly. The

first is the standard k-fold CV estimate, as in (5.3). The second is a bias-corrected version. On this data set, the two estimates are very similar to each other.

5.3.4   The Bootstrap

We illustrate the use of the bootstrap in the simple example of Section 5.2, as well as on an example involving estimating the accuracy of the linear regression model on the Auto data set.

Estimating the Accuracy of a Statistic of Interest

One of the great advantages of the bootstrap approach is that it can be applied in almost all situations. No complicated mathematical calculations are required. Performing a bootstrap analysis in R entails only two steps. First, we must create a function that computes the statistic of interest. Second, we use the boot() function, which is part of the boot library, to perform the bootstrap by repeatedly sampling observations from the data set with replacement.

boot()

The Portfolio data set in the ISLR package is described in Section 5.2. To illustrate the use of the bootstrap on this data, we must first create a function, alpha.fn(), which takes as input the (X, Y) data as well as a vector indicating which observations should be used to estimate α. The function then outputs the estimate for α based on the selected observations.

```
> alpha.fn=function(data,index){
+ X=data$X[index]
+ Y=data$Y[index]
+ return((var(Y)-cov(X,Y))/(var(X)+var(Y)-2*cov(X,Y)))
+ }
```

This function *returns*, or outputs, an estimate for α based on applying (5.7) to the observations indexed by the argument index. For instance, the following command tells R to estimate α using all 100 observations.

```
> alpha.fn(Portfolio,1:100)
[1] 0.576
```

The next command uses the sample() function to randomly select 100 observations from the range 1 to 100, with replacement. This is equivalent to constructing a new bootstrap data set and recomputing $\hat{\alpha}$ based on the new data set.

```
> set.seed(1)
> alpha.fn(Portfolio,sample(100,100,replace=T))
[1] 0.596
```

We can implement a bootstrap analysis by performing this command many times, recording all of the corresponding estimates for α, and computing the resulting standard deviation. However, the `boot()` function automates this approach. Below we produce $R = 1,000$ bootstrap estimates for α.

```
> boot(Portfolio,alpha.fn,R=1000)

ORDINARY NONPARAMETRIC BOOTSTRAP


Call:
boot(data = Portfolio, statistic = alpha.fn, R = 1000)


Bootstrap Statistics :
      original        bias       std. error
t1*   0.5758     -7.315e-05     0.0886
```

The final output shows that using the original data, $\hat{\alpha} = 0.5758$, and that the bootstrap estimate for $SE(\hat{\alpha})$ is 0.0886.

Estimating the Accuracy of a Linear Regression Model

The bootstrap approach can be used to assess the variability of the coefficient estimates and predictions from a statistical learning method. Here we use the bootstrap approach in order to assess the variability of the estimates for β_0 and β_1, the intercept and slope terms for the linear regression model that uses horsepower to predict mpg in the Auto data set. We will compare the estimates obtained using the bootstrap to those obtained using the formulas for $SE(\hat{\beta}_0)$ and $SE(\hat{\beta}_1)$ described in Section 3.1.2.

We first create a simple function, `boot.fn()`, which takes in the Auto data set as well as a set of indices for the observations, and returns the intercept and slope estimates for the linear regression model. We then apply this function to the full set of 392 observations in order to compute the estimates of β_0 and β_1 on the entire data set using the usual linear regression coefficient estimate formulas from Chapter 3. Note that we do not need the { and } at the beginning and end of the function because it is only one line long.

```
> boot.fn=function(data,index)
+  return(coef(lm(mpg~horsepower,data=data,subset=index)))
> boot.fn(Auto,1:392)
(Intercept) horsepower
    39.936     -0.158
```

The `boot.fn()` function can also be used in order to create bootstrap estimates for the intercept and slope terms by randomly sampling from among the observations with replacement. Here we give two examples.

```
> set.seed(1)
> boot.fn(Auto,sample(392,392,replace=T))
(Intercept) horsepower
    38.739     -0.148
> boot.fn(Auto,sample(392,392,replace=T))
(Intercept) horsepower
    40.038     -0.160
```

Next, we use the `boot()` function to compute the standard errors of 1,000 bootstrap estimates for the intercept and slope terms.

```
> boot(Auto,boot.fn,1000)

ORDINARY NONPARAMETRIC BOOTSTRAP

Call:
boot(data = Auto, statistic = boot.fn, R = 1000)

Bootstrap Statistics :
      original       bias    std. error
t1* 39.936       0.0297    0.8600
t2* -0.158      -0.0003    0.0074
```

This indicates that the bootstrap estimate for $SE(\hat{\beta}_0)$ is 0.86, and that the bootstrap estimate for $SE(\hat{\beta}_1)$ is 0.0074. As discussed in Section 3.1.2, standard formulas can be used to compute the standard errors for the regression coefficients in a linear model. These can be obtained using the `summary()` function.

```
> summary(lm(mpg~horsepower,data=Auto))$coef
              Estimate Std. Error t value   Pr(>|t|)
(Intercept)    39.936    0.71750    55.7 1.22e-187
horsepower     -0.158    0.00645   -24.5  7.03e-81
```

The standard error estimates for $\hat{\beta}_0$ and $\hat{\beta}_1$ obtained using the formulas from Section 3.1.2 are 0.717 for the intercept and 0.0064 for the slope. Interestingly, these are somewhat different from the estimates obtained using the bootstrap. Does this indicate a problem with the bootstrap? In fact, it suggests the opposite. Recall that the standard formulas given in Equation 3.8 on page 66 rely on certain assumptions. For example, they depend on the unknown parameter σ^2, the noise variance. We then estimate σ^2 using the RSS. Now although the formula for the standard errors do not rely on the linear model being correct, the estimate for σ^2 does. We see in Figure 3.8 on page 91 that there is a non-linear relationship in the data, and so the residuals from a linear fit will be inflated, and so will $\hat{\sigma}^2$. Secondly, the standard formulas assume (somewhat unrealistically) that the x_i are fixed, and all the variability comes from the variation in the errors ϵ_i. The bootstrap approach does not rely on any of these assumptions, and so it is likely giving a more accurate estimate of the standard errors of $\hat{\beta}_0$ and $\hat{\beta}_1$ than is the `summary()` function.

Below we compute the bootstrap standard error estimates and the standard linear regression estimates that result from fitting the quadratic model to the data. Since this model provides a good fit to the data (Figure 3.8), there is now a better correspondence between the bootstrap estimates and the standard estimates of $SE(\hat{\beta}_0)$, $SE(\hat{\beta}_1)$ and $SE(\hat{\beta}_2)$.

```
> boot.fn=function(data,index)
+ coefficients(lm(mpg~horsepower+I(horsepower^2),data=data,
    subset=index))
> set.seed(1)
> boot(Auto,boot.fn,1000)

ORDINARY NONPARAMETRIC BOOTSTRAP


Call:
boot(data = Auto, statistic = boot.fn, R = 1000)


Bootstrap Statistics :
      original      bias      std. error
t1*   56.900    6.098e-03   2.0945
t2*   -0.466   -1.777e-04   0.0334
t3*    0.001    1.324e-06   0.0001

> summary(lm(mpg~horsepower+I(horsepower^2),data=Auto))$coef
                Estimate  Std. Error  t value  Pr(>|t|)
(Intercept)      56.9001    1.80043       32   1.7e-109
horsepower       -0.4662    0.03112      -15   2.3e-40
I(horsepower^2)   0.0012    0.00012       10   2.2e-21
```

5.4   Exercises

Conceptual

1. Using basic statistical properties of the variance, as well as single-variable calculus, derive (5.6). In other words, prove that α given by (5.6) does indeed minimize $\text{Var}(\alpha X + (1 - \alpha)Y)$.

2. We will now derive the probability that a given observation is part of a bootstrap sample. Suppose that we obtain a bootstrap sample from a set of n observations.

   (a) What is the probability that the first bootstrap observation is *not* the jth observation from the original sample? Justify your answer.

   (b) What is the probability that the second bootstrap observation is *not* the jth observation from the original sample?

   (c) Argue that the probability that the jth observation is *not* in the bootstrap sample is $(1 - 1/n)^n$.

   (d) When $n = 5$, what is the probability that the jth observation is in the bootstrap sample?

   (e) When $n = 100$, what is the probability that the jth observation is in the bootstrap sample?

(f) When $n = 10,000$, what is the probability that the jth observation is in the bootstrap sample?

(g) Create a plot that displays, for each integer value of n from 1 to $100,000$, the probability that the jth observation is in the bootstrap sample. Comment on what you observe.

(h) We will now investigate numerically the probability that a bootstrap sample of size $n = 100$ contains the jth observation. Here $j = 4$. We repeatedly create bootstrap samples, and each time we record whether or not the fourth observation is contained in the bootstrap sample.

```
> store=rep(NA, 10000)
> for(i in 1:10000){
    store[i]=sum(sample(1:100, rep=TRUE)==4)>0
}
> mean(store)
```

Comment on the results obtained.

3. We now review k-fold cross-validation.

(a) Explain how k-fold cross-validation is implemented.

(b) What are the advantages and disadvantages of k-fold cross-validation relative to:

i. The validation set approach?

ii. LOOCV?

4. Suppose that we use some statistical learning method to make a prediction for the response Y for a particular value of the predictor X. Carefully describe how we might estimate the standard deviation of our prediction.

Applied

5. In Chapter 4, we used logistic regression to predict the probability of default using income and balance on the Default data set. We will now estimate the test error of this logistic regression model using the validation set approach. Do not forget to set a random seed before beginning your analysis.

(a) Fit a logistic regression model that uses income and balance to predict default.

(b) Using the validation set approach, estimate the test error of this model. In order to do this, you must perform the following steps:

i. Split the sample set into a training set and a validation set.

      ii. Fit a multiple logistic regression model using only the training observations.

     iii. Obtain a prediction of default status for each individual in the validation set by computing the posterior probability of default for that individual, and classifying the individual to the default category if the posterior probability is greater than 0.5.

     iv. Compute the validation set error, which is the fraction of the observations in the validation set that are misclassified.

  (c) Repeat the process in (b) three times, using three different splits of the observations into a training set and a validation set. Comment on the results obtained.

  (d) Now consider a logistic regression model that predicts the probability of default using income, balance, and a dummy variable for student. Estimate the test error for this model using the validation set approach. Comment on whether or not including a dummy variable for student leads to a reduction in the test error rate.

6. We continue to consider the use of a logistic regression model to predict the probability of default using income and balance on the Default data set. In particular, we will now compute estimates for the standard errors of the income and balance logistic regression coefficients in two different ways: (1) using the bootstrap, and (2) using the standard formula for computing the standard errors in the glm() function. Do not forget to set a random seed before beginning your analysis.

  (a) Using the summary() and glm() functions, determine the estimated standard errors for the coefficients associated with income and balance in a multiple logistic regression model that uses both predictors.

  (b) Write a function, boot.fn(), that takes as input the Default data set as well as an index of the observations, and that outputs the coefficient estimates for income and balance in the multiple logistic regression model.

  (c) Use the boot() function together with your boot.fn() function to estimate the standard errors of the logistic regression coefficients for income and balance.

  (d) Comment on the estimated standard errors obtained using the glm() function and using your bootstrap function.

7. In Sections 5.3.2 and 5.3.3, we saw that the cv.glm() function can be used in order to compute the LOOCV test error estimate. Alternatively, one could compute those quantities using just the glm() and

`predict.glm()` functions, and a for loop. You will now take this approach in order to compute the LOOCV error for a simple logistic regression model on the `Weekly` data set. Recall that in the context of classification problems, the LOOCV error is given in (5.4).

(a) Fit a logistic regression model that predicts `Direction` using `Lag1` and `Lag2`.

(b) Fit a logistic regression model that predicts `Direction` using `Lag1` and `Lag2` *using all but the first observation.*

(c) Use the model from (b) to predict the direction of the first observation. You can do this by predicting that the first observation will go up if $P(\text{Direction="Up"}|\text{Lag1, Lag2}) > 0.5$. Was this observation correctly classified?

(d) Write a for loop from $i = 1$ to $i = n$, where n is the number of observations in the data set, that performs each of the following steps:

    i. Fit a logistic regression model using all but the ith observation to predict `Direction` using `Lag1` and `Lag2`.

    ii. Compute the posterior probability of the market moving up for the ith observation.

    iii. Use the posterior probability for the ith observation in order to predict whether or not the market moves up.

    iv. Determine whether or not an error was made in predicting the direction for the ith observation. If an error was made, then indicate this as a 1, and otherwise indicate it as a 0.

(e) Take the average of the n numbers obtained in (d)iv in order to obtain the LOOCV estimate for the test error. Comment on the results.

8. We will now perform cross-validation on a simulated data set.

(a) Generate a simulated data set as follows:

```
> set.seed(1)
> x=rnorm(100)
> y=x-2*x^2+rnorm(100)
```

In this data set, what is n and what is p? Write out the model used to generate the data in equation form.

(b) Create a scatterplot of X against Y. Comment on what you find.

(c) Set a random seed, and then compute the LOOCV errors that result from fitting the following four models using least squares:

    i. $Y = \beta_0 + \beta_1 X + \epsilon$

   ii. $Y = \beta_0 + \beta_1 X + \beta_2 X^2 + \epsilon$

  iii. $Y = \beta_0 + \beta_1 X + \beta_2 X^2 + \beta_3 X^3 + \epsilon$

  iv. $Y = \beta_0 + \beta_1 X + \beta_2 X^2 + \beta_3 X^3 + \beta_4 X^4 + \epsilon$.

Note you may find it helpful to use the data.frame() function to create a single data set containing both X and Y.

(d) Repeat (c) using another random seed, and report your results. Are your results the same as what you got in (c)? Why?

(e) Which of the models in (c) had the smallest LOOCV error? Is this what you expected? Explain your answer.

(f) Comment on the statistical significance of the coefficient estimates that results from fitting each of the models in (c) using least squares. Do these results agree with the conclusions drawn based on the cross-validation results?

9. We will now consider the Boston housing data set, from the MASS library.

(a) Based on this data set, provide an estimate for the population mean of medv. Call this estimate $\hat{\mu}$.

(b) Provide an estimate of the standard error of $\hat{\mu}$. Interpret this result.

Hint: We can compute the standard error of the sample mean by dividing the sample standard deviation by the square root of the number of observations.

(c) Now estimate the standard error of $\hat{\mu}$ using the bootstrap. How does this compare to your answer from (b)?

(d) Based on your bootstrap estimate from (c), provide a 95 % confidence interval for the mean of medv. Compare it to the results obtained using t.test(Boston$medv).

Hint: You can approximate a 95 % confidence interval using the formula $[\hat{\mu} - 2SE(\hat{\mu}), \hat{\mu} + 2SE(\hat{\mu})]$.

(e) Based on this data set, provide an estimate, $\hat{\mu}_{med}$, for the median value of medv in the population.

(f) We now would like to estimate the standard error of $\hat{\mu}_{med}$. Unfortunately, there is no simple formula for computing the standard error of the median. Instead, estimate the standard error of the median using the bootstrap. Comment on your findings.

(g) Based on this data set, provide an estimate for the tenth percentile of medv in Boston suburbs. Call this quantity $\hat{\mu}_{0.1}$. (You can use the quantile() function.)

(h) Use the bootstrap to estimate the standard error of $\hat{\mu}_{0.1}$. Comment on your findings.

6

Linear Model Selection
and Regularization

In the regression setting, the standard linear model

$$Y = \beta_0 + \beta_1 X_1 + \cdots + \beta_p X_p + \epsilon \qquad (6.1)$$

is commonly used to describe the relationship between a response Y and a set of variables $X_1, X_2, \ldots, X_p$. We have seen in Chapter 3 that one typically fits this model using least squares.

In the chapters that follow, we consider some approaches for extending the linear model framework. In Chapter 7 we generalize (6.1) in order to accommodate non-linear, but still additive, relationships, while in Chapter 8 we consider even more general non-linear models. However, the linear model has distinct advantages in terms of inference and, on real-world problems, is often surprisingly competitive in relation to non-linear methods. Hence, before moving to the non-linear world, we discuss in this chapter some ways in which the simple linear model can be improved, by replacing plain least squares fitting with some alternative fitting procedures.

Why might we want to use another fitting procedure instead of least squares? As we will see, alternative fitting procedures can yield better *prediction accuracy* and *model interpretability*.

- *Prediction Accuracy*: Provided that the true relationship between the response and the predictors is approximately linear, the least squares estimates will have low bias. If $n \gg p$—that is, if n, the number of observations, is much larger than p, the number of variables—then the least squares estimates tend to also have low variance, and hence will perform well on test observations. However, if n is not much larger

G. James et al., *An Introduction to Statistical Learning: with Applications in R*, 203
Springer Texts in Statistics, DOI 10.1007/978-1-4614-7138-7_6,
© Springer Science+Business Media New York 2013

than p, then there can be a lot of variability in the least squares fit, resulting in overfitting and consequently poor predictions on future observations not used in model training. And if $p > n$, then there is no longer a unique least squares coefficient estimate: the variance is *infinite* so the method cannot be used at all. By *constraining* or *shrinking* the estimated coefficients, we can often substantially reduce the variance at the cost of a negligible increase in bias. This can lead to substantial improvements in the accuracy with which we can predict the response for observations not used in model training.

- *Model Interpretability*: It is often the case that some or many of the variables used in a multiple regression model are in fact not associated with the response. Including such *irrelevant* variables leads to unnecessary complexity in the resulting model. By removing these variables—that is, by setting the corresponding coefficient estimates to zero—we can obtain a model that is more easily interpreted. Now least squares is extremely unlikely to yield any coefficient estimates that are exactly zero. In this chapter, we see some approaches for automatically performing *feature selection* or *variable selection*—that is, for excluding irrelevant variables from a multiple regression model.

feature selection

variable selection

There are many alternatives, both classical and modern, to using least squares to fit (6.1). In this chapter, we discuss three important classes of methods.

- *Subset Selection*. This approach involves identifying a subset of the p predictors that we believe to be related to the response. We then fit a model using least squares on the reduced set of variables.

- *Shrinkage*. This approach involves fitting a model involving all p predictors. However, the estimated coefficients are shrunken towards zero relative to the least squares estimates. This shrinkage (also known as *regularization*) has the effect of reducing variance. Depending on what type of shrinkage is performed, some of the coefficients may be estimated to be exactly zero. Hence, shrinkage methods can also perform variable selection.

- *Dimension Reduction*. This approach involves *projecting* the p predictors into a M-dimensional subspace, where $M < p$. This is achieved by computing M different *linear combinations*, or *projections*, of the variables. Then these M projections are used as predictors to fit a linear regression model by least squares.

In the following sections we describe each of these approaches in greater detail, along with their advantages and disadvantages. Although this chapter describes extensions and modifications to the linear model for regression seen in Chapter 3, the same concepts apply to other methods, such as the classification models seen in Chapter 4.

6.1 Subset Selection

In this section we consider some methods for selecting subsets of predictors. These include best subset and stepwise model selection procedures.

6.1.1 Best Subset Selection

To perform *best subset selection*, we fit a separate least squares regression for each possible combination of the p predictors. That is, we fit all p models that contain exactly one predictor, all $\binom{p}{2} = p(p-1)/2$ models that contain exactly two predictors, and so forth. We then look at all of the resulting models, with the goal of identifying the one that is *best*.

best subset
selection

    The problem of selecting the *best model* from among the 2^p possibilities considered by best subset selection is not trivial. This is usually broken up into two stages, as described in Algorithm 6.1.

Algorithm 6.1 *Best subset selection*

---

1. Let $\mathcal{M}_0$ denote the *null model*, which contains no predictors. This model simply predicts the sample mean for each observation.

2. For $k = 1, 2, \ldots p$:

    (a) Fit all $\binom{p}{k}$ models that contain exactly k predictors.

    (b) Pick the best among these $\binom{p}{k}$ models, and call it $\mathcal{M}_k$. Here *best* is defined as having the smallest RSS, or equivalently largest R^2.

3. Select a single best model from among $\mathcal{M}_0, \ldots, \mathcal{M}_p$ using cross-validated prediction error, C_p (AIC), BIC, or adjusted R^2.

In Algorithm 6.1, Step 2 identifies the best model (on the training data) for each subset size, in order to reduce the problem from one of 2^p possible models to one of $p + 1$ possible models. In Figure 6.1, these models form the lower frontier depicted in red.

    Now in order to select a single best model, we must simply choose among these $p + 1$ options. This task must be performed with care, because the RSS of these $p + 1$ models decreases monotonically, and the R^2 increases monotonically, as the number of features included in the models increases. Therefore, if we use these statistics to select the best model, then we will always end up with a model involving all of the variables. The problem is that a low RSS or a high R^2 indicates a model with a low *training* error, whereas we wish to choose a model that has a low *test* error. (As shown in Chapter 2 in Figures 2.9–2.11, training error tends to be quite a bit smaller than test error, and a low training error by no means guarantees a low test error.) Therefore, in Step 3, we use cross-validated prediction

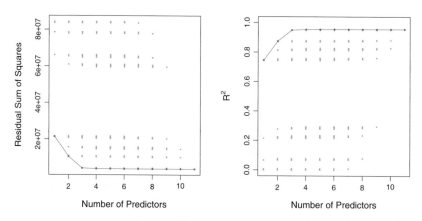

FIGURE 6.1. *For each possible model containing a subset of the ten predictors in the* Credit *data set, the RSS and R^2 are displayed. The red frontier tracks the best model for a given number of predictors, according to RSS and R^2. Though the data set contains only ten predictors, the x-axis ranges from 1 to 11, since one of the variables is categorical and takes on three values, leading to the creation of two dummy variables.*

error, C_p, BIC, or adjusted R^2 in order to select among $\mathcal{M}_0, \mathcal{M}_1, \ldots, \mathcal{M}_p$. These approaches are discussed in Section 6.1.3.

An application of best subset selection is shown in Figure 6.1. Each plotted point corresponds to a least squares regression model fit using a different subset of the 11 predictors in the Credit data set, discussed in Chapter 3. Here the variable ethnicity is a three-level qualitative variable, and so is represented by two dummy variables, which are selected separately in this case. We have plotted the RSS and R^2 statistics for each model, as a function of the number of variables. The red curves connect the best models for each model size, according to RSS or R^2. The figure shows that, as expected, these quantities improve as the number of variables increases; however, from the three-variable model on, there is little improvement in RSS and R^2 as a result of including additional predictors.

Although we have presented best subset selection here for least squares regression, the same ideas apply to other types of models, such as logistic regression. In the case of logistic regression, instead of ordering models by RSS in Step 2 of Algorithm 6.1, we instead use the *deviance*, a measure that plays the role of RSS for a broader class of models. The deviance is negative two times the maximized log-likelihood; the smaller the deviance, the better the fit.

deviance

While best subset selection is a simple and conceptually appealing approach, it suffers from computational limitations. The number of possible models that must be considered grows rapidly as p increases. In general, there are 2^p models that involve subsets of p predictors. So if $p = 10$, then there are approximately 1,000 possible models to be considered, and if

$p = 20$, then there are over one million possibilities! Consequently, best subset selection becomes computationally infeasible for values of p greater than around 40, even with extremely fast modern computers. There are computational shortcuts—so called branch-and-bound techniques—for eliminating some choices, but these have their limitations as p gets large. They also only work for least squares linear regression. We present computationally efficient alternatives to best subset selection next.

6.1.2 Stepwise Selection

For computational reasons, best subset selection cannot be applied with very large p. Best subset selection may also suffer from statistical problems when p is large. The larger the search space, the higher the chance of finding models that look good on the training data, even though they might not have any predictive power on future data. Thus an enormous search space can lead to overfitting and high variance of the coefficient estimates.

For both of these reasons, *stepwise* methods, which explore a far more restricted set of models, are attractive alternatives to best subset selection.

Forward Stepwise Selection

Forward stepwise selection is a computationally efficient alternative to best subset selection. While the best subset selection procedure considers all 2^p possible models containing subsets of the p predictors, forward stepwise considers a much smaller set of models. Forward stepwise selection begins with a model containing no predictors, and then adds predictors to the model, one-at-a-time, until all of the predictors are in the model. In particular, at each step the variable that gives the greatest *additional* improvement to the fit is added to the model. More formally, the forward stepwise selection procedure is given in Algorithm 6.2.

forward
stepwise
selection

Algorithm 6.2 *Forward stepwise selection*

---

1. Let $\mathcal{M}_0$ denote the *null* model, which contains no predictors.

2. For $k = 0, \ldots, p - 1$:

    (a) Consider all $p - k$ models that augment the predictors in $\mathcal{M}_k$ with one additional predictor.

    (b) Choose the *best* among these $p - k$ models, and call it $\mathcal{M}_{k+1}$. Here *best* is defined as having smallest RSS or highest R^2.

3. Select a single best model from among $\mathcal{M}_0, \ldots, \mathcal{M}_p$ using cross-validated prediction error, C_p (AIC), BIC, or adjusted R^2.

Unlike best subset selection, which involved fitting 2^p models, forward stepwise selection involves fitting one null model, along with $p - k$ models in the kth iteration, for $k = 0, \ldots, p - 1$. This amounts to a total of $1 + \sum_{k=0}^{p-1}(p-k) = 1+p(p+1)/2$ models. This is a substantial difference: when $p = 20$, best subset selection requires fitting 1,048,576 models, whereas forward stepwise selection requires fitting only 211 models.[1]

In Step 2(b) of Algorithm 6.2, we must identify the *best* model from among those $p-k$ that augment $\mathcal{M}_k$ with one additional predictor. We can do this by simply choosing the model with the lowest RSS or the highest R^2. However, in Step 3, we must identify the best model among a set of models with different numbers of variables. This is more challenging, and is discussed in Section 6.1.3.

Forward stepwise selection's computational advantage over best subset selection is clear. Though forward stepwise tends to do well in practice, it is not guaranteed to find the best possible model out of all 2^p models containing subsets of the p predictors. For instance, suppose that in a given data set with $p = 3$ predictors, the best possible one-variable model contains X_1, and the best possible two-variable model instead contains X_2 and X_3. Then forward stepwise selection will fail to select the best possible two-variable model, because $\mathcal{M}_1$ will contain X_1, so $\mathcal{M}_2$ must also contain X_1 together with one additional variable.

Table 6.1, which shows the first four selected models for best subset and forward stepwise selection on the `Credit` data set, illustrates this phenomenon. Both best subset selection and forward stepwise selection choose `rating` for the best one-variable model and then include `income` and `student` for the two- and three-variable models. However, best subset selection replaces `rating` by `cards` in the four-variable model, while forward stepwise selection must maintain `rating` in its four-variable model. In this example, Figure 6.1 indicates that there is not much difference between the three- and four-variable models in terms of RSS, so either of the four-variable models will likely be adequate.

Forward stepwise selection can be applied even in the high-dimensional setting where $n < p$; however, in this case, it is possible to construct submodels $\mathcal{M}_0, \ldots, \mathcal{M}_{n-1}$ only, since each submodel is fit using least squares, which will not yield a unique solution if $p \geq n$.

Backward Stepwise Selection

Like forward stepwise selection, *backward stepwise selection* provides an efficient alternative to best subset selection. However, unlike forward

backward stepwise selection

---

[1]Though forward stepwise selection considers $p(p+1)/2 + 1$ models, it performs a *guided* search over model space, and so the *effective* model space considered contains substantially more than $p(p+1)/2 + 1$ models.

# Variables	Best subset	Forward stepwise
One	rating	rating
Two	rating, income	rating, income
Three	rating, income, student	rating, income, student
Four	cards, income, student, limit	rating, income, student, limit

TABLE 6.1. *The first four selected models for best subset selection and forward stepwise selection on the* Credit *data set. The first three models are identical but the fourth models differ.*

stepwise selection, it begins with the full least squares model containing all p predictors, and then iteratively removes the least useful predictor, one-at-a-time. Details are given in Algorithm 6.3.

Algorithm 6.3 *Backward stepwise selection*

---

1. Let $\mathcal{M}_p$ denote the *full* model, which contains all p predictors.

2. For $k = p, p - 1, \ldots, 1$:

   (a) Consider all k models that contain all but one of the predictors in $\mathcal{M}_k$, for a total of $k - 1$ predictors.

   (b) Choose the *best* among these k models, and call it $\mathcal{M}_{k-1}$. Here *best* is defined as having smallest RSS or highest R^2.

3. Select a single best model from among $\mathcal{M}_0, \ldots, \mathcal{M}_p$ using cross-validated prediction error, C_p (AIC), BIC, or adjusted R^2.

Like forward stepwise selection, the backward selection approach searches through only $1 + p(p+1)/2$ models, and so can be applied in settings where p is too large to apply best subset selection.[2] Also like forward stepwise selection, backward stepwise selection is not guaranteed to yield the *best* model containing a subset of the p predictors.

Backward selection requires that the number of samples n is larger than the number of variables p (so that the full model can be fit). In contrast, forward stepwise can be used even when $n < p$, and so is the only viable subset method when p is very large.

---

[2]Like forward stepwise selection, backward stepwise selection performs a *guided* search over model space, and so effectively considers substantially more than $1 + p(p+1)/2$ models.

Hybrid Approaches

The best subset, forward stepwise, and backward stepwise selection approaches generally give similar but not identical models. As another alternative, hybrid versions of forward and backward stepwise selection are available, in which variables are added to the model sequentially, in analogy to forward selection. However, after adding each new variable, the method may also remove any variables that no longer provide an improvement in the model fit. Such an approach attempts to more closely mimic best subset selection while retaining the computational advantages of forward and backward stepwise selection.

6.1.3 *Choosing the Optimal Model*

Best subset selection, forward selection, and backward selection result in the creation of a set of models, each of which contains a subset of the p predictors. In order to implement these methods, we need a way to determine which of these models is *best*. As we discussed in Section 6.1.1, the model containing all of the predictors will always have the smallest RSS and the largest R^2, since these quantities are related to the training error. Instead, we wish to choose a model with a low test error. As is evident here, and as we show in Chapter 2, the training error can be a poor estimate of the test error. Therefore, RSS and R^2 are not suitable for selecting the best model among a collection of models with different numbers of predictors.

In order to select the best model with respect to test error, we need to estimate this test error. There are two common approaches:

1. We can indirectly estimate test error by making an *adjustment* to the training error to account for the bias due to overfitting.

2. We can *directly* estimate the test error, using either a validation set approach or a cross-validation approach, as discussed in Chapter 5.

We consider both of these approaches below.

C_p, AIC, BIC, and Adjusted R^2

We show in Chapter 2 that the training set MSE is generally an underestimate of the test MSE. (Recall that MSE $=$ RSS$/n$.) This is because when we fit a model to the training data using least squares, we specifically estimate the regression coefficients such that the training RSS (but not the test RSS) is as small as possible. In particular, the training error will decrease as more variables are included in the model, but the test error may not. Therefore, training set RSS and training set R^2 cannot be used to select from among a set of models with different numbers of variables.

However, a number of techniques for *adjusting* the training error for the model size are available. These approaches can be used to select among a set

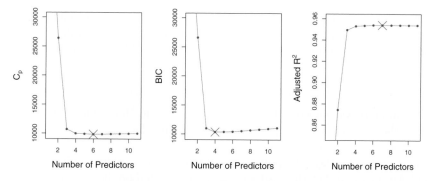

FIGURE 6.2. C_p, BIC, and adjusted R^2 are shown for the best models of each size for the Credit data set (the lower frontier in Figure 6.1). C_p and BIC are estimates of test MSE. In the middle plot we see that the BIC estimate of test error shows an increase after four variables are selected. The other two plots are rather flat after four variables are included.

of models with different numbers of variables. We now consider four such approaches: C_p, *Akaike information criterion* (AIC), *Bayesian information criterion* (BIC), and *adjusted R^2*. Figure 6.2 displays C_p, BIC, and adjusted R^2 for the best model of each size produced by best subset selection on the Credit data set.

For a fitted least squares model containing d predictors, the C_p estimate of test MSE is computed using the equation

$$C_p = \frac{1}{n}\left(\text{RSS} + 2d\hat{\sigma}^2\right),\qquad(6.2)$$

where $\hat{\sigma}^2$ is an estimate of the variance of the error ϵ associated with each response measurement in (6.1).[3] Typically $\hat{\sigma}^2$ is estimated using the full model containing all predictors. Essentially, the C_p statistic adds a penalty of $2d\hat{\sigma}^2$ to the training RSS in order to adjust for the fact that the training error tends to underestimate the test error. Clearly, the penalty increases as the number of predictors in the model increases; this is intended to adjust for the corresponding decrease in training RSS. Though it is beyond the scope of this book, one can show that if $\hat{\sigma}^2$ is an unbiased estimate of σ^2 in (6.2), then C_p is an unbiased estimate of test MSE. As a consequence, the C_p statistic tends to take on a small value for models with a low test error, so when determining which of a set of models is best, we choose the model with the lowest C_p value. In Figure 6.2, C_p selects the six-variable model containing the predictors income, limit, rating, cards, age and student.

C_p

Akaike information criterion

Bayesian information criterion

adjusted R^2

---

[3]Mallow's C_p is sometimes defined as $C_p' = \text{RSS}/\hat{\sigma}^2 + 2d - n$. This is equivalent to the definition given above in the sense that $C_p = \frac{1}{n}\hat{\sigma}^2(C_p' + n)$, and so the model with smallest C_p also has smallest C_p'.

The AIC criterion is defined for a large class of models fit by maximum likelihood. In the case of the model (6.1) with Gaussian errors, maximum likelihood and least squares are the same thing. In this case AIC is given by

$$\text{AIC} = \frac{1}{n\hat{\sigma}^2}\left(\text{RSS} + 2d\hat{\sigma}^2\right),$$

where, for simplicity, we have omitted an additive constant. Hence for least squares models, C_p and AIC are proportional to each other, and so only C_p is displayed in Figure 6.2.

BIC is derived from a Bayesian point of view, but ends up looking similar to C_p (and AIC) as well. For the least squares model with d predictors, the BIC is, up to irrelevant constants, given by

$$\text{BIC} = \frac{1}{n\hat{\sigma}^2}\left(\text{RSS} + \log(n)d\hat{\sigma}^2\right). \tag{6.3}$$

Like C_p, the BIC will tend to take on a small value for a model with a low test error, and so generally we select the model that has the lowest BIC value. Notice that BIC replaces the $2d\hat{\sigma}^2$ used by C_p with a $\log(n)d\hat{\sigma}^2$ term, where n is the number of observations. Since $\log n > 2$ for any $n > 7$, the BIC statistic generally places a heavier penalty on models with many variables, and hence results in the selection of smaller models than C_p. In Figure 6.2, we see that this is indeed the case for the Credit data set; BIC chooses a model that contains only the four predictors income, limit, cards, and student. In this case the curves are very flat and so there does not appear to be much difference in accuracy between the four-variable and six-variable models.

The adjusted R^2 statistic is another popular approach for selecting among a set of models that contain different numbers of variables. Recall from Chapter 3 that the usual R^2 is defined as $1 - \text{RSS}/\text{TSS}$, where $\text{TSS} = \sum(y_i - \bar{y})^2$ is the *total sum of squares* for the response. Since RSS always decreases as more variables are added to the model, the R^2 always increases as more variables are added. For a least squares model with d variables, the adjusted R^2 statistic is calculated as

$$\text{Adjusted } R^2 = 1 - \frac{\text{RSS}/(n-d-1)}{\text{TSS}/(n-1)}. \tag{6.4}$$

Unlike C_p, AIC, and BIC, for which a *small* value indicates a model with a low test error, a *large* value of adjusted R^2 indicates a model with a small test error. Maximizing the adjusted R^2 is equivalent to minimizing $\frac{\text{RSS}}{n-d-1}$. While RSS always decreases as the number of variables in the model increases, $\frac{\text{RSS}}{n-d-1}$ may increase or decrease, due to the presence of d in the denominator.

The intuition behind the adjusted R^2 is that once all of the correct variables have been included in the model, adding additional *noise* variables will lead to only a very small decrease in RSS. Since adding noise variables leads to an increase in d, such variables will lead to an increase in $\frac{\text{RSS}}{n-d-1}$, and consequently a decrease in the adjusted R^2. Therefore, in theory, the model with the largest adjusted R^2 will have only correct variables and no noise variables. Unlike the R^2 statistic, the adjusted R^2 statistic *pays a price* for the inclusion of unnecessary variables in the model. Figure 6.2 displays the adjusted R^2 for the Credit data set. Using this statistic results in the selection of a model that contains seven variables, adding gender to the model selected by C_p and AIC.

C_p, AIC, and BIC all have rigorous theoretical justifications that are beyond the scope of this book. These justifications rely on asymptotic arguments (scenarios where the sample size n is very large). Despite its popularity, and even though it is quite intuitive, the adjusted R^2 is not as well motivated in statistical theory as AIC, BIC, and C_p. All of these measures are simple to use and compute. Here we have presented the formulas for AIC, BIC, and C_p in the case of a linear model fit using least squares; however, these quantities can also be defined for more general types of models.

Validation and Cross-Validation

As an alternative to the approaches just discussed, we can directly estimate the test error using the validation set and cross-validation methods discussed in Chapter 5. We can compute the validation set error or the cross-validation error for each model under consideration, and then select the model for which the resulting estimated test error is smallest. This procedure has an advantage relative to AIC, BIC, C_p, and adjusted R^2, in that it provides a direct estimate of the test error, and makes fewer assumptions about the true underlying model. It can also be used in a wider range of model selection tasks, even in cases where it is hard to pinpoint the model degrees of freedom (e.g. the number of predictors in the model) or hard to estimate the error variance σ^2.

In the past, performing cross-validation was computationally prohibitive for many problems with large p and/or large n, and so AIC, BIC, C_p, and adjusted R^2 were more attractive approaches for choosing among a set of models. However, nowadays with fast computers, the computations required to perform cross-validation are hardly ever an issue. Thus, cross-validation is a very attractive approach for selecting from among a number of models under consideration.

Figure 6.3 displays, as a function of d, the BIC, validation set errors, and cross-validation errors on the Credit data, for the best d-variable model. The validation errors were calculated by randomly selecting three-quarters of the observations as the training set, and the remainder as the validation set. The cross-validation errors were computed using $k = 10$ folds. In this case, the validation and cross-validation methods both result in a

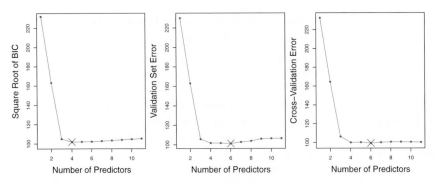

FIGURE 6.3. *For the* `Credit` *data set, three quantities are displayed for the best model containing d predictors, for d ranging from 1 to 11. The overall best model, based on each of these quantities, is shown as a blue cross. Left: Square root of BIC. Center: Validation set errors. Right: Cross-validation errors.*

six-variable model. However, all three approaches suggest that the four-, five-, and six-variable models are roughly equivalent in terms of their test errors.

In fact, the estimated test error curves displayed in the center and right-hand panels of Figure 6.3 are quite flat. While a three-variable model clearly has lower estimated test error than a two-variable model, the estimated test errors of the 3- to 11-variable models are quite similar. Furthermore, if we repeated the validation set approach using a different split of the data into a training set and a validation set, or if we repeated cross-validation using a different set of cross-validation folds, then the precise model with the lowest estimated test error would surely change. In this setting, we can select a model using the *one-standard-error rule*. We first calculate the standard error of the estimated test MSE for each model size, and then select the smallest model for which the estimated test error is within one standard error of the lowest point on the curve. The rationale here is that if a set of models appear to be more or less equally good, then we might as well choose the simplest model—that is, the model with the smallest number of predictors. In this case, applying the one-standard-error rule to the validation set or cross-validation approach leads to selection of the three-variable model.

one-standard-error rule

6.2 Shrinkage Methods

The subset selection methods described in Section 6.1 involve using least squares to fit a linear model that contains a subset of the predictors. As an alternative, we can fit a model containing all p predictors using a technique that *constrains* or *regularizes* the coefficient estimates, or equivalently, that *shrinks* the coefficient estimates towards zero. It may not be immediately obvious why such a constraint should improve the fit, but it turns out that shrinking the coefficient estimates can significantly reduce their variance. The two best-known techniques for shrinking the regression coefficients towards zero are *ridge regression* and the *lasso*.

6.2.1   Ridge Regression

Recall from Chapter 3 that the least squares fitting procedure estimates $\beta_0, \beta_1, \ldots, \beta_p$ using the values that minimize

$$\text{RSS} = \sum_{i=1}^{n} \left(y_i - \beta_0 - \sum_{j=1}^{p} \beta_j x_{ij} \right)^2 .$$

Ridge regression is very similar to least squares, except that the coefficients are estimated by minimizing a slightly different quantity. In particular, the ridge regression coefficient estimates $\hat{\beta}^R$ are the values that minimize

$$\sum_{i=1}^{n} \left(y_i - \beta_0 - \sum_{j=1}^{p} \beta_j x_{ij} \right)^2 + \lambda \sum_{j=1}^{p} \beta_j^2 = \text{RSS} + \lambda \sum_{j=1}^{p} \beta_j^2, \qquad (6.5)$$

where $\lambda \geq 0$ is a *tuning parameter*, to be determined separately. Equation 6.5 trades off two different criteria. As with least squares, ridge regression seeks coefficient estimates that fit the data well, by making the RSS small. However, the second term, $\lambda \sum_j \beta_j^2$, called a *shrinkage penalty*, is small when $\beta_1, \ldots, \beta_p$ are close to zero, and so it has the effect of *shrinking* the estimates of β_j towards zero. The tuning parameter λ serves to control the relative impact of these two terms on the regression coefficient estimates. When $\lambda = 0$, the penalty term has no effect, and ridge regression will produce the least squares estimates. However, as $\lambda \to \infty$, the impact of the shrinkage penalty grows, and the ridge regression coefficient estimates will approach zero. Unlike least squares, which generates only one set of coefficient estimates, ridge regression will produce a different set of coefficient estimates, $\hat{\beta}_\lambda^R$, for each value of λ. Selecting a good value for λ is critical; we defer this discussion to Section 6.2.3, where we use cross-validation.

Note that in (6.5), the shrinkage penalty is applied to $\beta_1, \ldots, \beta_p$, but not to the intercept β_0. We want to shrink the estimated association of each variable with the response; however, we do not want to shrink the intercept, which is simply a measure of the mean value of the response when $x_{i1} = x_{i2} = \ldots = x_{ip} = 0$. If we assume that the variables—that is, the columns of the data matrix $\mathbf{X}$—have been centered to have mean zero before ridge regression is performed, then the estimated intercept will take the form $\hat{\beta}_0 = \bar{y} = \sum_{i=1}^{n} y_i / n$.

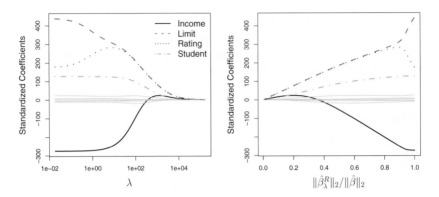

FIGURE 6.4. *The standardized ridge regression coefficients are displayed for the* Credit *data set, as a function of* λ *and* $\|\hat{\beta}_\lambda^R\|_2/\|\hat{\beta}\|_2$.

An Application to the Credit Data

In Figure 6.4, the ridge regression coefficient estimates for the Credit data set are displayed. In the left-hand panel, each curve corresponds to the ridge regression coefficient estimate for one of the ten variables, plotted as a function of λ. For example, the black solid line represents the ridge regression estimate for the income coefficient, as λ is varied. At the extreme left-hand side of the plot, λ is essentially zero, and so the corresponding ridge coefficient estimates are the same as the usual least squares estimates. But as λ increases, the ridge coefficient estimates shrink towards zero. When λ is extremely large, then all of the ridge coefficient estimates are basically zero; this corresponds to the *null model* that contains no predictors. In this plot, the income, limit, rating, and student variables are displayed in distinct colors, since these variables tend to have by far the largest coefficient estimates. While the ridge coefficient estimates tend to decrease in aggregate as λ increases, individual coefficients, such as rating and income, may occasionally increase as λ increases.

The right-hand panel of Figure 6.4 displays the same ridge coefficient estimates as the left-hand panel, but instead of displaying λ on the x-axis, we now display $\|\hat{\beta}_\lambda^R\|_2/\|\hat{\beta}\|_2$, where $\hat{\beta}$ denotes the vector of least squares coefficient estimates. The notation $\|\beta\|_2$ denotes the ℓ_2 *norm* (pronounced "ell 2") of a vector, and is defined as $\|\beta\|_2 = \sqrt{\sum_{j=1}^p \beta_j^2}$. It measures the distance of β from zero. As λ increases, the ℓ_2 norm of $\hat{\beta}_\lambda^R$ will *always* decrease, and so will $\|\hat{\beta}_\lambda^R\|_2/\|\hat{\beta}\|_2$. The latter quantity ranges from 1 (when $\lambda = 0$, in which case the ridge regression coefficient estimate is the same as the least squares estimate, and so their ℓ_2 norms are the same) to 0 (when $\lambda = \infty$, in which case the ridge regression coefficient estimate is a vector of zeros, with ℓ_2 norm equal to zero). Therefore, we can think of the x-axis in the right-hand panel of Figure 6.4 as the amount that the ridge

ℓ_2 norm

regression coefficient estimates have been shrunken towards zero; a small value indicates that they have been shrunken very close to zero.

The standard least squares coefficient estimates discussed in Chapter 3 are *scale equivariant*: multiplying X_j by a constant c simply leads to a scaling of the least squares coefficient estimates by a factor of $1/c$. In other words, regardless of how the jth predictor is scaled, $X_j \hat{\beta}_j$ will remain the same. In contrast, the ridge regression coefficient estimates can change *substantially* when multiplying a given predictor by a constant. For instance, consider the income variable, which is measured in dollars. One could reasonably have measured income in thousands of dollars, which would result in a reduction in the observed values of income by a factor of 1,000. Now due to the sum of squared coefficients term in the ridge regression formulation (6.5), such a change in scale will not simply cause the ridge regression coefficient estimate for income to change by a factor of 1,000. In other words, $X_j \hat{\beta}^R_{j,\lambda}$ will depend not only on the value of λ, but also on the scaling of the jth predictor. In fact, the value of $X_j \hat{\beta}^R_{j,\lambda}$ may even depend on the scaling of the *other* predictors! Therefore, it is best to apply ridge regression after *standardizing the predictors*, using the formula

$$\tilde{x}_{ij} = \frac{x_{ij}}{\sqrt{\frac{1}{n} \sum_{i=1}^{n} (x_{ij} - \overline{x}_j)^2}}, \tag{6.6}$$

so that they are all on the same scale. In (6.6), the denominator is the estimated standard deviation of the jth predictor. Consequently, all of the standardized predictors will have a standard deviation of one. As a result the final fit will not depend on the scale on which the predictors are measured. In Figure 6.4, the y-axis displays the standardized ridge regression coefficient estimates—that is, the coefficient estimates that result from performing ridge regression using standardized predictors.

Why Does Ridge Regression Improve Over Least Squares?

Ridge regression's advantage over least squares is rooted in the *bias-variance trade-off*. As λ increases, the flexibility of the ridge regression fit decreases, leading to decreased variance but increased bias. This is illustrated in the left-hand panel of Figure 6.5, using a simulated data set containing $p = 45$ predictors and $n = 50$ observations. The green curve in the left-hand panel of Figure 6.5 displays the variance of the ridge regression predictions as a function of λ. At the least squares coefficient estimates, which correspond to ridge regression with $\lambda = 0$, the variance is high but there is no bias. But as λ increases, the shrinkage of the ridge coefficient estimates leads to a substantial reduction in the variance of the predictions, at the expense of a slight increase in bias. Recall that the test mean squared error (MSE), plotted in purple, is a function of the variance plus the squared bias. For values

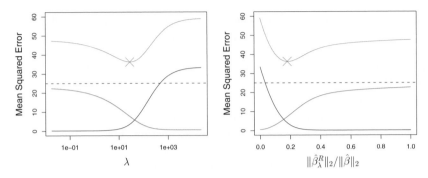

FIGURE 6.5. *Squared bias (black), variance (green), and test mean squared error (purple) for the ridge regression predictions on a simulated data set, as a function of λ and $\|\hat{\beta}_\lambda^R\|_2/\|\hat{\beta}\|_2$. The horizontal dashed lines indicate the minimum possible MSE. The purple crosses indicate the ridge regression models for which the MSE is smallest.*

of λ up to about 10, the variance decreases rapidly, with very little increase in bias, plotted in black. Consequently, the MSE drops considerably as λ increases from 0 to 10. Beyond this point, the decrease in variance due to increasing λ slows, and the shrinkage on the coefficients causes them to be significantly underestimated, resulting in a large increase in the bias. The minimum MSE is achieved at approximately $\lambda = 30$. Interestingly, because of its high variance, the MSE associated with the least squares fit, when $\lambda = 0$, is almost as high as that of the null model for which all coefficient estimates are zero, when $\lambda = \infty$. However, for an intermediate value of λ, the MSE is considerably lower.

The right-hand panel of Figure 6.5 displays the same curves as the left-hand panel, this time plotted against the ℓ_2 norm of the ridge regression coefficient estimates divided by the ℓ_2 norm of the least squares estimates. Now as we move from left to right, the fits become more flexible, and so the bias decreases and the variance increases.

In general, in situations where the relationship between the response and the predictors is close to linear, the least squares estimates will have low bias but may have high variance. This means that a small change in the training data can cause a large change in the least squares coefficient estimates. In particular, when the number of variables p is almost as large as the number of observations n, as in the example in Figure 6.5, the least squares estimates will be extremely variable. And if $p > n$, then the least squares estimates do not even have a unique solution, whereas ridge regression can still perform well by trading off a small increase in bias for a large decrease in variance. Hence, ridge regression works best in situations where the least squares estimates have high variance.

Ridge regression also has substantial computational advantages over best subset selection, which requires searching through 2^p models. As we discussed previously, even for moderate values of p, such a search can be computationally infeasible. In contrast, for any fixed value of λ, ridge regression only fits a single model, and the model-fitting procedure can be performed quite quickly. In fact, one can show that the computations required to solve (6.5), *simultaneously for all values of λ*, are almost identical to those for fitting a model using least squares.

6.2.2 The Lasso

Ridge regression does have one obvious disadvantage. Unlike best subset, forward stepwise, and backward stepwise selection, which will generally select models that involve just a subset of the variables, ridge regression will include all p predictors in the final model. The penalty $\lambda \sum \beta_j^2$ in (6.5) will shrink all of the coefficients towards zero, but it will not set any of them exactly to zero (unless $\lambda = \infty$). This may not be a problem for prediction accuracy, but it can create a challenge in model interpretation in settings in which the number of variables p is quite large. For example, in the Credit data set, it appears that the most important variables are income, limit, rating, and student. So we might wish to build a model including just these predictors. However, ridge regression will always generate a model involving all ten predictors. Increasing the value of λ will tend to reduce the magnitudes of the coefficients, but will not result in exclusion of any of the variables.

The *lasso* is a relatively recent alternative to ridge regression that over-comes this disadvantage. The lasso coefficients, $\hat{\beta}_\lambda^L$, minimize the quantity   lasso

$$\sum_{i=1}^{n} \left(y_i - \beta_0 - \sum_{j=1}^{p} \beta_j x_{ij} \right)^2 + \lambda \sum_{j=1}^{p} |\beta_j| = \text{RSS} + \lambda \sum_{j=1}^{p} |\beta_j|. \qquad (6.7)$$

Comparing (6.7) to (6.5), we see that the lasso and ridge regression have similar formulations. The only difference is that the β_j^2 term in the ridge regression penalty (6.5) has been replaced by $|\beta_j|$ in the lasso penalty (6.7). In statistical parlance, the lasso uses an ℓ_1 (pronounced "ell 1") penalty instead of an ℓ_2 penalty. The ℓ_1 norm of a coefficient vector β is given by $\|\beta\|_1 = \sum |\beta_j|$.

As with ridge regression, the lasso shrinks the coefficient estimates towards zero. However, in the case of the lasso, the ℓ_1 penalty has the effect of forcing some of the coefficient estimates to be exactly equal to zero when the tuning parameter λ is sufficiently large. Hence, much like best subset selection, the lasso performs *variable selection*. As a result, models generated from the lasso are generally much easier to interpret than those produced by ridge regression. We say that the lasso yields *sparse* models—that is,   sparse
models that involve only a subset of the variables. As in ridge regression, selecting a good value of λ for the lasso is critical; we defer this discussion to Section 6.2.3, where we use cross-validation.

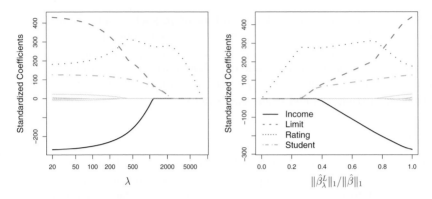

FIGURE 6.6. *The standardized lasso coefficients on the* Credit *data set are shown as a function of λ and $\|\hat{\beta}_\lambda^L\|_1/\|\hat{\beta}\|_1$.*

As an example, consider the coefficient plots in Figure 6.6, which are generated from applying the lasso to the Credit data set. When $\lambda = 0$, then the lasso simply gives the least squares fit, and when λ becomes sufficiently large, the lasso gives the null model in which all coefficient estimates equal zero. However, in between these two extremes, the ridge regression and lasso models are quite different from each other. Moving from left to right in the right-hand panel of Figure 6.6, we observe that at first the lasso results in a model that contains only the rating predictor. Then student and limit enter the model almost simultaneously, shortly followed by income. Eventually, the remaining variables enter the model. Hence, depending on the value of λ, the lasso can produce a model involving any number of variables. In contrast, ridge regression will always include all of the variables in the model, although the magnitude of the coefficient estimates will depend on λ.

Another Formulation for Ridge Regression and the Lasso

One can show that the lasso and ridge regression coefficient estimates solve the problems

$$\underset{\beta}{\text{minimize}} \left\{ \sum_{i=1}^{n} \left(y_i - \beta_0 - \sum_{j=1}^{p} \beta_j x_{ij} \right)^2 \right\} \quad \text{subject to} \quad \sum_{j=1}^{p} |\beta_j| \le s \tag{6.8}$$

and

$$\underset{\beta}{\text{minimize}} \left\{ \sum_{i=1}^{n} \left(y_i - \beta_0 - \sum_{j=1}^{p} \beta_j x_{ij} \right)^2 \right\} \quad \text{subject to} \quad \sum_{j=1}^{p} \beta_j^2 \le s, \tag{6.9}$$

respectively. In other words, for every value of λ, there is some s such that the Equations (6.7) and (6.8) will give the same lasso coefficient estimates. Similarly, for every value of λ there is a corresponding s such that Equations (6.5) and (6.9) will give the same ridge regression coefficient estimates. When $p = 2$, then (6.8) indicates that the lasso coefficient estimates have the smallest RSS out of all points that lie within the diamond defined by $|\beta_1| + |\beta_2| \leq s$. Similarly, the ridge regression estimates have the smallest RSS out of all points that lie within the circle defined by $\beta_1^2 + \beta_2^2 \leq s$.

We can think of (6.8) as follows. When we perform the lasso we are trying to find the set of coefficient estimates that lead to the smallest RSS, subject to the constraint that there is a *budget* s for how large $\sum_{j=1}^{p} |\beta_j|$ can be. When s is extremely large, then this budget is not very restrictive, and so the coefficient estimates can be large. In fact, if s is large enough that the least squares solution falls within the budget, then (6.8) will simply yield the least squares solution. In contrast, if s is small, then $\sum_{j=1}^{p} |\beta_j|$ must be small in order to avoid violating the budget. Similarly, (6.9) indicates that when we perform ridge regression, we seek a set of coefficient estimates such that the RSS is as small as possible, subject to the requirement that $\sum_{j=1}^{p} \beta_j^2$ not exceed the budget s.

The formulations (6.8) and (6.9) reveal a close connection between the lasso, ridge regression, and best subset selection. Consider the problem

$$
\underset{\beta}{\text{minimize}} \left\{ \sum_{i=1}^{n} \left(y_i - \beta_0 - \sum_{j=1}^{p} \beta_j x_{ij} \right)^2 \right\} \quad \text{subject to} \quad \sum_{j=1}^{p} I(\beta_j \neq 0) \leq s.
$$

(6.10)

Here $I(\beta_j \neq 0)$ is an indicator variable: it takes on a value of 1 if $\beta_j \neq 0$, and equals zero otherwise. Then (6.10) amounts to finding a set of coefficient estimates such that RSS is as small as possible, subject to the constraint that no more than s coefficients can be nonzero. The problem (6.10) is equivalent to best subset selection. Unfortunately, solving (6.10) is computationally infeasible when p is large, since it requires considering all $\binom{p}{s}$ models containing s predictors. Therefore, we can interpret ridge regression and the lasso as computationally feasible alternatives to best subset selection that replace the intractable form of the budget in (6.10) with forms that are much easier to solve. Of course, the lasso is much more closely related to best subset selection, since only the lasso performs feature selection for s sufficiently small in (6.8).

The Variable Selection Property of the Lasso

Why is it that the lasso, unlike ridge regression, results in coefficient estimates that are exactly equal to zero? The formulations (6.8) and (6.9) can be used to shed light on the issue. Figure 6.7 illustrates the situation. The least squares solution is marked as $\hat{\beta}$, while the blue diamond and

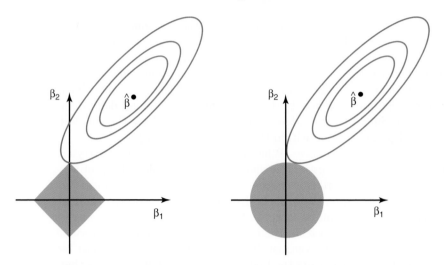

FIGURE 6.7. *Contours of the error and constraint functions for the lasso (left) and ridge regression (right). The solid blue areas are the constraint regions, $|\beta_1| + |\beta_2| \leq s$ and $\beta_1^2 + \beta_2^2 \leq s$, while the red ellipses are the contours of the RSS.*

circle represent the lasso and ridge regression constraints in (6.8) and (6.9), respectively. If s is sufficiently large, then the constraint regions will contain $\hat{\beta}$, and so the ridge regression and lasso estimates will be the same as the least squares estimates. (Such a large value of s corresponds to $\lambda = 0$ in (6.5) and (6.7).) However, in Figure 6.7 the least squares estimates lie outside of the diamond and the circle, and so the least squares estimates are not the same as the lasso and ridge regression estimates.

The ellipses that are centered around $\hat{\beta}$ represent regions of constant RSS. In other words, all of the points on a given ellipse share a common value of the RSS. As the ellipses expand away from the least squares coefficient estimates, the RSS increases. Equations (6.8) and (6.9) indicate that the lasso and ridge regression coefficient estimates are given by the first point at which an ellipse contacts the constraint region. Since ridge regression has a circular constraint with no sharp points, this intersection will not generally occur on an axis, and so the ridge regression coefficient estimates will be exclusively non-zero. However, the lasso constraint has *corners* at each of the axes, and so the ellipse will often intersect the constraint region at an axis. When this occurs, one of the coefficients will equal zero. In higher dimensions, many of the coefficient estimates may equal zero simultaneously. In Figure 6.7, the intersection occurs at $\beta_1 = 0$, and so the resulting model will only include β_2.

In Figure 6.7, we considered the simple case of $p = 2$. When $p = 3$, then the constraint region for ridge regression becomes a sphere, and the constraint region for the lasso becomes a polyhedron. When $p > 3$, the

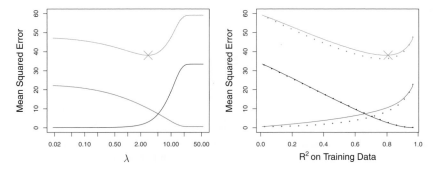

FIGURE 6.8. Left: *Plots of squared bias (black), variance (green), and test MSE (purple) for the lasso on a simulated data set.* Right: *Comparison of squared bias, variance and test MSE between lasso (solid) and ridge (dotted). Both are plotted against their R^2 on the training data, as a common form of indexing. The crosses in both plots indicate the lasso model for which the MSE is smallest.*

constraint for ridge regression becomes a hypersphere, and the constraint for the lasso becomes a polytope. However, the key ideas depicted in Figure 6.7 still hold. In particular, the lasso leads to feature selection when $p > 2$ due to the sharp corners of the polyhedron or polytope.

Comparing the Lasso and Ridge Regression

It is clear that the lasso has a major advantage over ridge regression, in that it produces simpler and more interpretable models that involve only a subset of the predictors. However, which method leads to better prediction accuracy? Figure 6.8 displays the variance, squared bias, and test MSE of the lasso applied to the same simulated data as in Figure 6.5. Clearly the lasso leads to qualitatively similar behavior to ridge regression, in that as λ increases, the variance decreases and the bias increases. In the right-hand panel of Figure 6.8, the dotted lines represent the ridge regression fits. Here we plot both against their R^2 on the training data. This is another useful way to index models, and can be used to compare models with different types of regularization, as is the case here. In this example, the lasso and ridge regression result in almost identical biases. However, the variance of ridge regression is slightly lower than the variance of the lasso. Consequently, the minimum MSE of ridge regression is slightly smaller than that of the lasso.

However, the data in Figure 6.8 were generated in such a way that all 45 predictors were related to the response—that is, none of the true coefficients $\beta_1, \ldots, \beta_{45}$ equaled zero. The lasso implicitly assumes that a number of the coefficients truly equal zero. Consequently, it is not surprising that ridge regression outperforms the lasso in terms of prediction error in this setting. Figure 6.9 illustrates a similar situation, except that now the response is a

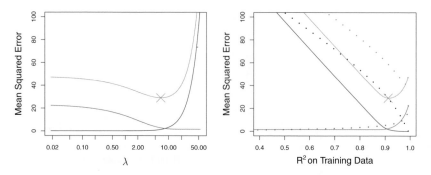

FIGURE 6.9. Left: *Plots of squared bias (black), variance (green), and test MSE (purple) for the lasso. The simulated data is similar to that in Figure 6.8, except that now only two predictors are related to the response.* Right: *Comparison of squared bias, variance and test MSE between lasso (solid) and ridge (dotted). Both are plotted against their R^2 on the training data, as a common form of indexing. The crosses in both plots indicate the lasso model for which the MSE is smallest.*

function of only 2 out of 45 predictors. Now the lasso tends to outperform ridge regression in terms of bias, variance, and MSE.

These two examples illustrate that neither ridge regression nor the lasso will universally dominate the other. In general, one might expect the lasso to perform better in a setting where a relatively small number of predictors have substantial coefficients, and the remaining predictors have coefficients that are very small or that equal zero. Ridge regression will perform better when the response is a function of many predictors, all with coefficients of roughly equal size. However, the number of predictors that is related to the response is never known *a priori* for real data sets. A technique such as cross-validation can be used in order to determine which approach is better on a particular data set.

As with ridge regression, when the least squares estimates have excessively high variance, the lasso solution can yield a reduction in variance at the expense of a small increase in bias, and consequently can generate more accurate predictions. Unlike ridge regression, the lasso performs variable selection, and hence results in models that are easier to interpret.

There are very efficient algorithms for fitting both ridge and lasso models; in both cases the entire coefficient paths can be computed with about the same amount of work as a single least squares fit. We will explore this further in the lab at the end of this chapter.

A Simple Special Case for Ridge Regression and the Lasso

In order to obtain a better intuition about the behavior of ridge regression and the lasso, consider a simple special case with $n = p$, and $\mathbf{X}$ a diagonal matrix with 1's on the diagonal and 0's in all off-diagonal elements. To simplify the problem further, assume also that we are performing regression without an intercept. With these assumptions, the usual least squares problem simplifies to finding $\beta_1, \ldots, \beta_p$ that minimize

$$\sum_{j=1}^{p} (y_j - \beta_j)^2. \tag{6.11}$$

In this case, the least squares solution is given by

$$\hat{\beta}_j = y_j.$$

And in this setting, ridge regression amounts to finding $\beta_1, \ldots, \beta_p$ such that

$$\sum_{j=1}^{p} (y_j - \beta_j)^2 + \lambda \sum_{j=1}^{p} \beta_j^2 \tag{6.12}$$

is minimized, and the lasso amounts to finding the coefficients such that

$$\sum_{j=1}^{p} (y_j - \beta_j)^2 + \lambda \sum_{j=1}^{p} |\beta_j| \tag{6.13}$$

is minimized. One can show that in this setting, the ridge regression estimates take the form

$$\hat{\beta}_j^R = y_j/(1 + \lambda), \tag{6.14}$$

and the lasso estimates take the form

$$\hat{\beta}_j^L = \begin{cases} y_j - \lambda/2 & \text{if } y_j > \lambda/2; \\ y_j + \lambda/2 & \text{if } y_j < -\lambda/2; \\ 0 & \text{if } |y_j| \le \lambda/2. \end{cases} \tag{6.15}$$

Figure 6.10 displays the situation. We can see that ridge regression and the lasso perform two very different types of shrinkage. In ridge regression, each least squares coefficient estimate is shrunken by the same proportion. In contrast, the lasso shrinks each least squares coefficient towards zero by a constant amount, $\lambda/2$; the least squares coefficients that are less than $\lambda/2$ in absolute value are shrunken entirely to zero. The type of shrinkage performed by the lasso in this simple setting (6.15) is known as *soft-thresholding*. The fact that some lasso coefficients are shrunken entirely to zero explains why the lasso performs feature selection.

soft-
thresholding

In the case of a more general data matrix $\mathbf{X}$, the story is a little more complicated than what is depicted in Figure 6.10, but the main ideas still hold approximately: ridge regression more or less shrinks every dimension of the data by the same proportion, whereas the lasso more or less shrinks all coefficients toward zero by a similar amount, and sufficiently small coefficients are shrunken all the way to zero.

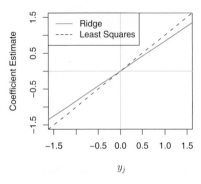
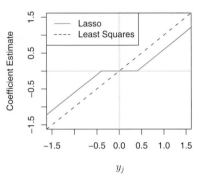

FIGURE 6.10. *The ridge regression and lasso coefficient estimates for a simple setting with $n = p$ and* **X** *a diagonal matrix with 1's on the diagonal. Left: The ridge regression coefficient estimates are shrunken proportionally towards zero, relative to the least squares estimates. Right: The lasso coefficient estimates are soft-thresholded towards zero.*

Bayesian Interpretation for Ridge Regression and the Lasso

We now show that one can view ridge regression and the lasso through a Bayesian lens. A Bayesian viewpoint for regression assumes that the coefficient vector β has some *prior* distribution, say $p(\beta)$, where $\beta = (\beta_0, \beta_1, \ldots, \beta_p)^T$. The likelihood of the data can be written as $f(Y|X, \beta)$, where $X = (X_1, \ldots, X_p)$. Multiplying the prior distribution by the likelihood gives us (up to a proportionality constant) the *posterior distribution*, which takes the form

posterior
distribution

$$p(\beta|X, Y) \propto f(Y|X, \beta)p(\beta|X) = f(Y|X, \beta)p(\beta),$$

where the proportionality above follows from Bayes' theorem, and the equality above follows from the assumption that X is fixed.

We assume the usual linear model,

$$Y = \beta_0 + X_1\beta_1 + \ldots + X_p\beta_p + \epsilon,$$

and suppose that the errors are independent and drawn from a normal distribution. Furthermore, assume that $p(\beta) = \prod_{j=1}^{p} g(\beta_j)$, for some density function g. It turns out that ridge regression and the lasso follow naturally from two special cases of g:

- If g is a Gaussian distribution with mean zero and standard deviation a function of λ, then it follows that the *posterior mode* for β—that is, the most likely value for β, given the data—is given by the ridge regression solution. (In fact, the ridge regression solution is also the posterior mean.)

posterior
mode

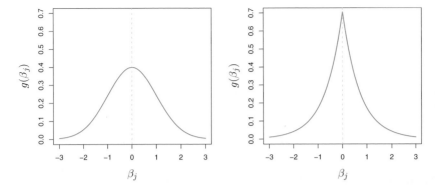

FIGURE 6.11. Left: *Ridge regression is the posterior mode for β under a Gaussian prior.* Right: *The lasso is the posterior mode for β under a double-exponential prior.*

- If g is a double-exponential (Laplace) distribution with mean zero and scale parameter a function of λ, then it follows that the posterior mode for β is the lasso solution. (However, the lasso solution is *not* the posterior mean, and in fact, the posterior mean does not yield a sparse coefficient vector.)

The Gaussian and double-exponential priors are displayed in Figure 6.11. Therefore, from a Bayesian viewpoint, ridge regression and the lasso follow directly from assuming the usual linear model with normal errors, together with a simple prior distribution for β. Notice that the lasso prior is steeply peaked at zero, while the Gaussian is flatter and fatter at zero. Hence, the lasso expects a priori that many of the coefficients are (exactly) zero, while ridge assumes the coefficients are randomly distributed about zero.

6.2.3 Selecting the Tuning Parameter

Just as the subset selection approaches considered in Section 6.1 require a method to determine which of the models under consideration is best, implementing ridge regression and the lasso requires a method for selecting a value for the tuning parameter λ in (6.5) and (6.7), or equivalently, the value of the constraint s in (6.9) and (6.8). Cross-validation provides a simple way to tackle this problem. We choose a grid of λ values, and compute the cross-validation error for each value of λ, as described in Chapter 5. We then select the tuning parameter value for which the cross-validation error is smallest. Finally, the model is re-fit using all of the available observations and the selected value of the tuning parameter.

   Figure 6.12 displays the choice of λ that results from performing leave-one-out cross-validation on the ridge regression fits from the Credit data set. The dashed vertical lines indicate the selected value of λ. In this case the value is relatively small, indicating that the optimal fit only involves a

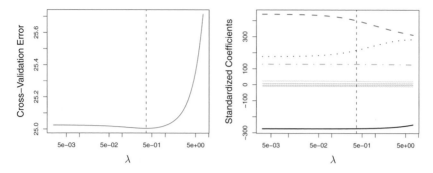

FIGURE 6.12. Left: *Cross-validation errors that result from applying ridge regression to the* Credit *data set with various value of* λ. Right: *The coefficient estimates as a function of* λ. *The vertical dashed lines indicate the value of* λ *selected by cross-validation.*

small amount of shrinkage relative to the least squares solution. In addition, the dip is not very pronounced, so there is rather a wide range of values that would give very similar error. In a case like this we might simply use the least squares solution.

Figure 6.13 provides an illustration of ten-fold cross-validation applied to the lasso fits on the sparse simulated data from Figure 6.9. The left-hand panel of Figure 6.13 displays the cross-validation error, while the right-hand panel displays the coefficient estimates. The vertical dashed lines indicate the point at which the cross-validation error is smallest. The two colored lines in the right-hand panel of Figure 6.13 represent the two predictors that are related to the response, while the grey lines represent the unrelated predictors; these are often referred to as *signal* and *noise* variables, respectively. Not only has the lasso correctly given much larger coefficient estimates to the two signal predictors, but also the minimum cross-validation error corresponds to a set of coefficient estimates for which only the signal variables are non-zero. Hence cross-validation together with the lasso has correctly identified the two signal variables in the model, even though this is a challenging setting, with $p = 45$ variables and only $n = 50$ observations. In contrast, the least squares solution—displayed on the far right of the right-hand panel of Figure 6.13—assigns a large coefficient estimate to only one of the two signal variables.

signal

6.3   Dimension Reduction Methods

The methods that we have discussed so far in this chapter have controlled variance in two different ways, either by using a subset of the original variables, or by shrinking their coefficients toward zero. All of these methods

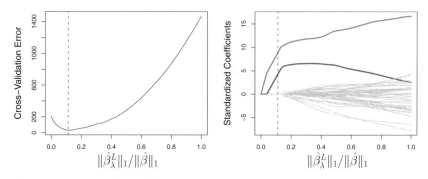

FIGURE 6.13. *Left: Ten-fold cross-validation MSE for the lasso, applied to the sparse simulated data set from Figure 6.9. Right: The corresponding lasso coefficient estimates are displayed. The vertical dashed lines indicate the lasso fit for which the cross-validation error is smallest.*

are defined using the original predictors, $X_1, X_2, \ldots, X_p$. We now explore a class of approaches that *transform* the predictors and then fit a least squares model using the transformed variables. We will refer to these techniques as *dimension reduction* methods.

Let $Z_1, Z_2, \ldots, Z_M$ represent $M < p$ *linear combinations* of our original p predictors. That is,

$$Z_m = \sum_{j=1}^{p} \phi_{jm} X_j \qquad (6.16)$$

for some constants $\phi_{1m}, \phi_{2m} \ldots, \phi_{pm}$, $m = 1, \ldots, M$. We can then fit the linear regression model

$$y_i = \theta_0 + \sum_{m=1}^{M} \theta_m z_{im} + \epsilon_i, \quad i = 1, \ldots, n, \qquad (6.17)$$

using least squares. Note that in (6.17), the regression coefficients are given by $\theta_0, \theta_1, \ldots, \theta_M$. If the constants $\phi_{1m}, \phi_{2m}, \ldots, \phi_{pm}$ are chosen wisely, then such dimension reduction approaches can often outperform least squares regression. In other words, fitting (6.17) using least squares can lead to better results than fitting (6.1) using least squares.

The term *dimension reduction* comes from the fact that this approach reduces the problem of estimating the $p+1$ coefficients $\beta_0, \beta_1, \ldots, \beta_p$ to the simpler problem of estimating the $M+1$ coefficients $\theta_0, \theta_1, \ldots, \theta_M$, where $M < p$. In other words, the dimension of the problem has been reduced from $p+1$ to $M+1$.

Notice that from (6.16),

$$\sum_{m=1}^{M} \theta_m z_{im} = \sum_{m=1}^{M} \theta_m \sum_{j=1}^{p} \phi_{jm} x_{ij} = \sum_{j=1}^{p} \sum_{m=1}^{M} \theta_m \phi_{jm} x_{ij} = \sum_{j=1}^{p} \beta_j x_{ij},$$

dimension
reduction

linear
combination

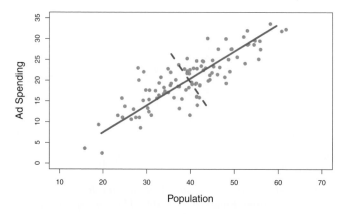

FIGURE 6.14. *The population size* (pop) *and ad spending* (ad) *for* 100 *different cities are shown as purple circles. The green solid line indicates the first principal component, and the blue dashed line indicates the second principal component.*

where

$$\beta_j = \sum_{m=1}^{M} \theta_m \phi_{jm}. \tag{6.18}$$

Hence (6.17) can be thought of as a special case of the original linear regression model given by (6.1). Dimension reduction serves to constrain the estimated β_j coefficients, since now they must take the form (6.18). This constraint on the form of the coefficients has the potential to bias the coefficient estimates. However, in situations where p is large relative to n, selecting a value of $M \ll p$ can significantly reduce the variance of the fitted coefficients. If $M = p$, and all the Z_m are linearly independent, then (6.18) poses no constraints. In this case, no dimension reduction occurs, and so fitting (6.17) is equivalent to performing least squares on the original p predictors.

All dimension reduction methods work in two steps. First, the trans-formed predictors $Z_1, Z_2, \ldots, Z_M$ are obtained. Second, the model is fit using these M predictors. However, the choice of $Z_1, Z_2, \ldots, Z_M$, or equiv-alently, the selection of the ϕ_{jm}'s, can be achieved in different ways. In this chapter, we will consider two approaches for this task: *principal components* and *partial least squares*.

6.3.1 Principal Components Regression

Principal components analysis (PCA) is a popular approach for deriving a low-dimensional set of features from a large set of variables. PCA is discussed in greater detail as a tool for *unsupervised learning* in Chapter 10. Here we describe its use as a dimension reduction technique for regression.

principal components analysis

An Overview of Principal Components Analysis

PCA is a technique for reducing the dimension of a $n \times p$ data matrix $\mathbf{X}$. The *first principal component* direction of the data is that along which the observations *vary the most*. For instance, consider Figure 6.14, which shows population size (pop) in tens of thousands of people, and ad spending for a particular company (ad) in thousands of dollars, for 100 cities. The green solid line represents the first principal component direction of the data. We can see by eye that this is the direction along which there is the greatest variability in the data. That is, if we *projected* the 100 observations onto this line (as shown in the left-hand panel of Figure 6.15), then the resulting projected observations would have the largest possible variance; projecting the observations onto any other line would yield projected observations with lower variance. Projecting a point onto a line simply involves finding the location on the line which is closest to the point.

The first principal component is displayed graphically in Figure 6.14, but how can it be summarized mathematically? It is given by the formula

$$Z_1 = 0.839 \times (\text{pop} - \overline{\text{pop}}) + 0.544 \times (\text{ad} - \overline{\text{ad}}). \tag{6.19}$$

Here $\phi_{11} = 0.839$ and $\phi_{21} = 0.544$ are the principal component loadings, which define the direction referred to above. In (6.19), $\overline{\text{pop}}$ indicates the mean of all pop values in this data set, and $\overline{\text{ad}}$ indicates the mean of all advertising spending. The idea is that out of every possible *linear combination* of pop and ad such that $\phi_{11}^2 + \phi_{21}^2 = 1$, this particular linear combination yields the highest variance: i.e. this is the linear combination for which $\text{Var}(\phi_{11} \times (\text{pop} - \overline{\text{pop}}) + \phi_{21} \times (\text{ad} - \overline{\text{ad}}))$ is maximized. It is necessary to consider only linear combinations of the form $\phi_{11}^2 + \phi_{21}^2 = 1$, since otherwise we could increase ϕ_{11} and ϕ_{21} arbitrarily in order to blow up the variance. In (6.19), the two loadings are both positive and have similar size, and so Z_1 is almost an *average* of the two variables.

Since $n = 100$, pop and ad are vectors of length 100, and so is Z_1 in (6.19). For instance,

$$z_{i1} = 0.839 \times (\text{pop}_i - \overline{\text{pop}}) + 0.544 \times (\text{ad}_i - \overline{\text{ad}}). \tag{6.20}$$

The values of $z_{11}, \ldots, z_{n1}$ are known as the *principal component scores*, and can be seen in the right-hand panel of Figure 6.15.

There is also another interpretation for PCA: the first principal component vector defines the line that is *as close as possible* to the data. For instance, in Figure 6.14, the first principal component line minimizes the sum of the squared perpendicular distances between each point and the line. These distances are plotted as dashed line segments in the left-hand panel of Figure 6.15, in which the crosses represent the *projection* of each point onto the first principal component line. The first principal component has been chosen so that the projected observations are *as close as possible* to the original observations.

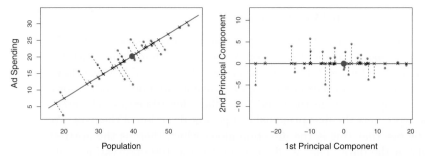

FIGURE 6.15. *A subset of the advertising data. The mean* pop *and* ad *budgets are indicated with a blue circle. Left: The first principal component direction is shown in green. It is the dimension along which the data vary the most, and it also defines the line that is closest to all n of the observations. The distances from each observation to the principal component are represented using the black dashed line segments. The blue dot represents* $(\overline{\text{pop}}, \overline{\text{ad}})$. *Right: The left-hand panel has been rotated so that the first principal component direction coincides with the x-axis.*

In the right-hand panel of Figure 6.15, the left-hand panel has been rotated so that the first principal component direction coincides with the x-axis. It is possible to show that the *first principal component score* for the ith observation, given in (6.20), is the distance in the x-direction of the ith cross from zero. So for example, the point in the bottom-left corner of the left-hand panel of Figure 6.15 has a large negative principal component score, $z_{i1} = -26.1$, while the point in the top-right corner has a large positive score, $z_{i1} = 18.7$. These scores can be computed directly using (6.20).

We can think of the values of the principal component Z_1 as single-number summaries of the joint pop and ad budgets for each location. In this example, if $z_{i1} = 0.839 \times (\text{pop}_i - \overline{\text{pop}}) + 0.544 \times (\text{ad}_i - \overline{\text{ad}}) < 0$, then this indicates a city with below-average population size and below-average ad spending. A positive score suggests the opposite. How well can a single number represent both pop and ad? In this case, Figure 6.14 indicates that pop and ad have approximately a linear relationship, and so we might expect that a single-number summary will work well. Figure 6.16 displays z_{i1} versus both pop and ad.[4] The plots show a strong relationship between the first principal component and the two features. In other words, the first principal component appears to capture most of the information contained in the pop and ad predictors.

So far we have concentrated on the first principal component. In general, one can construct up to p distinct principal components. The second principal component Z_2 is a linear combination of the variables that is un-correlated with Z_1, and has largest variance subject to this constraint. The second principal component direction is illustrated as a dashed blue line in Figure 6.14. It turns out that the zero correlation condition of Z_1 with Z_2

---

[4]The principal components were calculated after first standardizing both pop and ad, a common approach. Hence, the x-axes on Figures 6.15 and 6.16 are not on the same scale.

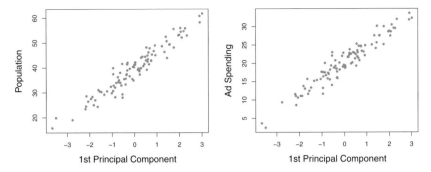

FIGURE 6.16. *Plots of the first principal component scores* z_{i1} *versus* pop *and* ad. *The relationships are strong.*

is equivalent to the condition that the direction must be *perpendicular*, or *orthogonal*, to the first principal component direction. The second principal component is given by the formula

$$Z_2 = 0.544 \times (\text{pop} - \overline{\text{pop}}) - 0.839 \times (\text{ad} - \overline{\text{ad}}).$$

Since the advertising data has two predictors, the first two principal components contain all of the information that is in pop and ad. However, by construction, the first component will contain the most information. Consider, for example, the much larger variability of z_{i1} (the x-axis) versus z_{i2} (the y-axis) in the right-hand panel of Figure 6.15. The fact that the second principal component scores are much closer to zero indicates that this component captures far less information. As another illustration, Figure 6.17 displays z_{i2} versus pop and ad. There is little relationship between the second principal component and these two predictors, again suggesting that in this case, one only needs the first principal component in order to accurately represent the pop and ad budgets.

With two-dimensional data, such as in our advertising example, we can construct at most two principal components. However, if we had other predictors, such as population age, income level, education, and so forth, then additional components could be constructed. They would successively maximize variance, subject to the constraint of being uncorrelated with the preceding components.

The Principal Components Regression Approach

The *principal components regression* (PCR) approach involves constructing the first M principal components, $Z_1, \dots, Z_M$, and then using these components as the predictors in a linear regression model that is fit using least squares. The key idea is that often a small number of principal components suffice to explain most of the variability in the data, as well as the relationship with the response. In other words, we assume that *the directions in which $X_1, \dots, X_p$ show the most variation are the directions that are associated with Y*. While this assumption is not guaranteed

perpendicular

orthogonal

principal components regression

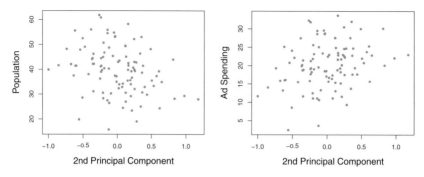

FIGURE 6.17. *Plots of the second principal component scores z_{i2} versus* pop *and* ad. *The relationships are weak.*

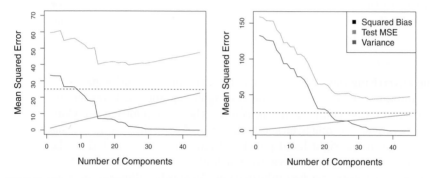

FIGURE 6.18. *PCR was applied to two simulated data sets.* Left: *Simulated data from Figure 6.8.* Right: *Simulated data from Figure 6.9.*

to be true, it often turns out to be a reasonable enough approximation to give good results.

If the assumption underlying PCR holds, then fitting a least squares model to $Z_1, \ldots, Z_M$ will lead to better results than fitting a least squares model to $X_1, \ldots, X_p$, since most or all of the information in the data that relates to the response is contained in $Z_1, \ldots, Z_M$, and by estimating only $M \ll p$ coefficients we can mitigate overfitting. In the advertising data, the first principal component explains most of the variance in both pop and ad, so a principal component regression that uses this single variable to predict some response of interest, such as sales, will likely perform quite well.

Figure 6.18 displays the PCR fits on the simulated data sets from Figures 6.8 and 6.9. Recall that both data sets were generated using $n = 50$ observations and $p = 45$ predictors. However, while the response in the first data set was a function of all the predictors, the response in the second data set was generated using only two of the predictors. The curves are plotted as a function of M, the number of principal components used as predictors in the regression model. As more principal components are used in

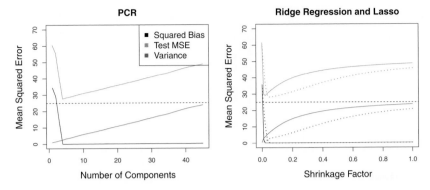

FIGURE 6.19. *PCR, ridge regression, and the lasso were applied to a simulated data set in which the first five principal components of X contain all the information about the response Y. In each panel, the irreducible error Var(ϵ) is shown as a horizontal dashed line. Left: Results for PCR. Right: Results for lasso (solid) and ridge regression (dotted). The x-axis displays the shrinkage factor of the coefficient estimates, defined as the ℓ_2 norm of the shrunken coefficient estimates divided by the ℓ_2 norm of the least squares estimate.*

the regression model, the bias decreases, but the variance increases. This results in a typical U-shape for the mean squared error. When $M = p = 45$, then PCR amounts simply to a least squares fit using all of the original predictors. The figure indicates that performing PCR with an appropriate choice of M can result in a substantial improvement over least squares, especially in the left-hand panel. However, by examining the ridge regression and lasso results in Figures 6.5, 6.8, and 6.9, we see that PCR does not perform as well as the two shrinkage methods in this example.

The relatively worse performance of PCR in Figure 6.18 is a consequence of the fact that the data were generated in such a way that many principal components are required in order to adequately model the response. In contrast, PCR will tend to do well in cases when the first few principal components are sufficient to capture most of the variation in the predictors as well as the relationship with the response. The left-hand panel of Figure 6.19 illustrates the results from another simulated data set designed to be more favorable to PCR. Here the response was generated in such a way that it depends exclusively on the first five principal components. Now the bias drops to zero rapidly as M, the number of principal components used in PCR, increases. The mean squared error displays a clear minimum at $M = 5$. The right-hand panel of Figure 6.19 displays the results on these data using ridge regression and the lasso. All three methods offer a significant improvement over least squares. However, PCR and ridge regression slightly outperform the lasso.

We note that even though PCR provides a simple way to perform regression using $M < p$ predictors, it is *not* a feature selection method. This is because each of the M principal components used in the regression

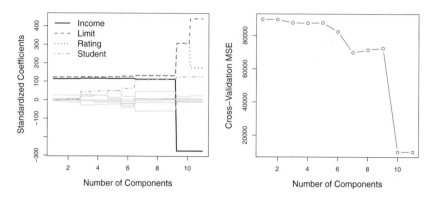

FIGURE 6.20. *Left: PCR standardized coefficient estimates on the* Credit *data set for different values of M. Right: The ten-fold cross validation MSE obtained using PCR, as a function of M.*

is a linear combination of all p of the *original* features. For instance, in (6.19), Z_1 was a linear combination of both pop and ad. Therefore, while PCR often performs quite well in many practical settings, it does not result in the development of a model that relies upon a small set of the original features. In this sense, PCR is more closely related to ridge regression than to the lasso. In fact, one can show that PCR and ridge regression are very closely related. One can even think of ridge regression as a continuous version of PCR![5]

In PCR, the number of principal components, M, is typically chosen by cross-validation. The results of applying PCR to the Credit data set are shown in Figure 6.20; the right-hand panel displays the cross-validation errors obtained, as a function of M. On these data, the lowest cross-validation error occurs when there are $M = 10$ components; this corresponds to almost no dimension reduction at all, since PCR with $M = 11$ is equivalent to simply performing least squares.

When performing PCR, we generally recommend *standardizing* each predictor, using (6.6), prior to generating the principal components. This standardization ensures that all variables are on the same scale. In the absence of standardization, the high-variance variables will tend to play a larger role in the principal components obtained, and the scale on which the variables are measured will ultimately have an effect on the final PCR model. However, if the variables are all measured in the same units (say, kilograms, or inches), then one might choose not to standardize them.

---

[5]More details can be found in Section 3.5 of *Elements of Statistical Learning* by Hastie, Tibshirani, and Friedman.

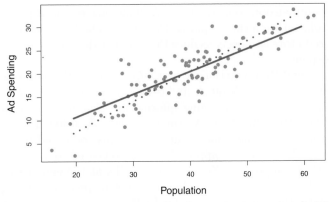

FIGURE 6.21. *For the advertising data, the first PLS direction (solid line) and first PCR direction (dotted line) are shown.*

6.3.2 Partial Least Squares

The PCR approach that we just described involves identifying linear combinations, or *directions*, that best represent the predictors $X_1, \ldots, X_p$. These directions are identified in an *unsupervised* way, since the response Y is not used to help determine the principal component directions. That is, the response does not *supervise* the identification of the principal components. Consequently, PCR suffers from a drawback: there is no guarantee that the directions that best explain the predictors will also be the best directions to use for predicting the response. Unsupervised methods are discussed further in Chapter 10.

We now present *partial least squares* (PLS), a *supervised* alternative to PCR. Like PCR, PLS is a dimension reduction method, which first identifies a new set of features $Z_1, \ldots, Z_M$ that are linear combinations of the original features, and then fits a linear model via least squares using these M new features. But unlike PCR, PLS identifies these new features in a supervised way—that is, it makes use of the response Y in order to identify new features that not only approximate the old features well, but also that *are related to the response*. Roughly speaking, the PLS approach attempts to find directions that help explain both the response and the predictors.

We now describe how the first PLS direction is computed. After standardizing the p predictors, PLS computes the first direction Z_1 by setting each ϕ_{j1} in (6.16) equal to the coefficient from the simple linear regression of Y onto X_j. One can show that this coefficient is proportional to the correlation between Y and X_j. Hence, in computing $Z_1 = \sum_{j=1}^{p} \phi_{j1} X_j$, PLS places the highest weight on the variables that are most strongly related to the response.

Figure 6.21 displays an example of PLS on a synthetic dataset with Sales in each of 100 regions as the response, and two predictors; Population Size and Advertising Spending.[6] The solid green line indicates the first PLS direction, while the dotted line shows the first principal component direction. PLS has chosen a direction that has less change in the ad dimension per unit

partial least squares

---

[6]This dataset is distinct from the Advertising data discussed in Chapter 3.

change in the pop dimension, relative to PCA. This suggests that pop is more highly correlated with the response than is ad. The PLS direction does not fit the predictors as closely as does PCA, but it does a better job explaining the response.

To identify the second PLS direction we first *adjust* each of the variables for Z_1, by regressing each variable on Z_1 and taking *residuals*. These residuals can be interpreted as the remaining information that has not been explained by the first PLS direction. We then compute Z_2 using this *orthogonalized* data in exactly the same fashion as Z_1 was computed based on the original data. This iterative approach can be repeated M times to identify multiple PLS components $Z_1, \ldots, Z_M$. Finally, at the end of this procedure, we use least squares to fit a linear model to predict Y using $Z_1, \ldots, Z_M$ in exactly the same fashion as for PCR.

As with PCR, the number M of partial least squares directions used in PLS is a tuning parameter that is typically chosen by cross-validation. We generally standardize the predictors and response before performing PLS.

PLS is popular in the field of chemometrics, where many variables arise from digitized spectrometry signals. In practice it often performs no better than ridge regression or PCR. While the supervised dimension reduction of PLS can reduce bias, it also has the potential to increase variance, so that the overall benefit of PLS relative to PCR is a wash.

6.4 Considerations in High Dimensions

6.4.1 High-Dimensional Data

Most traditional statistical techniques for regression and classification are intended for the *low-dimensional* setting in which n, the number of observations, is much greater than p, the number of features. This is due in part to the fact that throughout most of the field's history, the bulk of scientific problems requiring the use of statistics have been low-dimensional. For instance, consider the task of developing a model to predict a patient's blood pressure on the basis of his or her age, gender, and body mass index (BMI). There are three predictors, or four if an intercept is included in the model, and perhaps several thousand patients for whom blood pressure and age, gender, and BMI are available. Hence $n \gg p$, and so the problem is low-dimensional. (By dimension here we are referring to the size of p.)

In the past 20 years, new technologies have changed the way that data are collected in fields as diverse as finance, marketing, and medicine. It is now commonplace to collect an almost unlimited number of feature measurements (p very large). While p can be extremely large, the number of observations n is often limited due to cost, sample availability, or other considerations. Two examples are as follows:

1. Rather than predicting blood pressure on the basis of just age, gender, and BMI, one might also collect measurements for half a million

low-
dimensional

single nucleotide polymorphisms (SNPs; these are individual DNA mutations that are relatively common in the population) for inclusion in the predictive model. Then $n \approx 200$ and $p \approx 500,000$.

2. A marketing analyst interested in understanding people's online shopping patterns could treat as features all of the search terms entered by users of a search engine. This is sometimes known as the "bag-of-words" model. The same researcher might have access to the search histories of only a few hundred or a few thousand search engine users who have consented to share their information with the researcher. For a given user, each of the p search terms is scored present (0) or absent (1), creating a large binary feature vector. Then $n \approx 1,000$ and p is much larger.

Data sets containing more features than observations are often referred to as *high-dimensional*. Classical approaches such as least squares linear regression are not appropriate in this setting. Many of the issues that arise in the analysis of high-dimensional data were discussed earlier in this book, since they apply also when $n > p$: these include the role of the bias-variance trade-off and the danger of overfitting. Though these issues are always relevant, they can become particularly important when the number of features is very large relative to the number of observations.

high-
dimensional

We have defined the *high-dimensional setting* as the case where the number of features p is larger than the number of observations n. But the considerations that we will now discuss certainly also apply if p is slightly smaller than n, and are best always kept in mind when performing supervised learning.

6.4.2 What Goes Wrong in High Dimensions?

In order to illustrate the need for extra care and specialized techniques for regression and classification when $p > n$, we begin by examining what can go wrong if we apply a statistical technique not intended for the high-dimensional setting. For this purpose, we examine least squares regression. But the same concepts apply to logistic regression, linear discriminant analysis, and other classical statistical approaches.

When the number of features p is as large as, or larger than, the number of observations n, least squares as described in Chapter 3 cannot (or rather, *should not*) be performed. The reason is simple: regardless of whether or not there truly is a relationship between the features and the response, least squares will yield a set of coefficient estimates that result in a perfect fit to the data, such that the residuals are zero.

An example is shown in Figure 6.22 with $p = 1$ feature (plus an intercept) in two cases: when there are 20 observations, and when there are only two observations. When there are 20 observations, $n > p$ and the least

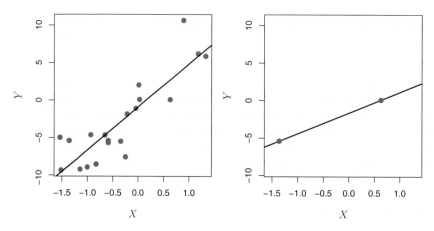

FIGURE 6.22. Left: *Least squares regression in the low-dimensional setting.* Right: *Least squares regression with n = 2 observations and two parameters to be estimated (an intercept and a coefficient).*

squares regression line does not perfectly fit the data; instead, the regression line seeks to approximate the 20 observations as well as possible. On the other hand, when there are only two observations, then regardless of the values of those observations, the regression line will fit the data exactly. This is problematic because this perfect fit will almost certainly lead to overfitting of the data. In other words, though it is possible to perfectly fit the training data in the high-dimensional setting, the resulting linear model will perform extremely poorly on an independent test set, and therefore does not constitute a useful model. In fact, we can see that this happened in Figure 6.22: the least squares line obtained in the right-hand panel will perform very poorly on a test set comprised of the observations in the left-hand panel. The problem is simple: when $p > n$ or $p \approx n$, a simple least squares regression line is too *flexible* and hence overfits the data.

Figure 6.23 further illustrates the risk of carelessly applying least squares when the number of features p is large. Data were simulated with $n = 20$ observations, and regression was performed with between 1 and 20 features, each of which was completely unrelated to the response. As shown in the figure, the model R^2 increases to 1 as the number of features included in the model increases, and correspondingly the training set MSE decreases to 0 as the number of features increases, *even though the features are completely unrelated to the response.* On the other hand, the MSE on an *independent test set* becomes extremely large as the number of features included in the model increases, because including the additional predictors leads to a vast increase in the variance of the coefficient estimates. Looking at the test set MSE, it is clear that the best model contains at most a few variables. However, someone who carelessly examines only the R^2 or the training set MSE might erroneously conclude that the model with the greatest number of variables is best. This indicates the importance of applying extra care

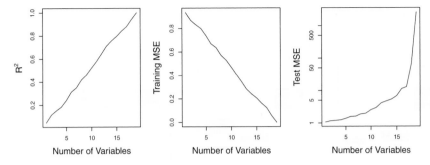

FIGURE 6.23. *On a simulated example with* $n = 20$ *training observations, features that are completely unrelated to the outcome are added to the model. Left: The* R^2 *increases to 1 as more features are included. Center: The training set MSE decreases to 0 as more features are included. Right: The test set MSE increases as more features are included.*

when analyzing data sets with a large number of variables, and of always evaluating model performance on an independent test set.

In Section 6.1.3, we saw a number of approaches for adjusting the training set RSS or R^2 in order to account for the number of variables used to fit a least squares model. Unfortunately, the C_p, AIC, and BIC approaches are not appropriate in the high-dimensional setting, because estimating $\hat{\sigma}^2$ is problematic. (For instance, the formula for $\hat{\sigma}^2$ from Chapter 3 yields an estimate $\hat{\sigma}^2 = 0$ in this setting.) Similarly, problems arise in the application of adjusted R^2 in the high-dimensional setting, since one can easily obtain a model with an adjusted R^2 value of 1. Clearly, alternative approaches that are better-suited to the high-dimensional setting are required.

6.4.3   Regression in High Dimensions

It turns out that many of the methods seen in this chapter for fitting *less flexible* least squares models, such as forward stepwise selection, ridge regression, the lasso, and principal components regression, are particularly useful for performing regression in the high-dimensional setting. Essentially, these approaches avoid overfitting by using a less flexible fitting approach than least squares.

Figure 6.24 illustrates the performance of the lasso in a simple simulated example. There are $p = 20$, 50, or 2,000 features, of which 20 are truly associated with the outcome. The lasso was performed on $n = 100$ training observations, and the mean squared error was evaluated on an independent test set. As the number of features increases, the test set error increases. When $p = 20$, the lowest validation set error was achieved when λ in (6.7) was small; however, when p was larger then the lowest validation set error was achieved using a larger value of λ. In each boxplot, rather than reporting the values of λ used, the *degrees of freedom* of the resulting

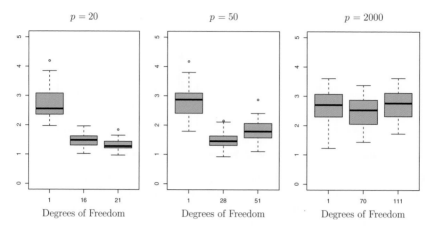

FIGURE 6.24. *The lasso was performed with* $n = 100$ *observations and three values of p, the number of features. Of the p features, 20 were associated with the response. The boxplots show the test MSEs that result using three different values of the tuning parameter* λ *in (6.7). For ease of interpretation, rather than reporting* λ, *the degrees of freedom are reported; for the lasso this turns out to be simply the number of estimated non-zero coefficients. When* $p = 20$, *the lowest test MSE was obtained with the smallest amount of regularization. When* $p = 50$, *the lowest test MSE was achieved when there is a substantial amount of regularization. When* $p = 2,000$ *the lasso performed poorly regardless of the amount of regularization, due to the fact that only 20 of the 2,000 features truly are associated with the outcome.*

lasso solution is displayed; this is simply the number of non-zero coefficient estimates in the lasso solution, and is a measure of the flexibility of the lasso fit. Figure 6.24 highlights three important points: (1) regularization or shrinkage plays a key role in high-dimensional problems, (2) appropriate tuning parameter selection is crucial for good predictive performance, and (3) the test error tends to increase as the dimensionality of the problem (i.e. the number of features or predictors) increases, unless the additional features are truly associated with the response.

The third point above is in fact a key principle in the analysis of high-dimensional data, which is known as the *curse of dimensionality*. One might think that as the number of features used to fit a model increases, the quality of the fitted model will increase as well. However, comparing the left-hand and right-hand panels in Figure 6.24, we see that this is not necessarily the case: in this example, the test set MSE almost doubles as p increases from 20 to 2,000. In general, *adding additional signal features that are truly associated with the response will improve the fitted model*, in the sense of leading to a reduction in test set error. However, adding noise features that are not truly associated with the response will lead to a deterioration in the fitted model, and consequently an increased test set error. This is because noise features increase the dimensionality of the problem, exacerbating the risk of overfitting (since noise features may be assigned nonzero coefficients due to chance associations with the response on the training set) without any potential upside in terms of improved test set error. Thus, we see that new technologies that allow for the collection of measurements for thousands or millions of features are a double-edged sword: they can lead to improved predictive models if these features are in fact relevant to the problem at hand, but will lead to worse results if the features are not relevant. Even if they are relevant, the variance incurred in fitting their coefficients may outweigh the reduction in bias that they bring.

6.4.4 Interpreting Results in High Dimensions

When we perform the lasso, ridge regression, or other regression procedures in the high-dimensional setting, we must be quite cautious in the way that we report the results obtained. In Chapter 3, we learned about *multi-collinearity*, the concept that the variables in a regression might be correlated with each other. In the high-dimensional setting, the multicollinearity problem is extreme: any variable in the model can be written as a linear combination of all of the other variables in the model. Essentially, this means that we can never know exactly which variables (if any) truly are predictive of the outcome, and we can never identify the *best* coefficients for use in the regression. At most, we can hope to assign large regression coefficients to variables that are correlated with the variables that truly are predictive of the outcome.

For instance, suppose that we are trying to predict blood pressure on the basis of half a million SNPs, and that forward stepwise selection indicates that 17 of those SNPs lead to a good predictive model on the training data. It would be incorrect to conclude that these 17 SNPs predict blood pressure more effectively than the other SNPs not included in the model. There are likely to be many sets of 17 SNPs that would predict blood pressure just as well as the selected model. If we were to obtain an independent data set and perform forward stepwise selection on that data set, we would likely obtain a model containing a different, and perhaps even non-overlapping, set of SNPs. This does not detract from the value of the model obtained—for instance, the model might turn out to be very effective in predicting blood pressure on an independent set of patients, and might be clinically useful for physicians. But we must be careful not to overstate the results obtained, and to make it clear that what we have identified is simply *one of many possible models* for predicting blood pressure, and that it must be further validated on independent data sets.

It is also important to be particularly careful in reporting errors and measures of model fit in the high-dimensional setting. We have seen that when $p > n$, it is easy to obtain a useless model that has zero residuals. Therefore, one should *never* use sum of squared errors, p-values, R^2

statistics, or other traditional measures of model fit on the training data as evidence of a good model fit in the high-dimensional setting. For instance, as we saw in Figure 6.23, one can easily obtain a model with $R^2 = 1$ when $p > n$. Reporting this fact might mislead others into thinking that a statistically valid and useful model has been obtained, whereas in fact this provides absolutely no evidence of a compelling model. It is important to instead report results on an independent test set, or cross-validation errors. For instance, the MSE or R^2 on an independent test set is a valid measure of model fit, but the MSE on the training set certainly is not.

6.5   Lab 1: Subset Selection Methods

6.5.1   Best Subset Selection

Here we apply the best subset selection approach to the Hitters data. We wish to predict a baseball player's Salary on the basis of various statistics associated with performance in the previous year.

First of all, we note that the Salary variable is missing for some of the players. The is.na() function can be used to identify the missing observations. It returns a vector of the same length as the input vector, with a TRUE for any elements that are missing, and a FALSE for non-missing elements. The sum() function can then be used to count all of the missing elements.

is.na()

sum()

```
> library(ISLR)
> fix(Hitters)
> names(Hitters)
 [1] "AtBat"     "Hits"      "HmRun"     "Runs"      "RBI"
 [6] "Walks"     "Years"     "CAtBat"    "CHits"     "CHmRun"
[11] "CRuns"     "CRBI"      "CWalks"    "League"    "Division"
[16] "PutOuts"   "Assists"   "Errors"    "Salary"    "NewLeague"
> dim(Hitters)
[1] 322  20
> sum(is.na(Hitters$Salary))
[1] 59
```

Hence we see that Salary is missing for 59 players. The na.omit() function removes all of the rows that have missing values in any variable.

```
> Hitters=na.omit(Hitters)
> dim(Hitters)
[1] 263  20
> sum(is.na(Hitters))
[1] 0
```

The regsubsets() function (part of the leaps library) performs best subset selection by identifying the best model that contains a given number of predictors, where *best* is quantified using RSS. The syntax is the same as for lm(). The summary() command outputs the best set of variables for each model size.

regsubsets()

```
> library(leaps)
> regfit.full=regsubsets(Salary~.,Hitters)
> summary(regfit.full)
Subset selection object
Call: regsubsets.formula(Salary ~ ., Hitters)
19 Variables  (and intercept)
...
1 subsets of each size up to 8
Selection Algorithm: exhaustive
          AtBat Hits HmRun Runs RBI Walks Years CAtBat CHits
1  ( 1 )  " "   " "  " "   " "  " " " "   " "   " "    " "
2  ( 1 )  " "   "*"  " "   " "  " " " "   " "   " "    " "
3  ( 1 )  " "   "*"  " "   " "  " " " "   " "   " "    " "
4  ( 1 )  " "   "*"  " "   " "  " " " "   " "   " "    " "
5  ( 1 )  "*"   "*"  " "   " "  " " " "   " "   " "    " "
6  ( 1 )  "*"   "*"  " "   " "  " " "*"   " "   " "    " "
7  ( 1 )  " "   "*"  " "   " "  " " "*"   " "   "*"    "*"
8  ( 1 )  "*"   "*"  " "   " "  " " "*"   " "   " "    " "
          CHmRun CRuns CRBI CWalks LeagueN DivisionW PutOuts
1  ( 1 )  " "    " "   "*"  " "    " "     " "       " "
2  ( 1 )  " "    " "   "*"  " "    " "     " "       " "
3  ( 1 )  " "    " "   "*"  " "    " "     " "       "*"
4  ( 1 )  " "    " "   "*"  " "    " "     "*"       "*"
5  ( 1 )  " "    " "   "*"  " "    " "     "*"       "*"
6  ( 1 )  " "    " "   "*"  " "    " "     "*"       "*"
7  ( 1 )  "*"    " "   " "  " "    " "     "*"       "*"
8  ( 1 )  "*"    "*"   " "  "*"    " "     "*"       "*"
          Assists Errors NewLeagueN
1  ( 1 )  " "     " "    " "
2  ( 1 )  " "     " "    " "
3  ( 1 )  " "     " "    " "
4  ( 1 )  " "     " "    " "
5  ( 1 )  " "     " "    " "
6  ( 1 )  " "     " "    " "
7  ( 1 )  " "     " "    " "
8  ( 1 )  " "     " "    " "
```

An asterisk indicates that a given variable is included in the corresponding model. For instance, this output indicates that the best two-variable model contains only Hits and CRBI. By default, regsubsets() only reports results up to the best eight-variable model. But the nvmax option can be used in order to return as many variables as are desired. Here we fit up to a 19-variable model.

```
> regfit.full=regsubsets(Salary~.,data=Hitters,nvmax=19)
> reg.summary=summary(regfit.full)
```

The summary() function also returns R^2, RSS, adjusted R^2, C_p, and BIC. We can examine these to try to select the *best* overall model.

```
> names(reg.summary)
[1] "which"  "rsq"   "rss"    "adjr2"  "cp"    "bic"
[7] "outmat" "obj"
```

For instance, we see that the R^2 statistic increases from 32 %, when only one variable is included in the model, to almost 55 %, when all variables are included. As expected, the R^2 statistic increases monotonically as more variables are included.

```
> reg.summary$rsq
 [1] 0.321 0.425 0.451 0.475 0.491 0.509 0.514 0.529 0.535
[10] 0.540 0.543 0.544 0.544 0.545 0.545 0.546 0.546 0.546
[19] 0.546
```

Plotting RSS, adjusted R^2, C_p, and BIC for all of the models at once will help us decide which model to select. Note the type="l" option tells R to connect the plotted points with lines.

```
> par(mfrow=c(2,2))
> plot(reg.summary$rss,xlab="Number of Variables",ylab="RSS",
    type="l")
> plot(reg.summary$adjr2,xlab="Number of Variables",
    ylab="Adjusted RSq",type="l")
```

The points() command works like the plot() command, except that it puts points on a plot that has already been created, instead of creating a new plot. The which.max() function can be used to identify the location of the maximum point of a vector. We will now plot a red dot to indicate the model with the largest adjusted R^2 statistic.

points()

```
> which.max(reg.summary$adjr2)
[1] 11
> points(11,reg.summary$adjr2[11], col="red",cex=2,pch=20)
```

In a similar fashion we can plot the C_p and BIC statistics, and indicate the models with the smallest statistic using which.min().

which.min()

```
> plot(reg.summary$cp,xlab="Number of Variables",ylab="Cp",
    type='l')
> which.min(reg.summary$cp)
[1] 10
> points(10,reg.summary$cp[10],col="red",cex=2,pch=20)
> which.min(reg.summary$bic)
[1] 6
> plot(reg.summary$bic,xlab="Number of Variables",ylab="BIC",
    type='l')
> points(6,reg.summary$bic[6],col="red",cex=2,pch=20)
```

The regsubsets() function has a built-in plot() command which can be used to display the selected variables for the best model with a given number of predictors, ranked according to the BIC, C_p, adjusted R^2, or AIC. To find out more about this function, type ?plot.regsubsets.

```
> plot(regfit.full,scale="r2")
> plot(regfit.full,scale="adjr2")
> plot(regfit.full,scale="Cp")
> plot(regfit.full,scale="bic")
```

The top row of each plot contains a black square for each variable selected according to the optimal model associated with that statistic. For instance, we see that several models share a BIC close to −150. However, the model with the lowest BIC is the six-variable model that contains only AtBat, Hits, Walks, CRBI, DivisionW, and PutOuts. We can use the coef() function to see the coefficient estimates associated with this model.

```
> coef(regfit.full,6)
(Intercept)         AtBat          Hits         Walks          CRBI
     91.512        -1.869         7.604         3.698         0.643
   DivisionW       PutOuts
    -122.952         0.264
```

6.5.2 Forward and Backward Stepwise Selection

We can also use the regsubsets() function to perform forward stepwise or backward stepwise selection, using the argument method="forward" or method="backward".

```
> regfit.fwd=regsubsets(Salary~.,data=Hitters,nvmax=19,
    method="forward")
> summary(regfit.fwd)
> regfit.bwd=regsubsets(Salary~.,data=Hitters,nvmax=19,
    method="backward")
> summary(regfit.bwd)
```

For instance, we see that using forward stepwise selection, the best one-variable model contains only CRBI, and the best two-variable model additionally includes Hits. For this data, the best one-variable through six-variable models are each identical for best subset and forward selection. However, the best seven-variable models identified by forward stepwise selection, backward stepwise selection, and best subset selection are different.

```
> coef(regfit.full,7)
(Intercept)          Hits         Walks        CAtBat         CHits
     79.451         1.283         3.227        -0.375         1.496
     CHmRun      DivisionW       PutOuts
      1.442      -129.987         0.237
> coef(regfit.fwd,7)
(Intercept)         AtBat          Hits         Walks          CRBI
    109.787        -1.959         7.450         4.913         0.854
     CWalks      DivisionW       PutOuts
     -0.305      -127.122         0.253
> coef(regfit.bwd,7)
(Intercept)         AtBat          Hits         Walks          CRuns
    105.649        -1.976         6.757         6.056         1.129
     CWalks      DivisionW       PutOuts
     -0.716      -116.169         0.303
```

6.5.3 Choosing Among Models Using the Validation Set Approach and Cross-Validation

We just saw that it is possible to choose among a set of models of different sizes using C_p, BIC, and adjusted R^2. We will now consider how to do this using the validation set and cross-validation approaches.

In order for these approaches to yield accurate estimates of the test error, we must use *only the training observations* to perform all aspects of model-fitting—including variable selection. Therefore, the determination of which model of a given size is best must be made using *only the training observations*. This point is subtle but important. If the full data set is used to perform the best subset selection step, the validation set errors and cross-validation errors that we obtain will not be accurate estimates of the test error.

In order to use the validation set approach, we begin by splitting the observations into a training set and a test set. We do this by creating a random vector, `train`, of elements equal to TRUE if the corresponding observation is in the training set, and FALSE otherwise. The vector `test` has a TRUE if the observation is in the test set, and a FALSE otherwise. Note the ! in the command to create `test` causes TRUEs to be switched to FALSEs and vice versa. We also set a random seed so that the user will obtain the same training set/test set split.

```
> set.seed(1)
> train=sample(c(TRUE,FALSE), nrow(Hitters),rep=TRUE)
> test=(!train)
```

Now, we apply `regsubsets()` to the training set in order to perform best subset selection.

```
> regfit.best=regsubsets(Salary~.,data=Hitters[train,],
    nvmax=19)
```

Notice that we subset the `Hitters` data frame directly in the call in order to access only the training subset of the data, using the expression `Hitters[train,]`. We now compute the validation set error for the best model of each model size. We first make a model matrix from the test data.

```
    test.mat=model.matrix(Salary~.,data=Hitters[test,])
```

The `model.matrix()` function is used in many regression packages for building an "X" matrix from data. Now we run a loop, and for each size i, we extract the coefficients from `regfit.best` for the best model of that size, multiply them into the appropriate columns of the test model matrix to form the predictions, and compute the test MSE.

`model.matrix()`

```
> val.errors=rep(NA,19)
> for(i in 1:19){
+    coefi=coef(regfit.best,id=i)
```

```
+    pred=test.mat[,names(coefi)]%*%coefi
+    val.errors[i]=mean((Hitters$Salary[test]-pred)^2)
}
```

We find that the best model is the one that contains ten variables.

```
> val.errors
 [1]  220968 169157 178518 163426 168418 171271 162377 157909
 [9]  154056 148162 151156 151742 152214 157359 158541 158743
[17]  159973 159860 160106
> which.min(val.errors)
[1] 10
> coef(regfit.best,10)
(Intercept)         AtBat          Hits         Walks        CAtBat
    -80.275        -1.468         7.163         3.643        -0.186
      CHits        CHmRun        CWalks       LeagueN      DivisionW
      1.105         1.384        -0.748        84.558       -53.029
    PutOuts
      0.238
```

This was a little tedious, partly because there is no `predict()` method for `regsubsets()`. Since we will be using this function again, we can capture our steps above and write our own predict method.

```
>    predict.regsubsets=function(object,newdata,id,...){
+    form=as.formula(object$call[[2]])
+    mat=model.matrix(form,newdata)
+    coefi=coef(object,id=id)
+    xvars=names(coefi)
+    mat[,xvars]%*%coefi
+    }
```

Our function pretty much mimics what we did above. The only complex part is how we extracted the formula used in the call to `regsubsets()`. We demonstrate how we use this function below, when we do cross-validation.

Finally, we perform best subset selection on the full data set, and select the best ten-variable model. It is important that we make use of the full data set in order to obtain more accurate coefficient estimates. Note that we perform best subset selection on the full data set and select the best ten-variable model, rather than simply using the variables that were obtained from the training set, because the best ten-variable model on the full data set may differ from the corresponding model on the training set.

```
> regfit.best=regsubsets(Salary~.,data=Hitters,nvmax=19)
> coef(regfit.best,10)
(Intercept)         AtBat          Hits         Walks        CAtBat
    162.535        -2.169         6.918         5.773        -0.130
      CRuns          CRBI        CWalks     DivisionW       PutOuts
      1.408         0.774        -0.831      -112.380         0.297
    Assists
      0.283
```

In fact, we see that the best ten-variable model on the full data set has a different set of variables than the best ten-variable model on the training set.

We now try to choose among the models of different sizes using cross-validation. This approach is somewhat involved, as we must perform best subset selection *within each of the k training sets*. Despite this, we see that with its clever subsetting syntax, R makes this job quite easy. First, we create a vector that allocates each observation to one of $k = 10$ folds, and we create a matrix in which we will store the results.

```
> k=10
> set.seed(1)
> folds=sample(1:k,nrow(Hitters),replace=TRUE)
> cv.errors=matrix(NA,k,19, dimnames=list(NULL, paste(1:19)))
```

Now we write a for loop that performs cross-validation. In the *j*th fold, the elements of folds that equal j are in the test set, and the remainder are in the training set. We make our predictions for each model size (using our new predict() method), compute the test errors on the appropriate subset, and store them in the appropriate slot in the matrix cv.errors.

```
> for(j in 1:k){
+   best.fit=regsubsets(Salary~.,data=Hitters[folds!=j,],
        nvmax=19)
+   for(i in 1:19){
+     pred=predict(best.fit,Hitters[folds==j,],id=i)
+     cv.errors[j,i]=mean( (Hitters$Salary[folds==j]-pred)^2)
+     }
+   }
```

This has given us a 10×19 matrix, of which the (i, j)th element corresponds to the test MSE for the *i*th cross-validation fold for the best *j*-variable model. We use the apply() function to average over the columns of this matrix in order to obtain a vector for which the *j*th element is the cross-validation error for the *j*-variable model.   apply()

```
> mean.cv.errors=apply(cv.errors,2,mean)
> mean.cv.errors
 [1] 160093 140197 153117 151159 146841 138303 144346 130208
 [9] 129460 125335 125154 128274 133461 133975 131826 131883
[17] 132751 133096 132805
> par(mfrow=c(1,1))
> plot(mean.cv.errors,type='b')
```

We see that cross-validation selects an 11-variable model. We now perform best subset selection on the full data set in order to obtain the 11-variable model.

```
> reg.best=regsubsets(Salary~.,data=Hitters, nvmax=19)
> coef(reg.best,11)
(Intercept)       AtBat         Hits        Walks       CAtBat
    135.751      -2.128        6.924        5.620       -0.139
```

CRuns	CRBI	CWalks	LeagueN	DivisionW
1.455	0.785	-0.823	43.112	-111.146
PutOuts	Assists			
0.289	0.269			

6.6 Lab 2: Ridge Regression and the Lasso

We will use the `glmnet` package in order to perform ridge regression and the lasso. The main function in this package is `glmnet()`, which can be used to fit ridge regression models, lasso models, and more. This function has slightly different syntax from other model-fitting functions that we have encountered thus far in this book. In particular, we must pass in an x matrix as well as a y vector, and we do not use the y ~ x syntax. We will now perform ridge regression and the lasso in order to predict `Salary` on the `Hitters` data. Before proceeding ensure that the missing values have been removed from the data, as described in Section 6.5.

`glmnet()`

```
> x=model.matrix(Salary~.,Hitters)[,-1]
> y=Hitters$Salary
```

The `model.matrix()` function is particularly useful for creating x; not only does it produce a matrix corresponding to the 19 predictors but it also automatically transforms any qualitative variables into dummy variables. The latter property is important because `glmnet()` can only take numerical, quantitative inputs.

6.6.1 Ridge Regression

The `glmnet()` function has an `alpha` argument that determines what type of model is fit. If `alpha=0` then a ridge regression model is fit, and if `alpha=1` then a lasso model is fit. We first fit a ridge regression model.

```
> library(glmnet)
> grid=10^seq(10,-2,length=100)
> ridge.mod=glmnet(x,y,alpha=0,lambda=grid)
```

By default the `glmnet()` function performs ridge regression for an automatically selected range of λ values. However, here we have chosen to implement the function over a grid of values ranging from $\lambda = 10^{10}$ to $\lambda = 10^{-2}$, essentially covering the full range of scenarios from the null model containing only the intercept, to the least squares fit. As we will see, we can also compute model fits for a particular value of λ that is not one of the original `grid` values. Note that by default, the `glmnet()` function standardizes the variables so that they are on the same scale. To turn off this default setting, use the argument `standardize=FALSE`.

Associated with each value of λ is a vector of ridge regression coefficients, stored in a matrix that can be accessed by `coef()`. In this case, it is a 20×100

matrix, with 20 rows (one for each predictor, plus an intercept) and 100 columns (one for each value of λ).

```
> dim(coef(ridge.mod))
[1]  20 100
```

We expect the coefficient estimates to be much smaller, in terms of ℓ_2 norm, when a large value of λ is used, as compared to when a small value of λ is used. These are the coefficients when $\lambda = 11{,}498$, along with their ℓ_2 norm:

```
> ridge.mod$lambda[50]
[1] 11498
> coef(ridge.mod)[,50]
(Intercept)        AtBat          Hits         HmRun          Runs
    407.356        0.037         0.138         0.525         0.231
        RBI        Walks         Years        CAtBat         CHits
      0.240        0.290         1.108         0.003         0.012
     CHmRun        CRuns          CRBI        CWalks       LeagueN
      0.088        0.023         0.024         0.025         0.085
  DivisionW      PutOuts       Assists        Errors    NewLeagueN
     -6.215        0.016         0.003        -0.021         0.301
> sqrt(sum(coef(ridge.mod)[-1,50]^2))
[1] 6.36
```

In contrast, here are the coefficients when $\lambda = 705$, along with their ℓ_2 norm. Note the much larger ℓ_2 norm of the coefficients associated with this smaller value of λ.

```
> ridge.mod$lambda[60]
[1] 705
> coef(ridge.mod)[,60]
(Intercept)        AtBat          Hits         HmRun          Runs
     54.325        0.112         0.656         1.180         0.938
        RBI        Walks         Years        CAtBat         CHits
      0.847        1.320         2.596         0.011         0.047
     CHmRun        CRuns          CRBI        CWalks       LeagueN
      0.338        0.094         0.098         0.072        13.684
  DivisionW      PutOuts       Assists        Errors    NewLeagueN
    -54.659        0.119         0.016        -0.704         8.612
> sqrt(sum(coef(ridge.mod)[-1,60]^2))
[1] 57.1
```

We can use the predict() function for a number of purposes. For instance, we can obtain the ridge regression coefficients for a new value of λ, say 50:

```
> predict(ridge.mod,s=50,type="coefficients")[1:20,]
(Intercept)        AtBat          Hits         HmRun          Runs
     48.766       -0.358         1.969        -1.278         1.146
        RBI        Walks         Years        CAtBat         CHits
      0.804        2.716        -6.218         0.005         0.106
     CHmRun        CRuns          CRBI        CWalks       LeagueN
      0.624        0.221         0.219        -0.150        45.926
  DivisionW      PutOuts       Assists        Errors    NewLeagueN
   -118.201        0.250         0.122        -3.279        -9.497
```

We now split the samples into a training set and a test set in order to estimate the test error of ridge regression and the lasso. There are two common ways to randomly split a data set. The first is to produce a random vector of TRUE, FALSE elements and select the observations corresponding to TRUE for the training data. The second is to randomly choose a subset of numbers between 1 and n; these can then be used as the indices for the training observations. The two approaches work equally well. We used the former method in Section 6.5.3. Here we demonstrate the latter approach.

We first set a random seed so that the results obtained will be reproducible.

```
> set.seed(1)
> train=sample(1:nrow(x), nrow(x)/2)
> test=(-train)
> y.test=y[test]
```

Next we fit a ridge regression model on the training set, and evaluate its MSE on the test set, using $\lambda = 4$. Note the use of the predict() function again. This time we get predictions for a test set, by replacing type="coefficients" with the newx argument.

```
> ridge.mod=glmnet(x[train,],y[train],alpha=0,lambda=grid,
    thresh=1e-12)
> ridge.pred=predict(ridge.mod,s=4,newx=x[test,])
> mean((ridge.pred-y.test)^2)
[1] 101037
```

The test MSE is 101037. Note that if we had instead simply fit a model with just an intercept, we would have predicted each test observation using the mean of the training observations. In that case, we could compute the test set MSE like this:

```
> mean((mean(y[train])-y.test)^2)
[1] 193253
```

We could also get the same result by fitting a ridge regression model with a *very* large value of λ. Note that 1e10 means 10^{10}.

```
> ridge.pred=predict(ridge.mod,s=1e10,newx=x[test,])
> mean((ridge.pred-y.test)^2)
[1] 193253
```

So fitting a ridge regression model with $\lambda = 4$ leads to a much lower test MSE than fitting a model with just an intercept. We now check whether there is any benefit to performing ridge regression with $\lambda = 4$ instead of just performing least squares regression. Recall that least squares is simply ridge regression with $\lambda = 0$.[7]

---

[7] In order for glmnet() to yield the exact least squares coefficients when $\lambda = 0$, we use the argument exact=T when calling the predict() function. Otherwise, the predict() function will interpolate over the grid of λ values used in fitting the

```
> ridge.pred=predict(ridge.mod,s=0,newx=x[test,],exact=T)
> mean((ridge.pred-y.test)^2)
[1] 114783
> lm(y~x, subset=train)
> predict(ridge.mod,s=0,exact=T,type="coefficients")[1:20,]
```

In general, if we want to fit a (unpenalized) least squares model, then
we should use the lm() function, since that function provides more useful
outputs, such as standard errors and p-values for the coefficients.

In general, instead of arbitrarily choosing $\lambda = 4$, it would be better to
use cross-validation to choose the tuning parameter λ. We can do this using
the built-in cross-validation function, cv.glmnet(). By default, the function
performs ten-fold cross-validation, though this can be changed using the
argument nfolds. Note that we set a random seed first so our results will
be reproducible, since the choice of the cross-validation folds is random.

cv.glmnet()

```
> set.seed(1)
> cv.out=cv.glmnet(x[train,],y[train],alpha=0)
> plot(cv.out)
> bestlam=cv.out$lambda.min
> bestlam
[1] 212
```

Therefore, we see that the value of λ that results in the smallest cross-
validation error is 212. What is the test MSE associated with this value of
λ?

```
> ridge.pred=predict(ridge.mod,s=bestlam,newx=x[test,])
> mean((ridge.pred-y.test)^2)
[1] 96016
```

This represents a further improvement over the test MSE that we got using
$\lambda = 4$. Finally, we refit our ridge regression model on the full data set,
using the value of λ chosen by cross-validation, and examine the coefficient
estimates.

```
> out=glmnet(x,y,alpha=0)
> predict(out,type="coefficients",s=bestlam)[1:20,]
(Intercept)        AtBat         Hits        HmRun         Runs
    9.8849       0.0314       1.0088       0.1393       1.1132
       RBI        Walks        Years       CAtBat        CHits
    0.8732       1.8041       0.1307       0.0111       0.0649
     CHmRun        CRuns         CRBI       CWalks      LeagueN
    0.4516       0.1290       0.1374       0.0291      27.1823
  DivisionW       PutOuts      Assists       Errors   NewLeagueN
  -91.6341       0.1915       0.0425      -1.8124       7.2121
```

---

glmnet() model, yielding approximate results. When we use exact=T, there remains
a slight discrepancy in the third decimal place between the output of glmnet() when
$\lambda = 0$ and the output of lm(); this is due to numerical approximation on the part of
glmnet().

As expected, none of the coefficients are zero—ridge regression does not perform variable selection!

6.6.2   The Lasso

We saw that ridge regression with a wise choice of λ can outperform least squares as well as the null model on the Hitters data set. We now ask whether the lasso can yield either a more accurate or a more interpretable model than ridge regression. In order to fit a lasso model, we once again use the glmnet() function; however, this time we use the argument alpha=1. Other than that change, we proceed just as we did in fitting a ridge model.

```
> lasso.mod=glmnet(x[train,],y[train],alpha=1,lambda=grid)
> plot(lasso.mod)
```

We can see from the coefficient plot that depending on the choice of tuning parameter, some of the coefficients will be exactly equal to zero. We now perform cross-validation and compute the associated test error.

```
> set.seed(1)
> cv.out=cv.glmnet(x[train,],y[train],alpha=1)
> plot(cv.out)
> bestlam=cv.out$lambda.min
> lasso.pred=predict(lasso.mod,s=bestlam,newx=x[test,])
> mean((lasso.pred-y.test)^2)
[1] 100743
```

This is substantially lower than the test set MSE of the null model and of least squares, and very similar to the test MSE of ridge regression with λ chosen by cross-validation.

However, the lasso has a substantial advantage over ridge regression in that the resulting coefficient estimates are sparse. Here we see that 12 of the 19 coefficient estimates are exactly zero. So the lasso model with λ chosen by cross-validation contains only seven variables.

```
> out=glmnet(x,y,alpha=1,lambda=grid)
> lasso.coef=predict(out,type="coefficients",s=bestlam)[1:20,]
> lasso.coef
(Intercept)        AtBat          Hits        HmRun          Runs
     18.539        0.000         1.874        0.000         0.000
        RBI        Walks         Years       CAtBat         CHits
      0.000        2.218         0.000        0.000         0.000
      CHmRun        CRuns          CRBI       CWalks       LeagueN
       0.000        0.207         0.413        0.000         3.267
   DivisionW      PutOuts       Assists       Errors    NewLeagueN
    -103.485        0.220         0.000        0.000         0.000
> lasso.coef[lasso.coef!=0]
(Intercept)         Hits         Walks        CRuns          CRBI
     18.539        1.874         2.218        0.207         0.413
    LeagueN    DivisionW       PutOuts
      3.267     -103.485         0.220
```

6.7   Lab 3: PCR and PLS Regression

6.7.1   Principal Components Regression

Principal components regression (PCR) can be performed using the pcr()
function, which is part of the pls library. We now apply PCR to the Hitters
data, in order to predict Salary. Again, ensure that the missing values have
been removed from the data, as described in Section 6.5.

pcr()

```
> library(pls)
> set.seed(2)
> pcr.fit=pcr(Salary~., data=Hitters,scale=TRUE,
    validation="CV")
```

The syntax for the pcr() function is similar to that for lm(), with a few
additional options. Setting scale=TRUE has the effect of *standardizing* each
predictor, using (6.6), prior to generating the principal components, so that
the scale on which each variable is measured will not have an effect. Setting
validation="CV" causes pcr() to compute the ten-fold cross-validation error
for each possible value of M, the number of principal components used. The
resulting fit can be examined using summary().

```
> summary(pcr.fit)
Data:    X dimension: 263 19
     Y dimension: 263 1
Fit method: svdpc
Number of components considered: 19

VALIDATION: RMSEP
Cross-validated using 10 random segments.
          (Intercept)  1 comps  2 comps  3 comps  4 comps
CV             452       348.9    352.2    353.5    352.8
adjCV          452       348.7    351.8    352.9    352.1
...


TRAINING: % variance explained
        1 comps  2 comps  3 comps  4 comps  5 comps  6 comps
X         38.31    60.16    70.84    79.03    84.29    88.63
Salary    40.63    41.58    42.17    43.22    44.90    46.48
...
```

The CV score is provided for each possible number of components, ranging
from $M = 0$ onwards. (We have printed the CV output only up to $M = 4$.)
Note that pcr() reports the *root mean squared error*; in order to obtain
the usual MSE, we must square this quantity. For instance, a root mean
squared error of 352.8 corresponds to an MSE of $352.8^2 = 124{,}468$.

One can also plot the cross-validation scores using the validationplot()
function. Using val.type="MSEP" will cause the cross-validation MSE to be
plotted.

validation plot()

```
> validationplot(pcr.fit,val.type="MSEP")
```

We see that the smallest cross-validation error occurs when $M = 16$ components are used. This is barely fewer than $M = 19$, which amounts to simply performing least squares, because when all of the components are used in PCR no dimension reduction occurs. However, from the plot we also see that the cross-validation error is roughly the same when only one component is included in the model. This suggests that a model that uses just a small number of components might suffice.

The summary() function also provides the *percentage of variance explained* in the predictors and in the response using different numbers of components. This concept is discussed in greater detail in Chapter 10. Briefly, we can think of this as the amount of information about the predictors or the response that is captured using M principal components. For example, setting $M = 1$ only captures 38.31 % of all the variance, or information, in the predictors. In contrast, using $M = 6$ increases the value to 88.63 %. If we were to use all $M = p = 19$ components, this would increase to 100 %.

We now perform PCR on the training data and evaluate its test set performance.

```
> set.seed(1)
> pcr.fit=pcr(Salary~., data=Hitters,subset=train,scale=TRUE,
    validation="CV")
> validationplot(pcr.fit,val.type="MSEP")
```

Now we find that the lowest cross-validation error occurs when $M = 7$ component are used. We compute the test MSE as follows.

```
> pcr.pred=predict(pcr.fit,x[test,],ncomp=7)
> mean((pcr.pred-y.test)^2)
[1] 96556
```

This test set MSE is competitive with the results obtained using ridge regression and the lasso. However, as a result of the way PCR is implemented, the final model is more difficult to interpret because it does not perform any kind of variable selection or even directly produce coefficient estimates.

Finally, we fit PCR on the full data set, using $M = 7$, the number of components identified by cross-validation.

```
> pcr.fit=pcr(y~x,scale=TRUE,ncomp=7)
> summary(pcr.fit)
Data:    X dimension: 263 19
         Y dimension: 263 1
Fit method: svdpc
Number of components considered: 7
TRAINING: % variance explained
       1 comps  2 comps  3 comps  4 comps  5 comps  6 comps
X        38.31    60.16    70.84    79.03    84.29    88.63
y        40.63    41.58    42.17    43.22    44.90    46.48
       7 comps
X        92.26
y        46.69
```

6.7.2 Partial Least Squares

We implement partial least squares (PLS) using the plsr() function, also
in the pls library. The syntax is just like that of the pcr() function.

plsr()

```
> set.seed(1)
> pls.fit=plsr(Salary~., data=Hitters,subset=train,scale=TRUE,
    validation="CV")
> summary(pls.fit)
Data:   X dimension: 131 19
        Y dimension: 131 1
Fit method: kernelpls
Number of components considered: 19

VALIDATION: RMSEP
Cross-validated using 10 random segments.
        (Intercept)  1 comps  2 comps  3 comps  4 comps
CV           464.6     394.2    391.5    393.1    395.0
adjCV        464.6     393.4    390.2    391.1    392.9
...


TRAINING: % variance explained
          1 comps  2 comps  3 comps  4 comps  5 comps  6 comps
X           38.12    53.46    66.05    74.49    79.33    84.56
Salary      33.58    38.96    41.57    42.43    44.04    45.59
...
> validationplot(pls.fit,val.type="MSEP")
```

The lowest cross-validation error occurs when only $M = 2$ partial least
squares directions are used. We now evaluate the corresponding test set
MSE.

```
> pls.pred=predict(pls.fit,x[test,],ncomp=2)
> mean((pls.pred-y.test)^2)
[1] 101417
```

The test MSE is comparable to, but slightly higher than, the test MSE
obtained using ridge regression, the lasso, and PCR.

  Finally, we perform PLS using the full data set, using $M = 2$, the number
of components identified by cross-validation.

```
> pls.fit=plsr(Salary~., data=Hitters,scale=TRUE,ncomp=2)
> summary(pls.fit)
Data:   X dimension: 263 19
        Y dimension: 263 1
Fit method: kernelpls
Number of components considered: 2
TRAINING: % variance explained
          1 comps  2 comps
X           38.08    51.03
Salary      43.05    46.40
```

Notice that the percentage of variance in Salary that the two-component
PLS fit explains, 46.40 %, is almost as much as that explained using the

final seven-component model PCR fit, 46.69 %. This is because PCR only attempts to maximize the amount of variance explained in the predictors, while PLS searches for directions that explain variance in both the predictors and the response.

6.8 Exercises

Conceptual

1. We perform best subset, forward stepwise, and backward stepwise selection on a single data set. For each approach, we obtain $p + 1$ models, containing $0, 1, 2, \ldots, p$ predictors. Explain your answers:

    (a) Which of the three models with k predictors has the smallest *training* RSS?

    (b) Which of the three models with k predictors has the smallest *test* RSS?

    (c) True or False:

        i. The predictors in the k-variable model identified by forward stepwise are a subset of the predictors in the $(k+1)$-variable model identified by forward stepwise selection.

        ii. The predictors in the k-variable model identified by backward stepwise are a subset of the predictors in the $(k + 1)$-variable model identified by backward stepwise selection.

        iii. The predictors in the k-variable model identified by backward stepwise are a subset of the predictors in the $(k + 1)$-variable model identified by forward stepwise selection.

        iv. The predictors in the k-variable model identified by forward stepwise are a subset of the predictors in the $(k+1)$-variable model identified by backward stepwise selection.

        v. The predictors in the k-variable model identified by best subset are a subset of the predictors in the $(k + 1)$-variable model identified by best subset selection.

2. For parts (a) through (c), indicate which of i. through iv. is correct. Justify your answer.

    (a) The lasso, relative to least squares, is:

        i. More flexible and hence will give improved prediction accuracy when its increase in bias is less than its decrease in variance.

        ii. More flexible and hence will give improved prediction accuracy when its increase in variance is less than its decrease in bias.

iii. Less flexible and hence will give improved prediction accuracy when its increase in bias is less than its decrease in
variance.

iv. Less flexible and hence will give improved prediction accuracy when its increase in variance is less than its decrease
in bias.

(b) Repeat (a) for ridge regression relative to least squares.

(c) Repeat (a) for non-linear methods relative to least squares.

3. Suppose we estimate the regression coefficients in a linear regression
model by minimizing

$$\sum_{i=1}^{n} \left(y_i - \beta_0 - \sum_{j=1}^{p} \beta_j x_{ij} \right)^2 \quad \text{subject to} \quad \sum_{j=1}^{p} |\beta_j| \leq s$$

for a particular value of s. For parts (a) through (e), indicate which
of i. through v. is correct. Justify your answer.

(a) As we increase s from 0, the training RSS will:

i. Increase initially, and then eventually start decreasing in an
inverted U shape.

ii. Decrease initially, and then eventually start increasing in a
U shape.

iii. Steadily increase.

iv. Steadily decrease.

v. Remain constant.

(b) Repeat (a) for test RSS.

(c) Repeat (a) for variance.

(d) Repeat (a) for (squared) bias.

(e) Repeat (a) for the irreducible error.

4. Suppose we estimate the regression coefficients in a linear regression
model by minimizing

$$\sum_{i=1}^{n} \left(y_i - \beta_0 - \sum_{j=1}^{p} \beta_j x_{ij} \right)^2 + \lambda \sum_{j=1}^{p} \beta_j^2$$

for a particular value of λ. For parts (a) through (e), indicate which
of i. through v. is correct. Justify your answer.

(a) As we increase λ from 0, the training RSS will:

  i. Increase initially, and then eventually start decreasing in an inverted U shape.

  ii. Decrease initially, and then eventually start increasing in a U shape.

  iii. Steadily increase.

  iv. Steadily decrease.

  v. Remain constant.

(b) Repeat (a) for test RSS.

(c) Repeat (a) for variance.

(d) Repeat (a) for (squared) bias.

(e) Repeat (a) for the irreducible error.

5. It is well-known that ridge regression tends to give similar coefficient values to correlated variables, whereas the lasso may give quite different coefficient values to correlated variables. We will now explore this property in a very simple setting.

  Suppose that $n = 2$, $p = 2$, $x_{11} = x_{12}$, $x_{21} = x_{22}$. Furthermore, suppose that $y_1 + y_2 = 0$ and $x_{11} + x_{21} = 0$ and $x_{12} + x_{22} = 0$, so that the estimate for the intercept in a least squares, ridge regression, or lasso model is zero: $\hat{\beta}_0 = 0$.

(a) Write out the ridge regression optimization problem in this setting.

(b) Argue that in this setting, the ridge coefficient estimates satisfy $\hat{\beta}_1 = \hat{\beta}_2$.

(c) Write out the lasso optimization problem in this setting.

(d) Argue that in this setting, the lasso coefficients $\hat{\beta}_1$ and $\hat{\beta}_2$ are not unique—in other words, there are many possible solutions to the optimization problem in (c). Describe these solutions.

6. We will now explore (6.12) and (6.13) further.

(a) Consider (6.12) with $p = 1$. For some choice of y_1 and $\lambda > 0$, plot (6.12) as a function of β_1. Your plot should confirm that (6.12) is solved by (6.14).

(b) Consider (6.13) with $p = 1$. For some choice of y_1 and $\lambda > 0$, plot (6.13) as a function of β_1. Your plot should confirm that (6.13) is solved by (6.15).

7. We will now derive the Bayesian connection to the lasso and ridge regression discussed in Section 6.2.2.

  (a) Suppose that $y_i = \beta_0 + \sum_{j=1}^{p} x_{ij}\beta_j + \epsilon_i$ where $\epsilon_1, \ldots, \epsilon_n$ are independent and identically distributed from a $N(0, \sigma^2)$ distribution. Write out the likelihood for the data.

  (b) Assume the following prior for β: $\beta_1, \ldots, \beta_p$ are independent and identically distributed according to a double-exponential distribution with mean 0 and common scale parameter b: i.e. $p(\beta) = \frac{1}{2b}\exp(-|\beta|/b)$. Write out the posterior for β in this setting.

  (c) Argue that the lasso estimate is the *mode* for β under this posterior distribution.

  (d) Now assume the following prior for β: $\beta_1, \ldots, \beta_p$ are independent and identically distributed according to a normal distribution with mean zero and variance c. Write out the posterior for β in this setting.

  (e) Argue that the ridge regression estimate is both the *mode* and the *mean* for β under this posterior distribution.

Applied

8. In this exercise, we will generate simulated data, and will then use this data to perform best subset selection.

  (a) Use the `rnorm()` function to generate a predictor X of length $n = 100$, as well as a noise vector ϵ of length $n = 100$.

  (b) Generate a response vector Y of length $n = 100$ according to the model

$$Y = \beta_0 + \beta_1 X + \beta_2 X^2 + \beta_3 X^3 + \epsilon,$$

  where β_0, β_1, β_2, and β_3 are constants of your choice.

  (c) Use the `regsubsets()` function to perform best subset selection in order to choose the best model containing the predictors $X, X^2, \ldots, X^{10}$. What is the best model obtained according to C_p, BIC, and adjusted R^2? Show some plots to provide evidence for your answer, and report the coefficients of the best model obtained. Note you will need to use the `data.frame()` function to create a single data set containing both X and Y.

(d) Repeat (c), using forward stepwise selection and also using backwards stepwise selection. How does your answer compare to the results in (c)?

(e) Now fit a lasso model to the simulated data, again using $X, X^2, \ldots, X^{10}$ as predictors. Use cross-validation to select the optimal value of λ. Create plots of the cross-validation error as a function of λ. Report the resulting coefficient estimates, and discuss the results obtained.

(f) Now generate a response vector Y according to the model

$$Y = \beta_0 + \beta_7 X^7 + \epsilon,$$

and perform best subset selection and the lasso. Discuss the results obtained.

9. In this exercise, we will predict the number of applications received using the other variables in the College data set.

(a) Split the data set into a training set and a test set.

(b) Fit a linear model using least squares on the training set, and report the test error obtained.

(c) Fit a ridge regression model on the training set, with λ chosen by cross-validation. Report the test error obtained.

(d) Fit a lasso model on the training set, with λ chosen by cross-validation. Report the test error obtained, along with the number of non-zero coefficient estimates.

(e) Fit a PCR model on the training set, with M chosen by cross-validation. Report the test error obtained, along with the value of M selected by cross-validation.

(f) Fit a PLS model on the training set, with M chosen by cross-validation. Report the test error obtained, along with the value of M selected by cross-validation.

(g) Comment on the results obtained. How accurately can we predict the number of college applications received? Is there much difference among the test errors resulting from these five approaches?

10. We have seen that as the number of features used in a model increases, the training error will necessarily decrease, but the test error may not. We will now explore this in a simulated data set.

(a) Generate a data set with $p = 20$ features, $n = 1,000$ observations, and an associated quantitative response vector generated according to the model

$$Y = X\beta + \epsilon,$$

where β has some elements that are exactly equal to zero.

(b) Split your data set into a training set containing 100 observations and a test set containing 900 observations.

(c) Perform best subset selection on the training set, and plot the training set MSE associated with the best model of each size.

(d) Plot the test set MSE associated with the best model of each size.

(e) For which model size does the test set MSE take on its minimum value? Comment on your results. If it takes on its minimum value for a model containing only an intercept or a model containing all of the features, then play around with the way that you are generating the data in (a) until you come up with a scenario in which the test set MSE is minimized for an intermediate model size.

(f) How does the model at which the test set MSE is minimized compare to the true model used to generate the data? Comment on the coefficient values.

(g) Create a plot displaying $\sqrt{\sum_{j=1}^{p}(\beta_j - \hat{\beta}_j^r)^2}$ for a range of values of r, where $\hat{\beta}_j^r$ is the jth coefficient estimate for the best model containing r coefficients. Comment on what you observe. How does this compare to the test MSE plot from (d)?

11. We will now try to predict per capita crime rate in the `Boston` data set.

(a) Try out some of the regression methods explored in this chapter, such as best subset selection, the lasso, ridge regression, and PCR. Present and discuss results for the approaches that you consider.

(b) Propose a model (or set of models) that seem to perform well on this data set, and justify your answer. Make sure that you are evaluating model performance using validation set error, cross-validation, or some other reasonable alternative, as opposed to using training error.

(c) Does your chosen model involve all of the features in the data set? Why or why not?

7

Moving Beyond Linearity

So far in this book, we have mostly focused on linear models. Linear models are relatively simple to describe and implement, and have advantages over other approaches in terms of interpretation and inference. However, standard linear regression can have significant limitations in terms of predictive power. This is because the linearity assumption is almost always an approximation, and sometimes a poor one. In Chapter 6 we see that we can improve upon least squares using ridge regression, the lasso, principal components regression, and other techniques. In that setting, the improvement is obtained by reducing the complexity of the linear model, and hence the variance of the estimates. But we are still using a linear model, which can only be improved so far! In this chapter we relax the linearity assumption while still attempting to maintain as much interpretability as possible. We do this by examining very simple extensions of linear models like polynomial regression and step functions, as well as more sophisticated approaches such as splines, local regression, and generalized additive models.

- *Polynomial regression* extends the linear model by adding extra predictors, obtained by raising each of the original predictors to a power. For example, a *cubic* regression uses three variables, X, X^2, and X^3, as predictors. This approach provides a simple way to provide a non-linear fit to data.

- *Step functions* cut the range of a variable into K distinct regions in order to produce a qualitative variable. This has the effect of fitting a piecewise constant function.

G. James et al., *An Introduction to Statistical Learning: with Applications in R*, 265
Springer Texts in Statistics, DOI 10.1007/978-1-4614-7138-7_7,
© Springer Science+Business Media New York 2013

- *Regression splines* are more flexible than polynomials and step functions, and in fact are an extension of the two. They involve dividing the range of X into K distinct regions. Within each region, a polynomial function is fit to the data. However, these polynomials are constrained so that they join smoothly at the region boundaries, or *knots*. Provided that the interval is divided into enough regions, this can produce an extremely flexible fit.

- *Smoothing splines* are similar to regression splines, but arise in a slightly different situation. Smoothing splines result from minimizing a residual sum of squares criterion subject to a smoothness penalty.

- *Local regression* is similar to splines, but differs in an important way. The regions are allowed to overlap, and indeed they do so in a very smooth way.

- *Generalized additive models* allow us to extend the methods above to deal with multiple predictors.

In Sections 7.1–7.6, we present a number of approaches for modeling the relationship between a response Y and a single predictor X in a flexible way. In Section 7.7, we show that these approaches can be seamlessly integrated in order to model a response Y as a function of several predictors $X_1, \ldots, X_p$.

7.1 Polynomial Regression

Historically, the standard way to extend linear regression to settings in which the relationship between the predictors and the response is non-linear has been to replace the standard linear model

$$y_i = \beta_0 + \beta_1 x_i + \epsilon_i$$

with a polynomial function

$$y_i = \beta_0 + \beta_1 x_i + \beta_2 x_i^2 + \beta_3 x_i^3 + \ldots + \beta_d x_i^d + \epsilon_i, \qquad (7.1)$$

where ϵ_i is the error term. This approach is known as *polynomial regression*, and in fact we saw an example of this method in Section 3.3.2. For large enough degree d, a polynomial regression allows us to produce an extremely non-linear curve. Notice that the coefficients in (7.1) can be easily estimated using least squares linear regression because this is just a standard linear model with predictors $x_i, x_i^2, x_i^3, \ldots, x_i^d$. Generally speaking, it is unusual to use d greater than 3 or 4 because for large values of d, the polynomial curve can become overly flexible and can take on some very strange shapes. This is especially true near the boundary of the X variable.

Degree-4 Polynomial

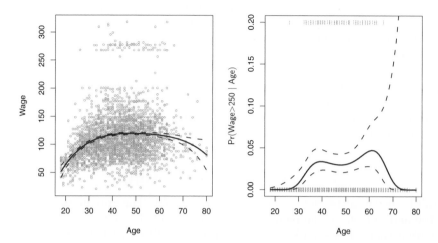

FIGURE 7.1. *The* Wage *data.* Left: *The solid blue curve is a degree-4 polynomial of* wage *(in thousands of dollars) as a function of* age, *fit by least squares. The dotted curves indicate an estimated 95 % confidence interval.* Right: *We model the binary event* wage>250 *using logistic regression, again with a degree-4 polynomial. The fitted posterior probability of* wage *exceeding $250,000 is shown in blue, along with an estimated 95 % confidence interval.*

The left-hand panel in Figure 7.1 is a plot of wage against age for the Wage data set, which contains income and demographic information for males who reside in the central Atlantic region of the United States. We see the results of fitting a degree-4 polynomial using least squares (solid blue curve). Even though this is a linear regression model like any other, the individual coefficients are not of particular interest. Instead, we look at the entire fitted function across a grid of 62 values for age from 18 to 80 in order to understand the relationship between age and wage.

In Figure 7.1, a pair of dotted curves accompanies the fit; these are (2×) standard error curves. Let's see how these arise. Suppose we have computed the fit at a particular value of age, x_0:

$$\hat{f}(x_0) = \hat{\beta}_0 + \hat{\beta}_1 x_0 + \hat{\beta}_2 x_0^2 + \hat{\beta}_3 x_0^3 + \hat{\beta}_4 x_0^4. \tag{7.2}$$

What is the variance of the fit, i.e. $\operatorname{Var} \hat{f}(x_0)$? Least squares returns variance estimates for each of the fitted coefficients $\hat{\beta}_j$, as well as the covariances between pairs of coefficient estimates. We can use these to compute the estimated variance of $\hat{f}(x_0)$.[1] The estimated *pointwise* standard error of $\hat{f}(x_0)$ is the square-root of this variance. This computation is repeated

---

[1]If $\hat{\mathbf{C}}$ is the 5×5 covariance matrix of the $\hat{\beta}_j$, and if $\ell_0^T = (1, x_0, x_0^2, x_0^3, x_0^4)$, then $\operatorname{Var}[\hat{f}(x_0)] = \ell_0^T \hat{\mathbf{C}} \ell_0$.

at each reference point x_0, and we plot the fitted curve, as well as twice the standard error on either side of the fitted curve. We plot twice the standard error because, for normally distributed error terms, this quantity corresponds to an approximate 95 % confidence interval.

It seems like the wages in Figure 7.1 are from two distinct populations: there appears to be a *high earners* group earning more than $250,000 per annum, as well as a *low earners* group. We can treat wage as a binary variable by splitting it into these two groups. Logistic regression can then be used to predict this binary response, using polynomial functions of age as predictors. In other words, we fit the model

$$\Pr(y_i > 250 | x_i) = \frac{\exp(\beta_0 + \beta_1 x_i + \beta_2 x_i^2 + \ldots + \beta_d x_i^d)}{1 + \exp(\beta_0 + \beta_1 x_i + \beta_2 x_i^2 + \ldots + \beta_d x_i^d)}. \tag{7.3}$$

The result is shown in the right-hand panel of Figure 7.1. The gray marks on the top and bottom of the panel indicate the ages of the high earners and the low earners. The solid blue curve indicates the fitted probabilities of being a high earner, as a function of age. The estimated 95 % confidence interval is shown as well. We see that here the confidence intervals are fairly wide, especially on the right-hand side. Although the sample size for this data set is substantial ($n = 3{,}000$), there are only 79 high earners, which results in a high variance in the estimated coefficients and consequently wide confidence intervals.

7.2 Step Functions

Using polynomial functions of the features as predictors in a linear model imposes a *global* structure on the non-linear function of X. We can instead use *step functions* in order to avoid imposing such a global structure. Here we break the range of X into *bins*, and fit a different constant in each bin. This amounts to converting a continuous variable into an *ordered categorical variable*.

step function

ordered categorical variable

In greater detail, we create cutpoints $c_1, c_2, \ldots, c_K$ in the range of X, and then construct $K + 1$ new variables

$$\begin{aligned}
C_0(X) &= I(X < c_1), \\
C_1(X) &= I(c_1 \leq X < c_2), \\
C_2(X) &= I(c_2 \leq X < c_3), \\
&\;\;\vdots \\
C_{K-1}(X) &= I(c_{K-1} \leq X < c_K), \\
C_K(X) &= I(c_K \leq X),
\end{aligned} \tag{7.4}$$

where $I(\cdot)$ is an *indicator function* that returns a 1 if the condition is true, and returns a 0 otherwise. For example, $I(c_K \leq X)$ equals 1 if $c_K \leq X$, and

indicator function

Piecewise Constant

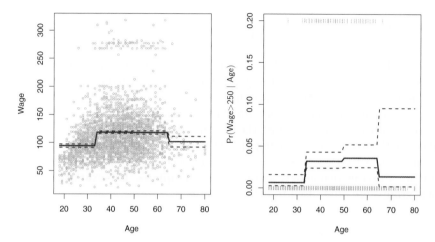

FIGURE 7.2. *The* Wage *data. Left: The solid curve displays the fitted value from a least squares regression of* wage *(in thousands of dollars) using step functions of* age. *The dotted curves indicate an estimated 95 % confidence interval. Right: We model the binary event* wage>250 *using logistic regression, again using step functions of* age. *The fitted posterior probability of* wage *exceeding $250,000 is shown, along with an estimated 95 % confidence interval.*

equals 0 otherwise. These are sometimes called *dummy* variables. Notice that for any value of X, $C_0(X) + C_1(X) + \ldots + C_K(X) = 1$, since X must be in exactly one of the $K + 1$ intervals. We then use least squares to fit a linear model using $C_1(X), C_2(X), \ldots, C_K(X)$ as predictors[2]:

$$y_i = \beta_0 + \beta_1 C_1(x_i) + \beta_2 C_2(x_i) + \ldots + \beta_K C_K(x_i) + \epsilon_i. \tag{7.5}$$

For a given value of X, at most one of $C_1, C_2, \ldots, C_K$ can be non-zero. Note that when $X < c_1$, all of the predictors in (7.5) are zero, so β_0 can be interpreted as the mean value of Y for $X < c_1$. By comparison, (7.5) predicts a response of $\beta_0 + \beta_j$ for $c_j \leq X < c_{j+1}$, so β_j represents the average increase in the response for X in $c_j \leq X < c_{j+1}$ relative to $X < c_1$.

An example of fitting step functions to the Wage data from Figure 7.1 is shown in the left-hand panel of Figure 7.2. We also fit the logistic regression model

---

[2]We exclude $C_0(X)$ as a predictor in (7.5) because it is redundant with the intercept. This is similar to the fact that we need only two dummy variables to code a qualitative variable with three levels, provided that the model will contain an intercept. The decision to exclude $C_0(X)$ instead of some other $C_k(X)$ in (7.5) is arbitrary. Alternatively, we could include $C_0(X), C_1(X), \ldots, C_K(X)$, and exclude the intercept.

$$\Pr(y_i > 250 | x_i) = \frac{\exp(\beta_0 + \beta_1 C_1(x_i) + \ldots + \beta_K C_K(x_i))}{1 + \exp(\beta_0 + \beta_1 C_1(x_i) + \ldots + \beta_K C_K(x_i))} \qquad (7.6)$$

in order to predict the probability that an individual is a high earner on the basis of age. The right-hand panel of Figure 7.2 displays the fitted posterior probabilities obtained using this approach.

Unfortunately, unless there are natural breakpoints in the predictors, piecewise-constant functions can miss the action. For example, in the left-hand panel of Figure 7.2, the first bin clearly misses the increasing trend of wage with age. Nevertheless, step function approaches are very popular in biostatistics and epidemiology, among other disciplines. For example, 5-year age groups are often used to define the bins.

7.3   Basis Functions

Polynomial and piecewise-constant regression models are in fact special cases of a *basis function* approach. The idea is to have at hand a family of functions or transformations that can be applied to a variable X: $b_1(X), b_2(X), \ldots, b_K(X)$. Instead of fitting a linear model in X, we fit the model

$$y_i = \beta_0 + \beta_1 b_1(x_i) + \beta_2 b_2(x_i) + \beta_3 b_3(x_i) + \ldots + \beta_K b_K(x_i) + \epsilon_i. \qquad (7.7)$$

Note that the basis functions $b_1(\cdot), b_2(\cdot), \ldots, b_K(\cdot)$ are fixed and known. (In other words, we choose the functions ahead of time.) For polynomial regression, the basis functions are $b_j(x_i) = x_i^j$, and for piecewise constant functions they are $b_j(x_i) = I(c_j \le x_i < c_{j+1})$. We can think of (7.7) as a standard linear model with predictors $b_1(x_i), b_2(x_i), \ldots, b_K(x_i)$. Hence, we can use least squares to estimate the unknown regression coefficients in (7.7). Importantly, this means that all of the inference tools for linear models that are discussed in Chapter 3, such as standard errors for the coefficient estimates and F-statistics for the model's overall significance, are available in this setting.

Thus far we have considered the use of polynomial functions and piecewise constant functions for our basis functions; however, many alternatives are possible. For instance, we can use wavelets or Fourier series to construct basis functions. In the next section, we investigate a very common choice for a basis function: *regression splines*.

basis function

regression spline

7.4   Regression Splines

Now we discuss a flexible class of basis functions that extends upon the polynomial regression and piecewise constant regression approaches that we have just seen.

7.4.1   Piecewise Polynomials

Instead of fitting a high-degree polynomial over the entire range of X, *piecewise polynomial regression* involves fitting separate low-degree polynomials over different regions of X. For example, a piecewise cubic polynomial works by fitting a cubic regression model of the form

$$y_i = \beta_0 + \beta_1 x_i + \beta_2 x_i^2 + \beta_3 x_i^3 + \epsilon_i, \tag{7.8}$$

piecewise polynomial regression

where the coefficients β_0, β_1, β_2, and β_3 differ in different parts of the range of X. The points where the coefficients change are called *knots*.

For example, a piecewise cubic with no knots is just a standard cubic polynomial, as in (7.1) with $d = 3$. A piecewise cubic polynomial with a single knot at a point c takes the form

knot

$$y_i = \begin{cases} \beta_{01} + \beta_{11} x_i + \beta_{21} x_i^2 + \beta_{31} x_i^3 + \epsilon_i & \text{if } x_i < c; \\ \beta_{02} + \beta_{12} x_i + \beta_{22} x_i^2 + \beta_{32} x_i^3 + \epsilon_i & \text{if } x_i \geq c. \end{cases}$$

In other words, we fit two different polynomial functions to the data, one on the subset of the observations with $x_i < c$, and one on the subset of the observations with $x_i \geq c$. The first polynomial function has coefficients $\beta_{01}, \beta_{11}, \beta_{21}, \beta_{31}$, and the second has coefficients $\beta_{02}, \beta_{12}, \beta_{22}, \beta_{32}$. Each of these polynomial functions can be fit using least squares applied to simple functions of the original predictor.

Using more knots leads to a more flexible piecewise polynomial. In general, if we place K different knots throughout the range of X, then we will end up fitting $K + 1$ different cubic polynomials. Note that we do not need to use a cubic polynomial. For example, we can instead fit piecewise linear functions. In fact, our piecewise constant functions of Section 7.2 are piecewise polynomials of degree 0!

The top left panel of Figure 7.3 shows a piecewise cubic polynomial fit to a subset of the Wage data, with a single knot at age=50. We immediately see a problem: the function is discontinuous and looks ridiculous! Since each polynomial has four parameters, we are using a total of eight *degrees of freedom* in fitting this piecewise polynomial model.

degrees of freedom

7.4.2   Constraints and Splines

The top left panel of Figure 7.3 looks wrong because the fitted curve is just too flexible. To remedy this problem, we can fit a piecewise polynomial

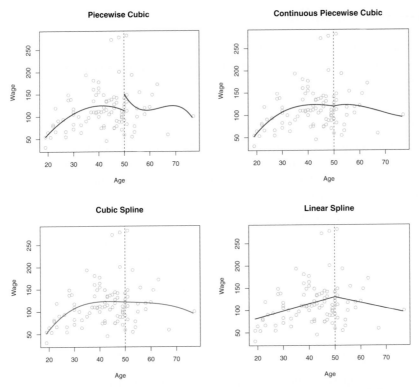

FIGURE 7.3. *Various piecewise polynomials are fit to a subset of the* Wage *data, with a knot at* age=50. *Top Left: The cubic polynomials are unconstrained. Top Right: The cubic polynomials are constrained to be continuous at* age=50. *Bottom Left: The cubic polynomials are constrained to be continuous, and to have continuous first and second derivatives. Bottom Right: A linear spline is shown, which is constrained to be continuous.*

under the *constraint* that the fitted curve must be continuous. In other words, there cannot be a jump when age=50. The top right plot in Figure 7.3 shows the resulting fit. This looks better than the top left plot, but the V-shaped join looks unnatural.

In the lower left plot, we have added two additional constraints: now both the first and second *derivatives* of the piecewise polynomials are continuous at age=50. In other words, we are requiring that the piecewise polynomial be not only continuous when age=50, but also very *smooth*. Each constraint that we impose on the piecewise cubic polynomials effectively frees up one degree of freedom, by reducing the complexity of the resulting piecewise polynomial fit. So in the top left plot, we are using eight degrees of freedom, but in the bottom left plot we imposed three constraints (continuity, continuity of the first derivative, and continuity of the second derivative) and so are left with five degrees of freedom. The curve in the bottom left

derivative

plot is called a *cubic spline*.[3] In general, a cubic spline with K knots uses a total of $4 + K$ degrees of freedom.

cubic spline

In Figure 7.3, the lower right plot is a *linear spline*, which is continuous at age=50. The general definition of a degree-d spline is that it is a piecewise degree-d polynomial, with continuity in derivatives up to degree $d - 1$ at each knot. Therefore, a linear spline is obtained by fitting a line in each region of the predictor space defined by the knots, requiring continuity at each knot.

linear spline

In Figure 7.3, there is a single knot at age=50. Of course, we could add more knots, and impose continuity at each.

7.4.3   The Spline Basis Representation

The regression splines that we just saw in the previous section may have seemed somewhat complex: how can we fit a piecewise degree-d polynomial under the constraint that it (and possibly its first $d - 1$ derivatives) be continuous? It turns out that we can use the basis model (7.7) to represent a regression spline. A cubic spline with K knots can be modeled as

$$y_i = \beta_0 + \beta_1 b_1(x_i) + \beta_2 b_2(x_i) + \cdots + \beta_{K+3} b_{K+3}(x_i) + \epsilon_i, \qquad (7.9)$$

for an appropriate choice of basis functions $b_1, b_2, \ldots, b_{K+3}$. The model (7.9) can then be fit using least squares.

Just as there were several ways to represent polynomials, there are also many equivalent ways to represent cubic splines using different choices of basis functions in (7.9). The most direct way to represent a cubic spline using (7.9) is to start off with a basis for a cubic polynomial—namely, x, x^2, x^3—and then add one *truncated power basis* function per knot. A truncated power basis function is defined as

truncated power basis

$$h(x, \xi) = (x - \xi)_+^3 = \begin{cases} (x - \xi)^3 & \text{if } x > \xi \\ 0 & \text{otherwise,} \end{cases} \qquad (7.10)$$

where ξ is the knot. One can show that adding a term of the form $\beta_4 h(x, \xi)$ to the model (7.8) for a cubic polynomial will lead to a discontinuity in only the third derivative at ξ; the function will remain continuous, with continuous first and second derivatives, at each of the knots.

In other words, in order to fit a cubic spline to a data set with K knots, we perform least squares regression with an intercept and $3 + K$ predictors, of the form $X, X^2, X^3, h(X, \xi_1), h(X, \xi_2), \ldots, h(X, \xi_K)$, where $\xi_1, \ldots, \xi_K$ are the knots. This amounts to estimating a total of $K + 4$ regression coefficients; for this reason, fitting a cubic spline with K knots uses $K+4$ degrees of freedom.

---

[3] Cubic splines are popular because most human eyes cannot detect the discontinuity at the knots.

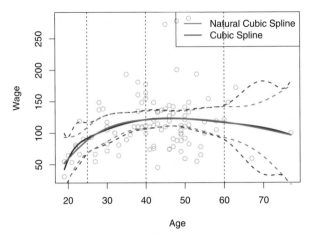

FIGURE 7.4. *A cubic spline and a natural cubic spline, with three knots, fit to a subset of the* Wage *data.*

Unfortunately, splines can have high variance at the outer range of the predictors—that is, when X takes on either a very small or very large value. Figure 7.4 shows a fit to the Wage data with three knots. We see that the confidence bands in the boundary region appear fairly wild. A *natural spline* is a regression spline with additional *boundary constraints*: the function is required to be linear at the boundary (in the region where X is smaller than the smallest knot, or larger than the largest knot). This additional constraint means that natural splines generally produce more stable estimates at the boundaries. In Figure 7.4, a natural cubic spline is also displayed as a red line. Note that the corresponding confidence intervals are narrower.

natural
spline

7.4.4   Choosing the Number and Locations of the Knots

When we fit a spline, where should we place the knots? The regression spline is most flexible in regions that contain a lot of knots, because in those regions the polynomial coefficients can change rapidly. Hence, one option is to place more knots in places where we feel the function might vary most rapidly, and to place fewer knots where it seems more stable. While this option can work well, in practice it is common to place knots in a uniform fashion. One way to do this is to specify the desired degrees of freedom, and then have the software automatically place the corresponding number of knots at uniform quantiles of the data.

Figure 7.5 shows an example on the Wage data. As in Figure 7.4, we have fit a natural cubic spline with three knots, except this time the knot locations were chosen automatically as the 25th, 50th, and 75th percentiles

Natural Cubic Spline

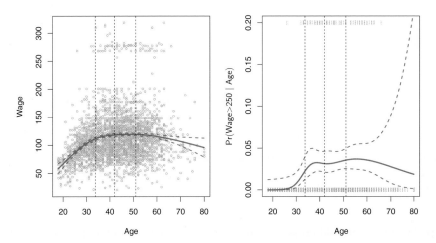

FIGURE 7.5. *A natural cubic spline function with four degrees of freedom is fit to the* Wage *data. Left: A spline is fit to* wage *(in thousands of dollars) as a function of* age. *Right: Logistic regression is used to model the binary event* wage>250 *as a function of* age. *The fitted posterior probability of* wage *exceeding* $250,000 *is shown.*

of age. This was specified by requesting four degrees of freedom. The argument by which four degrees of freedom leads to three interior knots is somewhat technical.[4]

How many knots should we use, or equivalently how many degrees of freedom should our spline contain? One option is to try out different numbers of knots and see which produces the best looking curve. A somewhat more objective approach is to use cross-validation, as discussed in Chapters 5 and 6. With this method, we remove a portion of the data (say 10 %), fit a spline with a certain number of knots to the remaining data, and then use the spline to make predictions for the held-out portion. We repeat this process multiple times until each observation has been left out once, and then compute the overall cross-validated RSS. This procedure can be repeated for different numbers of knots K. Then the value of K giving the smallest RSS is chosen.

---

[4]There are actually five knots, including the two boundary knots. A cubic spline with five knots would have nine degrees of freedom. But natural cubic splines have two additional *natural* constraints at each boundary to enforce linearity, resulting in $9-4=5$ degrees of freedom. Since this includes a constant, which is absorbed in the intercept, we count it as four degrees of freedom.

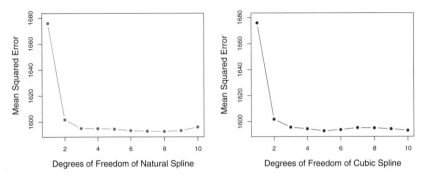

FIGURE 7.6. *Ten-fold cross-validated mean squared errors for selecting the degrees of freedom when fitting splines to the* Wage *data. The response is* wage *and the predictor* age. *Left: A natural cubic spline. Right: A cubic spline.*

Figure 7.6 shows ten-fold cross-validated mean squared errors for splines with various degrees of freedom fit to the Wage data. The left-hand panel corresponds to a natural spline and the right-hand panel to a cubic spline. The two methods produce almost identical results, with clear evidence that a one-degree fit (a linear regression) is not adequate. Both curves flatten out quickly, and it seems that three degrees of freedom for the natural spline and four degrees of freedom for the cubic spline are quite adequate.

In Section 7.7 we fit additive spline models simultaneously on several variables at a time. This could potentially require the selection of degrees of freedom for each variable. In cases like this we typically adopt a more pragmatic approach and set the degrees of freedom to a fixed number, say four, for all terms.

7.4.5 Comparison to Polynomial Regression

Regression splines often give superior results to polynomial regression. This is because unlike polynomials, which must use a high degree (exponent in the highest monomial term, e.g. X^{15}) to produce flexible fits, splines introduce flexibility by increasing the number of knots but keeping the degree fixed. Generally, this approach produces more stable estimates. Splines also allow us to place more knots, and hence flexibility, over regions where the function f seems to be changing rapidly, and fewer knots where f appears more stable. Figure 7.7 compares a natural cubic spline with 15 degrees of freedom to a degree-15 polynomial on the Wage data set. The extra flexibility in the polynomial produces undesirable results at the boundaries, while the natural cubic spline still provides a reasonable fit to the data.

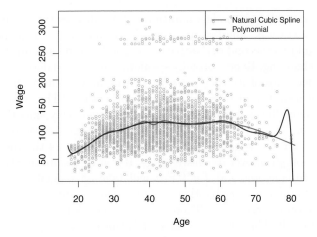

FIGURE 7.7. *On the* Wage *data set, a natural cubic spline with 15 degrees of freedom is compared to a degree-*15 *polynomial. Polynomials can show wild behavior, especially near the tails.*

7.5 Smoothing Splines

7.5.1 An Overview of Smoothing Splines

In the last section we discussed regression splines, which we create by specifying a set of knots, producing a sequence of basis functions, and then using least squares to estimate the spline coefficients. We now introduce a somewhat different approach that also produces a spline.

In fitting a smooth curve to a set of data, what we really want to do is find some function, say $g(x)$, that fits the observed data well: that is, we want RSS $= \sum_{i=1}^{n}(y_i - g(x_i))^2$ to be small. However, there is a problem with this approach. If we don't put any constraints on $g(x_i)$, then we can always make RSS zero simply by choosing g such that it *interpolates* all of the y_i. Such a function would woefully overfit the data—it would be far too flexible. What we really want is a function g that makes RSS small, but that is also *smooth*.

How might we ensure that g is smooth? There are a number of ways to do this. A natural approach is to find the function g that minimizes

$$\sum_{i=1}^{n}(y_i - g(x_i))^2 + \lambda \int g''(t)^2 dt \qquad (7.11)$$

where λ is a nonnegative *tuning parameter*. The function g that minimizes (7.11) is known as a *smoothing spline*.

What does (7.11) mean? Equation 7.11 takes the "Loss+Penalty" formulation that we encounter in the context of ridge regression and the lasso in Chapter 6. The term $\sum_{i=1}^{n}(y_i - g(x_i))^2$ is a *loss function* that encourages g to fit the data well, and the term $\lambda \int g''(t)^2 dt$ is a *penalty term*

smoothing spline

loss function

that penalizes the variability in g. The notation $g''(t)$ indicates the second derivative of the function g. The first derivative $g'(t)$ measures the slope of a function at t, and the second derivative corresponds to the amount by which the slope is changing. Hence, broadly speaking, the second derivative of a function is a measure of its *roughness*: it is large in absolute value if $g(t)$ is very wiggly near t, and it is close to zero otherwise. (The second derivative of a straight line is zero; note that a line is perfectly smooth.) The $\int$ notation is an *integral*, which we can think of as a summation over the range of t. In other words, $\int g''(t)^2 dt$ is simply a measure of the total change in the function $g'(t)$, over its entire range. If g is very smooth, then $g'(t)$ will be close to constant and $\int g''(t)^2 dt$ will take on a small value. Conversely, if g is jumpy and variable then $g'(t)$ will vary significantly and $\int g''(t)^2 dt$ will take on a large value. Therefore, in (7.11), $\lambda \int g''(t)^2 dt$ encourages g to be smooth. The larger the value of λ, the smoother g will be.

When $\lambda = 0$, then the penalty term in (7.11) has no effect, and so the function g will be very jumpy and will exactly interpolate the training observations. When $\lambda \to \infty$, g will be perfectly smooth—it will just be a straight line that passes as closely as possible to the training points. In fact, in this case, g will be the linear least squares line, since the loss function in (7.11) amounts to minimizing the residual sum of squares. For an intermediate value of λ, g will approximate the training observations but will be somewhat smooth. We see that λ controls the bias-variance trade-off of the smoothing spline.

The function $g(x)$ that minimizes (7.11) can be shown to have some special properties: it is a piecewise cubic polynomial with knots at the unique values of $x_1, \ldots, x_n$, and continuous first and second derivatives at each knot. Furthermore, it is linear in the region outside of the extreme knots. In other words, *the function $g(x)$ that minimizes (7.11) is a natural cubic spline with knots at $x_1, \ldots, x_n$!* However, it is not the same natural cubic spline that one would get if one applied the basis function approach described in Section 7.4.3 with knots at $x_1, \ldots, x_n$—rather, it is a *shrunken* version of such a natural cubic spline, where the value of the tuning parameter λ in (7.11) controls the level of shrinkage.

7.5.2 *Choosing the Smoothing Parameter λ*

We have seen that a smoothing spline is simply a natural cubic spline with knots at every unique value of x_i. It might seem that a smoothing spline will have far too many degrees of freedom, since a knot at each data point allows a great deal of flexibility. But the tuning parameter λ controls the roughness of the smoothing spline, and hence the *effective degrees of freedom*. It is possible to show that as λ increases from 0 to ∞, the effective degrees of freedom, which we write df_λ, decrease from n to 2.

effective degrees of freedom

In the context of smoothing splines, why do we discuss *effective* degrees of freedom instead of degrees of freedom? Usually degrees of freedom refer to the number of free parameters, such as the number of coefficients fit in a polynomial or cubic spline. Although a smoothing spline has n parameters and hence n nominal degrees of freedom, these n parameters are heavily constrained or shrunk down. Hence df_λ is a measure of the flexibility of the smoothing spline—the higher it is, the more flexible (and the lower-bias but higher-variance) the smoothing spline. The definition of effective degrees of freedom is somewhat technical. We can write

$$\hat{\mathbf{g}}_\lambda = \mathbf{S}_\lambda \mathbf{y}, \qquad (7.12)$$

where $\hat{\mathbf{g}}$ is the solution to (7.11) for a particular choice of λ—that is, it is a n-vector containing the fitted values of the smoothing spline at the training points $x_1, \ldots, x_n$. Equation 7.12 indicates that the vector of fitted values when applying a smoothing spline to the data can be written as a $n \times n$ matrix $\mathbf{S}_\lambda$ (for which there is a formula) times the response vector $\mathbf{y}$. Then the effective degrees of freedom is defined to be

$$df_\lambda = \sum_{i=1}^{n} \{\mathbf{S}_\lambda\}_{ii}, \qquad (7.13)$$

the sum of the diagonal elements of the matrix $\mathbf{S}_\lambda$.

In fitting a smoothing spline, we do not need to select the number or location of the knots—there will be a knot at each training observation, $x_1, \ldots, x_n$. Instead, we have another problem: we need to choose the value of λ. It should come as no surprise that one possible solution to this problem is cross-validation. In other words, we can find the value of λ that makes the cross-validated RSS as small as possible. It turns out that the *leave-one-out* cross-validation error (LOOCV) can be computed very efficiently for smoothing splines, with essentially the same cost as computing a single fit, using the following formula:

$$\text{RSS}_{cv}(\lambda) = \sum_{i=1}^{n} (y_i - \hat{g}_\lambda^{(-i)}(x_i))^2 = \sum_{i=1}^{n} \left[\frac{y_i - \hat{g}_\lambda(x_i)}{1 - \{\mathbf{S}_\lambda\}_{ii}} \right]^2 .$$

The notation $\hat{g}_\lambda^{(-i)}(x_i)$ indicates the fitted value for this smoothing spline evaluated at x_i, where the fit uses all of the training observations except for the ith observation (x_i, y_i). In contrast, $\hat{g}_\lambda(x_i)$ indicates the smoothing spline function fit to all of the training observations and evaluated at x_i. This remarkable formula says that we can compute each of these *leave-one-out* fits using only $\hat{g}_\lambda$, the original fit to *all* of the data![5] We have a very similar formula (5.2) on page 180 in Chapter 5 for least squares linear regression. Using (5.2), we can very quickly perform LOOCV for the regression splines discussed earlier in this chapter, as well as for least squares regression using arbitrary basis functions.

---

[5] The exact formulas for computing $\hat{g}(x_i)$ and $\mathbf{S}_\lambda$ are very technical; however, efficient algorithms are available for computing these quantities.

Smoothing Spline

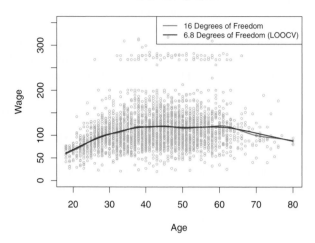

FIGURE 7.8. *Smoothing spline fits to the* Wage *data. The red curve results from specifying* 16 *effective degrees of freedom. For the blue curve,* λ *was found automatically by leave-one-out cross-validation, which resulted in* 6.8 *effective degrees of freedom.*

Figure 7.8 shows the results from fitting a smoothing spline to the Wage data. The red curve indicates the fit obtained from pre-specifying that we would like a smoothing spline with 16 effective degrees of freedom. The blue curve is the smoothing spline obtained when λ is chosen using LOOCV; in this case, the value of λ chosen results in 6.8 effective degrees of freedom (computed using (7.13)). For this data, there is little discernible difference between the two smoothing splines, beyond the fact that the one with 16 degrees of freedom seems slightly wigglier. Since there is little difference between the two fits, the smoothing spline fit with 6.8 degrees of freedom is preferable, since in general simpler models are better unless the data provides evidence in support of a more complex model.

7.6 Local Regression

Local regression is a different approach for fitting flexible non-linear functions, which involves computing the fit at a target point x_0 using only the nearby training observations. Figure 7.9 illustrates the idea on some simulated data, with one target point near 0.4, and another near the boundary at 0.05. In this figure the blue line represents the function $f(x)$ from which the data were generated, and the light orange line corresponds to the local regression estimate $\hat{f}(x)$. Local regression is described in Algorithm 7.1.

Note that in Step 3 of Algorithm 7.1, the weights K_{i0} will differ for each value of x_0. In other words, in order to obtain the local regression fit at a new point, we need to fit a new weighted least squares regression model by

local regression

Local Regression

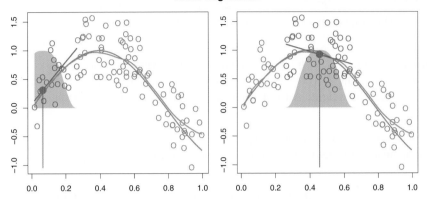

FIGURE 7.9. *Local regression illustrated on some simulated data, where the blue curve represents $f(x)$ from which the data were generated, and the light orange curve corresponds to the local regression estimate $\hat{f}(x)$. The orange colored points are local to the target point x_0, represented by the orange vertical line. The yellow bell-shape superimposed on the plot indicates weights assigned to each point, decreasing to zero with distance from the target point. The fit $\hat{f}(x_0)$ at x_0 is obtained by fitting a weighted linear regression (orange line segment), and using the fitted value at x_0 (orange solid dot) as the estimate $\hat{f}(x_0)$.*

minimizing (7.14) for a new set of weights. Local regression is sometimes referred to as a *memory-based* procedure, because like nearest-neighbors, we need all the training data each time we wish to compute a prediction. We will avoid getting into the technical details of local regression here—there are books written on the topic.

In order to perform local regression, there are a number of choices to be made, such as how to define the weighting function K, and whether to fit a linear, constant, or quadratic regression in Step 3 above. (Equation 7.14 corresponds to a linear regression.) While all of these choices make some difference, the most important choice is the *span s*, defined in Step 1 above. The span plays a role like that of the tuning parameter λ in smoothing splines: it controls the flexibility of the non-linear fit. The smaller the value of s, the more *local* and wiggly will be our fit; alternatively, a very large value of s will lead to a global fit to the data using all of the training observations. We can again use cross-validation to choose s, or we can specify it directly. Figure 7.10 displays local linear regression fits on the Wage data, using two values of s: 0.7 and 0.2. As expected, the fit obtained using $s = 0.7$ is smoother than that obtained using $s = 0.2$.

The idea of local regression can be generalized in many different ways. In a setting with multiple features $X_1, X_2, \ldots, X_p$, one very useful generalization involves fitting a multiple linear regression model that is global in some variables, but local in another, such as time. Such *varying coefficient*

Algorithm 7.1 *Local Regression At* $X = x_0$

---

1. Gather the fraction $s = k/n$ of training points whose x_i are closest to x_0.

2. Assign a weight $K_{i0} = K(x_i, x_0)$ to each point in this neighborhood, so that the point furthest from x_0 has weight zero, and the closest has the highest weight. All but these k nearest neighbors get weight zero.

3. Fit a *weighted least squares regression* of the y_i on the x_i using the aforementioned weights, by finding $\hat{\beta}_0$ and $\hat{\beta}_1$ that minimize

$$\sum_{i=1}^{n} K_{i0}(y_i - \beta_0 - \beta_1 x_i)^2. \tag{7.14}$$

4. The fitted value at x_0 is given by $\hat{f}(x_0) = \hat{\beta}_0 + \hat{\beta}_1 x_0$.

---

models are a useful way of adapting a model to the most recently gathered data. Local regression also generalizes very naturally when we want to fit models that are local in a pair of variables X_1 and X_2, rather than one. We can simply use two-dimensional neighborhoods, and fit bivariate linear regression models using the observations that are near each target point in two-dimensional space. Theoretically the same approach can be implemented in higher dimensions, using linear regressions fit to p-dimensional neighborhoods. However, local regression can perform poorly if p is much larger than about 3 or 4 because there will generally be very few training observations close to x_0. Nearest-neighbors regression, discussed in Chapter 3, suffers from a similar problem in high dimensions.

varying coefficient model

7.7 Generalized Additive Models

In Sections 7.1–7.6, we present a number of approaches for flexibly predicting a response Y on the basis of a single predictor X. These approaches can be seen as extensions of simple linear regression. Here we explore the problem of flexibly predicting Y on the basis of several predictors, $X_1, \ldots, X_p$. This amounts to an extension of multiple linear regression.

Generalized additive models (GAMs) provide a general framework for extending a standard linear model by allowing non-linear functions of each of the variables, while maintaining *additivity*. Just like linear models, GAMs can be applied with both quantitative and qualitative responses. We first examine GAMs for a quantitative response in Section 7.7.1, and then for a qualitative response in Section 7.7.2.

generalized additive model

additivity

Local Linear Regression

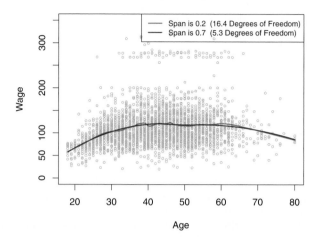

FIGURE 7.10. *Local linear fits to the* Wage *data. The span specifies the fraction of the data used to compute the fit at each target point.*

7.7.1   GAMs for Regression Problems

A natural way to extend the multiple linear regression model

$$y_i = \beta_0 + \beta_1 x_{i1} + \beta_2 x_{i2} + \cdots + \beta_p x_{ip} + \epsilon_i$$

in order to allow for non-linear relationships between each feature and the response is to replace each linear component $\beta_j x_{ij}$ with a (smooth) non-linear function $f_j(x_{ij})$. We would then write the model as

$$
\begin{aligned}
y_i &= \beta_0 + \sum_{j=1}^{p} f_j(x_{ij}) + \epsilon_i \\
&= \beta_0 + f_1(x_{i1}) + f_2(x_{i2}) + \cdots + f_p(x_{ip}) + \epsilon_i. \quad (7.15)
\end{aligned}
$$

This is an example of a GAM. It is called an *additive* model because we calculate a separate f_j for each X_j, and then add together all of their contributions.

In Sections 7.1–7.6, we discuss many methods for fitting functions to a single variable. The beauty of GAMs is that we can use these methods as building blocks for fitting an additive model. In fact, for most of the methods that we have seen so far in this chapter, this can be done fairly trivially. Take, for example, natural splines, and consider the task of fitting the model

$$\text{wage} = \beta_0 + f_1(\text{year}) + f_2(\text{age}) + f_3(\text{education}) + \epsilon \quad (7.16)$$

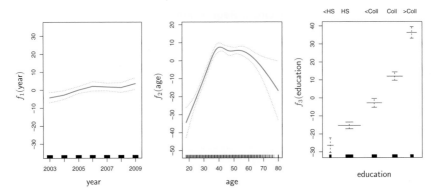

FIGURE 7.11. *For the* Wage *data, plots of the relationship between each feature and the response,* wage, *in the fitted model (7.16). Each plot displays the fitted function and pointwise standard errors. The first two functions are natural splines in* year *and* age, *with four and five degrees of freedom, respectively. The third function is a step function, fit to the qualitative variable* education.

on the Wage data. Here year and age are quantitative variables, and education is a qualitative variable with five levels: <HS, HS, <Coll, Coll, >Coll, referring to the amount of high school or college education that an individual has completed. We fit the first two functions using natural splines. We fit the third function using a separate constant for each level, via the usual dummy variable approach of Section 3.3.1.

Figure 7.11 shows the results of fitting the model (7.16) using least squares. This is easy to do, since as discussed in Section 7.4, natural splines can be constructed using an appropriately chosen set of basis functions. Hence the entire model is just a big regression onto spline basis variables and dummy variables, all packed into one big regression matrix.

Figure 7.11 can be easily interpreted. The left-hand panel indicates that holding age and education fixed, wage tends to increase slightly with year; this may be due to inflation. The center panel indicates that holding education and year fixed, wage tends to be highest for intermediate values of age, and lowest for the very young and very old. The right-hand panel indicates that holding year and age fixed, wage tends to increase with education: the more educated a person is, the higher their salary, on average. All of these findings are intuitive.

Figure 7.12 shows a similar triple of plots, but this time f_1 and f_2 are smoothing splines with four and five degrees of freedom, respectively. Fitting a GAM with a smoothing spline is not quite as simple as fitting a GAM with a natural spline, since in the case of smoothing splines, least squares cannot be used. However, standard software such as the gam() function in R can be used to fit GAMs using smoothing splines, via an approach known as *backfitting*. This method fits a model involving multiple predictors by

backfitting

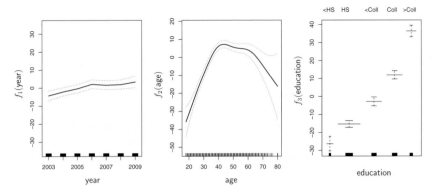

FIGURE 7.12. *Details are as in Figure 7.11, but now f_1 and f_2 are smoothing splines with four and five degrees of freedom, respectively.*

repeatedly updating the fit for each predictor in turn, holding the others fixed. The beauty of this approach is that each time we update a function, we simply apply the fitting method for that variable to a *partial residual*.[6]

The fitted functions in Figures 7.11 and 7.12 look rather similar. In most situations, the differences in the GAMs obtained using smoothing splines versus natural splines are small.

We do not have to use splines as the building blocks for GAMs: we can just as well use local regression, polynomial regression, or any combination of the approaches seen earlier in this chapter in order to create a GAM. GAMs are investigated in further detail in the lab at the end of this chapter.

Pros and Cons of GAMs

Before we move on, let us summarize the advantages and limitations of a GAM.

▲ GAMs allow us to fit a non-linear f_j to each X_j, so that we can automatically model non-linear relationships that standard linear regression will miss. This means that we do not need to manually try out many different transformations on each variable individually.

▲ The non-linear fits can potentially make more accurate predictions for the response Y.

▲ Because the model is additive, we can still examine the effect of each X_j on Y individually while holding all of the other variables fixed. Hence if we are interested in inference, GAMs provide a useful representation.

---

[6]A partial residual for X_3, for example, has the form $r_i = y_i - f_1(x_{i1}) - f_2(x_{i2})$. If we know f_1 and f_2, then we can fit f_3 by treating this residual as a response in a non-linear regression on X_3.

▲ The smoothness of the function f_j for the variable X_j can be summarized via degrees of freedom.

◆ The main limitation of GAMs is that the model is restricted to be additive. With many variables, important interactions can be missed. However, as with linear regression, we can manually add interaction terms to the GAM model by including additional predictors of the form $X_j \times X_k$. In addition we can add low-dimensional interaction functions of the form $f_{jk}(X_j, X_k)$ into the model; such terms can be fit using two-dimensional smoothers such as local regression, or two-dimensional splines (not covered here).

For fully general models, we have to look for even more flexible approaches such as random forests and boosting, described in Chapter 8. GAMs provide a useful compromise between linear and fully nonparametric models.

7.7.2 GAMs for Classification Problems

GAMs can also be used in situations where Y is qualitative. For simplicity, here we will assume Y takes on values zero or one, and let $p(X) = \Pr(Y = 1|X)$ be the conditional probability (given the predictors) that the response equals one. Recall the logistic regression model (4.6):

$$\log\left(\frac{p(X)}{1 - p(X)}\right) = \beta_0 + \beta_1 X_1 + \beta_2 X_2 + \cdots + \beta_p X_p. \qquad (7.17)$$

This *logit* is the log of the odds of $P(Y = 1|X)$ versus $P(Y = 0|X)$, which (7.17) represents as a linear function of the predictors. A natural way to extend (7.17) to allow for non-linear relationships is to use the model

$$\log\left(\frac{p(X)}{1 - p(X)}\right) = \beta_0 + f_1(X_1) + f_2(X_2) + \cdots + f_p(X_p). \qquad (7.18)$$

Equation 7.18 is a logistic regression GAM. It has all the same pros and cons as discussed in the previous section for quantitative responses.

We fit a GAM to the Wage data in order to predict the probability that an individual's income exceeds \$250,000 per year. The GAM that we fit takes the form

$$\log\left(\frac{p(X)}{1 - p(X)}\right) = \beta_0 + \beta_1 \times \texttt{year} + f_2(\texttt{age}) + f_3(\texttt{education}), \quad (7.19)$$

where

$$p(X) = \Pr(\texttt{wage} > 250|\texttt{year}, \texttt{age}, \texttt{education}).$$

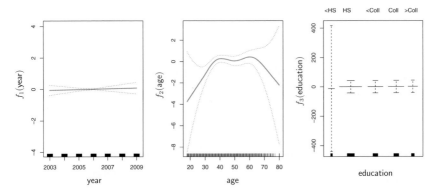

FIGURE 7.13. *For the* Wage *data, the logistic regression GAM given in (7.19) is fit to the binary response* I(wage>250). *Each plot displays the fitted function and pointwise standard errors. The first function is linear in* year, *the second function a smoothing spline with five degrees of freedom in* age, *and the third a step function for* education. *There are very wide standard errors for the first level* <HS *of* education.

Once again f_2 is fit using a smoothing spline with five degrees of freedom, and f_3 is fit as a step function, by creating dummy variables for each of the levels of education. The resulting fit is shown in Figure 7.13. The last panel looks suspicious, with very wide confidence intervals for level <HS. In fact, there are no ones for that category: no individuals with less than a high school education make more than \$250,000 per year. Hence we refit the GAM, excluding the individuals with less than a high school education. The resulting model is shown in Figure 7.14. As in Figures 7.11 and 7.12, all three panels have the same vertical scale. This allows us to visually assess the relative contributions of each of the variables. We observe that age and education have a much larger effect than year on the probability of being a high earner.

7.8   Lab: Non-linear Modeling

In this lab, we re-analyze the Wage data considered in the examples throughout this chapter, in order to illustrate the fact that many of the complex non-linear fitting procedures discussed can be easily implemented in R. We begin by loading the ISLR library, which contains the data.

```
> library(ISLR)
> attach(Wage)
```

288   7. Moving Beyond Linearity

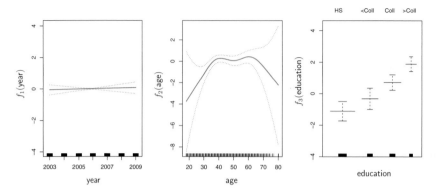

FIGURE 7.14. *The same model is fit as in Figure 7.13, this time excluding the observations for which* education *is* <HS. *Now we see that increased education tends to be associated with higher salaries.*

7.8.1 Polynomial Regression and Step Functions

We now examine how Figure 7.1 was produced. We first fit the model using the following command:

```
> fit=lm(wage~poly(age,4),data=Wage)
> coef(summary(fit))
                 Estimate  Std. Error  t value  Pr(>|t|)
(Intercept)      111.704      0.729    153.28   <2e-16
poly(age, 4)1    447.068     39.915     11.20   <2e-16
poly(age, 4)2   -478.316     39.915    -11.98   <2e-16
poly(age, 4)3    125.522     39.915      3.14   0.0017
poly(age, 4)4    -77.911     39.915     -1.95   0.0510
```

This syntax fits a linear model, using the lm() function, in order to predict wage using a fourth-degree polynomial in age: poly(age,4). The poly() command allows us to avoid having to write out a long formula with powers of age. The function returns a matrix whose columns are a basis of *orthogonal polynomials*, which essentially means that each column is a linear combination of the variables age, age^2, age^3 and age^4.

orthogonal polynomial

However, we can also use poly() to obtain age, age^2, age^3 and age^4 directly, if we prefer. We can do this by using the raw=TRUE argument to the poly() function. Later we see that this does not affect the model in a meaningful way—though the choice of basis clearly affects the coefficient estimates, it does not affect the fitted values obtained.

```
> fit2=lm(wage~poly(age,4,raw=T),data=Wage)
> coef(summary(fit2))
                            Estimate  Std. Error  t value  Pr(>|t|)
(Intercept)                -1.84e+02   6.00e+01    -3.07   0.002180
poly(age, 4, raw = T)1      2.12e+01   5.89e+00     3.61   0.000312
poly(age, 4, raw = T)2     -5.64e-01   2.06e-01    -2.74   0.006261
```

```
poly(age, 4, raw = T)3   6.81e-03    3.07e-03    2.22 0.026398
poly(age, 4, raw = T)4  -3.20e-05    1.64e-05   -1.95 0.051039
```

There are several other equivalent ways of fitting this model, which show-case the flexibility of the formula language in R. For example

```
> fit2a=lm(wage~age+I(age^2)+I(age^3)+I(age^4),data=Wage)
> coef(fit2a)
(Intercept)         age      I(age^2)     I(age^3)     I(age^4)
  -1.84e+02    2.12e+01    -5.64e-01     6.81e-03    -3.20e-05
```

This simply creates the polynomial basis functions on the fly, taking care to protect terms like age^2 via the *wrapper* function I() (the ^ symbol has a special meaning in formulas).   wrapper

```
> fit2b=lm(wage~cbind(age,age^2,age^3,age^4),data=Wage)
```

This does the same more compactly, using the cbind() function for building a matrix from a collection of vectors; any function call such as cbind() inside a formula also serves as a wrapper.

We now create a grid of values for age at which we want predictions, and then call the generic predict() function, specifying that we want standard errors as well.

```
> agelims=range(age)
> age.grid=seq(from=agelims[1],to=agelims[2])
> preds=predict(fit,newdata=list(age=age.grid),se=TRUE)
> se.bands=cbind(preds$fit+2*preds$se.fit,preds$fit-2*preds$se.
    fit)
```

Finally, we plot the data and add the fit from the degree-4 polynomial.

```
> par(mfrow=c(1,2),mar=c(4.5,4.5,1,1),oma=c(0,0,4,0))
> plot(age,wage,xlim=agelims,cex=.5,col="darkgrey")
> title("Degree-4 Polynomial",outer=T)
> lines(age.grid,preds$fit,lwd=2,col="blue")
> matlines(age.grid,se.bands,lwd=1,col="blue",lty=3)
```

Here the mar and oma arguments to par() allow us to control the margins of the plot, and the title() function creates a figure title that spans both subplots.   title()

We mentioned earlier that whether or not an orthogonal set of basis functions is produced in the poly() function will not affect the model obtained in a meaningful way. What do we mean by this? The fitted values obtained in either case are identical:

```
> preds2=predict(fit2,newdata=list(age=age.grid),se=TRUE)
> max(abs(preds$fit-preds2$fit))
[1] 7.39e-13
```

In performing a polynomial regression we must decide on the degree of the polynomial to use. One way to do this is by using hypothesis tests. We now fit models ranging from linear to a degree-5 polynomial and seek to determine the simplest model which is sufficient to explain the relationship between wage and age. We use the anova() function, which performs an *analysis of variance* (ANOVA, using an F-test) in order to test the null hypothesis that a model $\mathcal{M}_1$ is sufficient to explain the data against the alternative hypothesis that a more complex model $\mathcal{M}_2$ is required. In order to use the anova() function, $\mathcal{M}_1$ and $\mathcal{M}_2$ must be *nested* models: the predictors in $\mathcal{M}_1$ must be a subset of the predictors in $\mathcal{M}_2$. In this case, we fit five different models and sequentially compare the simpler model to the more complex model.

anova()

analysis of variance

```
> fit.1=lm(wage~age,data=Wage)
> fit.2=lm(wage~poly(age,2),data=Wage)
> fit.3=lm(wage~poly(age,3),data=Wage)
> fit.4=lm(wage~poly(age,4),data=Wage)
> fit.5=lm(wage~poly(age,5),data=Wage)
> anova(fit.1,fit.2,fit.3,fit.4,fit.5)
Analysis of Variance Table

Model 1: wage ~ age
Model 2: wage ~ poly(age, 2)
Model 3: wage ~ poly(age, 3)
Model 4: wage ~ poly(age, 4)
Model 5: wage ~ poly(age, 5)
  Res.Df     RSS Df Sum of Sq       F Pr(>F)
1   2998 5022216
2   2997 4793430  1    228786 143.59 <2e-16 ***
3   2996 4777674  1     15756   9.89 0.0017 **
4   2995 4771604  1      6070   3.81 0.0510 .
5   2994 4770322  1      1283   0.80 0.3697
---
Signif. codes:  0 '***' 0.001 '**' 0.01 '*' 0.05 '.' 0.1 ' ' 1
```

The p-value comparing the linear Model 1 to the quadratic Model 2 is essentially zero ($<10^{-15}$), indicating that a linear fit is not sufficient. Similarly the p-value comparing the quadratic Model 2 to the cubic Model 3 is very low (0.0017), so the quadratic fit is also insufficient. The p-value comparing the cubic and degree-4 polynomials, Model 3 and Model 4, is approximately 5 % while the degree-5 polynomial Model 5 seems unnecessary because its p-value is 0.37. Hence, either a cubic or a quartic polynomial appear to provide a reasonable fit to the data, but lower- or higher-order models are not justified.

In this case, instead of using the anova() function, we could have obtained these p-values more succinctly by exploiting the fact that poly() creates orthogonal polynomials.

```
> coef(summary(fit.5))
               Estimate Std. Error  t value  Pr(>|t|)
(Intercept)      111.70     0.7288 153.2780 0.000e+00
poly(age, 5)1    447.07    39.9161  11.2002 1.491e-28
poly(age, 5)2   -478.32    39.9161 -11.9830 2.368e-32
poly(age, 5)3    125.52    39.9161   3.1446 1.679e-03
```

```
poly(age, 5)4    -77.91    39.9161   -1.9519 5.105e-02
poly(age, 5)5    -35.81    39.9161   -0.8972 3.697e-01
```

Notice that the p-values are the same, and in fact the square of the t-statistics are equal to the F-statistics from the `anova()` function; for example:

```
> (-11.983)^2
[1] 143.6
```

However, the ANOVA method works whether or not we used orthogonal polynomials; it also works when we have other terms in the model as well. For example, we can use `anova()` to compare these three models:

```
> fit.1=lm(wage~education+age,data=Wage)
> fit.2=lm(wage~education+poly(age,2),data=Wage)
> fit.3=lm(wage~education+poly(age,3),data=Wage)
> anova(fit.1,fit.2,fit.3)
```

As an alternative to using hypothesis tests and ANOVA, we could choose the polynomial degree using cross-validation, as discussed in Chapter 5.

Next we consider the task of predicting whether an individual earns more than \$250,000 per year. We proceed much as before, except that first we create the appropriate response vector, and then apply the `glm()` function using `family="binomial"` in order to fit a polynomial logistic regression model.

```
> fit=glm(I(wage>250)~poly(age,4),data=Wage,family=binomial)
```

Note that we again use the wrapper `I()` to create this binary response variable on the fly. The expression `wage>250` evaluates to a logical variable containing `TRUE`s and `FALSE`s, which `glm()` coerces to binary by setting the `TRUE`s to 1 and the `FALSE`s to 0.

Once again, we make predictions using the `predict()` function.

```
> preds=predict(fit,newdata=list(age=age.grid),se=T)
```

However, calculating the confidence intervals is slightly more involved than in the linear regression case. The default prediction type for a `glm()` model is `type="link"`, which is what we use here. This means we get predictions for the *logit*: that is, we have fit a model of the form

$$\log\left(\frac{\Pr(Y=1|X)}{1-\Pr(Y=1|X)}\right) = X\beta,$$

and the predictions given are of the form $X\hat{\beta}$. The standard errors given are also of this form. In order to obtain confidence intervals for $\Pr(Y=1|X)$, we use the transformation

$$\Pr(Y=1|X) = \frac{\exp(X\beta)}{1+\exp(X\beta)}.$$

```
> pfit=exp(preds$fit)/(1+exp(preds$fit))
> se.bands.logit = cbind(preds$fit+2*preds$se.fit, preds$fit-2*
    preds$se.fit)
> se.bands = exp(se.bands.logit)/(1+exp(se.bands.logit))
```

Note that we could have directly computed the probabilities by selecting the `type="response"` option in the `predict()` function.

```
> preds=predict(fit,newdata=list(age=age.grid),type="response",
    se=T)
```

However, the corresponding confidence intervals would not have been sensible because we would end up with negative probabilities!

Finally, the right-hand plot from Figure 7.1 was made as follows:

```
> plot(age,I(wage>250),xlim=agelims,type="n",ylim=c(0,.2))
> points(jitter(age), I((wage>250)/5),cex=.5,pch="|",
    col="darkgrey")
> lines(age.grid,pfit,lwd=2, col="blue")
> matlines(age.grid,se.bands,lwd=1,col="blue",lty=3)
```

We have drawn the age values corresponding to the observations with `wage` values above 250 as gray marks on the top of the plot, and those with `wage` values below 250 are shown as gray marks on the bottom of the plot. We used the `jitter()` function to jitter the age values a bit so that observations with the same age value do not cover each other up. This is often called a *rug plot*.

`jitter()`

`rug plot`

In order to fit a step function, as discussed in Section 7.2, we use the `cut()` function.

`cut()`

```
> table(cut(age,4))
(17.9,33.5]    (33.5,49]     (49,64.5] (64.5,80.1]
       750         1399           779          72
> fit=lm(wage~cut(age,4),data=Wage)
> coef(summary(fit))
```

	Estimate	Std. Error	t value	Pr(>|t|)
(Intercept)	94.16	1.48	63.79	0.00e+00
cut(age, 4)(33.5,49]	24.05	1.83	13.15	1.98e-38
cut(age, 4)(49,64.5]	23.66	2.07	11.44	1.04e-29
cut(age, 4)(64.5,80.1]	7.64	4.99	1.53	1.26e-01

Here `cut()` automatically picked the cutpoints at 33.5, 49, and 64.5 years of age. We could also have specified our own cutpoints directly using the `breaks` option. The function `cut()` returns an ordered categorical variable; the `lm()` function then creates a set of dummy variables for use in the regression. The `age<33.5` category is left out, so the intercept coefficient of $94,160 can be interpreted as the average salary for those under 33.5 years of age, and the other coefficients can be interpreted as the average additional salary for those in the other age groups. We can produce predictions and plots just as we did in the case of the polynomial fit.

7.8.2 Splines

In order to fit regression splines in R, we use the splines library. In Section 7.4, we saw that regression splines can be fit by constructing an appropriate matrix of basis functions. The bs() function generates the entire matrix of basis functions for splines with the specified set of knots. By default, cubic splines are produced. Fitting wage to age using a regression spline is simple:

```
> library(splines)
> fit=lm(wage~bs(age,knots=c(25,40,60)),data=Wage)
> pred=predict(fit,newdata=list(age=age.grid),se=T)
> plot(age,wage,col="gray")
> lines(age.grid,pred$fit,lwd=2)
> lines(age.grid,pred$fit+2*pred$se,lty="dashed")
> lines(age.grid,pred$fit-2*pred$se,lty="dashed")
```

bs()

Here we have prespecified knots at ages 25, 40, and 60. This produces a spline with six basis functions. (Recall that a cubic spline with three knots has seven degrees of freedom; these degrees of freedom are used up by an intercept, plus six basis functions.) We could also use the df option to produce a spline with knots at uniform quantiles of the data.

```
> dim(bs(age,knots=c(25,40,60)))
[1] 3000    6
> dim(bs(age,df=6))
[1] 3000    6
> attr(bs(age,df=6),"knots")
  25%  50%  75%
33.8 42.0 51.0
```

In this case R chooses knots at ages 33.8, 42.0, and 51.0, which correspond to the 25th, 50th, and 75th percentiles of age. The function bs() also has a degree argument, so we can fit splines of any degree, rather than the default degree of 3 (which yields a cubic spline).

In order to instead fit a natural spline, we use the ns() function. Here we fit a natural spline with four degrees of freedom.

ns()

```
> fit2=lm(wage~ns(age,df=4),data=Wage)
> pred2=predict(fit2,newdata=list(age=age.grid),se=T)
> lines(age.grid, pred2$fit,col="red",lwd=2)
```

As with the bs() function, we could instead specify the knots directly using the knots option.

In order to fit a smoothing spline, we use the smooth.spline() function. Figure 7.8 was produced with the following code:

smooth.
spline()

```
> plot(age,wage,xlim=agelims,cex=.5,col="darkgrey")
> title("Smoothing Spline")
> fit=smooth.spline(age,wage,df=16)
> fit2=smooth.spline(age,wage,cv=TRUE)
> fit2$df
[1] 6.8
> lines(fit,col="red",lwd=2)
```

```
> lines(fit2,col="blue",lwd=2)
> legend("topright",legend=c("16 DF","6.8 DF"),
    col=c("red","blue"),lty=1,lwd=2,cex=.8)
```

Notice that in the first call to smooth.spline(), we specified df=16. The function then determines which value of λ leads to 16 degrees of freedom. In the second call to smooth.spline(), we select the smoothness level by cross-validation; this results in a value of λ that yields 6.8 degrees of freedom.

In order to perform local regression, we use the loess() function.

loess()

```
> plot(age,wage,xlim=agelims,cex=.5,col="darkgrey")
> title("Local Regression")
> fit=loess(wage~age,span=.2,data=Wage)
> fit2=loess(wage~age,span=.5,data=Wage)
> lines(age.grid,predict(fit,data.frame(age=age.grid)),
    col="red",lwd=2)
> lines(age.grid,predict(fit2,data.frame(age=age.grid)),
    col="blue",lwd=2)
> legend("topright",legend=c("Span=0.2","Span=0.5"),
    col=c("red","blue"),lty=1,lwd=2,cex=.8)
```

Here we have performed local linear regression using spans of 0.2 and 0.5: that is, each neighborhood consists of 20 % or 50 % of the observations. The larger the span, the smoother the fit. The locfit library can also be used for fitting local regression models in R.

7.8.3   GAMs

We now fit a GAM to predict wage using natural spline functions of year and age, treating education as a qualitative predictor, as in (7.16). Since this is just a big linear regression model using an appropriate choice of basis functions, we can simply do this using the lm() function.

```
> gam1=lm(wage~ns(year,4)+ns(age,5)+education,data=Wage)
```

We now fit the model (7.16) using smoothing splines rather than natural splines. In order to fit more general sorts of GAMs, using smoothing splines or other components that cannot be expressed in terms of basis functions and then fit using least squares regression, we will need to use the gam library in R.

The s() function, which is part of the gam library, is used to indicate that we would like to use a smoothing spline. We specify that the function of year should have 4 degrees of freedom, and that the function of age will have 5 degrees of freedom. Since education is qualitative, we leave it as is, and it is converted into four dummy variables. We use the gam() function in order to fit a GAM using these components. All of the terms in (7.16) are fit simultaneously, taking each other into account to explain the response.

s()

gam()

```
> library(gam)
> gam.m3=gam(wage~s(year,4)+s(age,5)+education,data=Wage)
```

In order to produce Figure 7.12, we simply call the `plot()` function:

```
> par(mfrow=c(1,3))
> plot(gam.m3, se=TRUE,col="blue")
```

The generic `plot()` function recognizes that gam.m3 is an object of class gam, and invokes the appropriate `plot.gam()` method. Conveniently, even though gam1 is not of class gam but rather of class lm, we can *still* use `plot.gam()` on it. Figure 7.11 was produced using the following expression:

`plot.gam()`

```
> plot.gam(gam1, se=TRUE, col="red")
```

Notice here we had to use `plot.gam()` rather than the *generic* `plot()` function.

In these plots, the function of year looks rather linear. We can perform a series of ANOVA tests in order to determine which of these three models is best: a GAM that excludes year ($\mathcal{M}_1$), a GAM that uses a linear function of year ($\mathcal{M}_2$), or a GAM that uses a spline function of year ($\mathcal{M}_3$).

```
> gam.m1=gam(wage~s(age,5)+education,data=Wage)
> gam.m2=gam(wage~year+s(age,5)+education,data=Wage)
> anova(gam.m1,gam.m2,gam.m3,test="F")
Analysis of Deviance Table

Model 1: wage ~ s(age, 5) + education
Model 2: wage ~ year + s(age, 5) + education
Model 3: wage ~ s(year, 4) + s(age, 5) + education
  Resid. Df Resid. Dev Df Deviance     F  Pr(>F)
1      2990    3711730
2      2989    3693841  1    17889  14.5 0.00014 ***
3      2986    3689770  3     4071   1.1 0.34857
---
Signif. codes:  0 '***' 0.001 '**' 0.01 '*' 0.05 '.' 0.1 ' ' 1
```

We find that there is compelling evidence that a GAM with a linear function of year is better than a GAM that does not include year at all (p-value $= 0.00014$). However, there is no evidence that a non-linear function of year is needed (p-value $= 0.349$). In other words, based on the results of this ANOVA, $\mathcal{M}_2$ is preferred.

The `summary()` function produces a summary of the gam fit.

```
> summary(gam.m3)

Call: gam(formula = wage ~ s(year, 4) + s(age, 5) + education,
    data = Wage)
Deviance Residuals:
    Min      1Q  Median      3Q     Max
-119.43  -19.70   -3.33   14.17  213.48

(Dispersion Parameter for gaussian family taken to be 1236)

    Null Deviance: 5222086 on 2999 degrees of freedom
Residual Deviance: 3689770 on 2986 degrees of freedom
```

```
AIC: 29888

Number of Local Scoring Iterations: 2

DF for Terms and F-values for Nonparametric Effects

            Df Npar Df Npar F  Pr(F)
(Intercept) 1
s(year, 4)  1       3    1.1    0.35
s(age, 5)   1       4   32.4  <2e-16 ***
education   4
---
Signif. codes:  0 '***' 0.001 '**' 0.01 '*' 0.05 '.' 0.1 ' ' 1
```

The p-values for year and age correspond to a null hypothesis of a linear
relationship versus the alternative of a non-linear relationship. The large
p-value for year reinforces our conclusion from the ANOVA test that a linear function is adequate for this term. However, there is very clear evidence
that a non-linear term is required for age.

We can make predictions from gam objects, just like from lm objects,
using the predict() method for the class gam. Here we make predictions on
the training set.

```
> preds=predict(gam.m2,newdata=Wage)
```

We can also use local regression fits as building blocks in a GAM, using
the lo() function.

lo()

```
> gam.lo=gam(wage~s(year,df=4)+lo(age,span=0.7)+education,
          data=Wage)
> plot.gam(gam.lo, se=TRUE, col="green")
```

Here we have used local regression for the age term, with a span of 0.7.
We can also use the lo() function to create interactions before calling the
gam() function. For example,

```
> gam.lo.i=gam(wage~lo(year,age,span=0.5)+education,
    data=Wage)
```

fits a two-term model, in which the first term is an interaction between
year and age, fit by a local regression surface. We can plot the resulting
two-dimensional surface if we first install the akima package.

```
> library(akima)
> plot(gam.lo.i)
```

In order to fit a logistic regression GAM, we once again use the I() function in constructing the binary response variable, and set family=binomial.

```
> gam.lr=gam(I(wage>250)~year+s(age,df=5)+education,
    family=binomial,data=Wage)
> par(mfrow=c(1,3))
> plot(gam.lr,se=T,col="green")
```

It is easy to see that there are no high earners in the <HS category:

```
> table(education,I(wage>250))
```

education	FALSE	TRUE
1. < HS Grad	268	0
2. HS Grad	966	5
3. Some College	643	7
4. College Grad	663	22
5. Advanced Degree	381	45

Hence, we fit a logistic regression GAM using all but this category. This provides more sensible results.

```
> gam.lr.s=gam(I(wage>250)~year+s(age,df=5)+education,family=
    binomial,data=Wage,subset=(education!="1. < HS Grad"))
> plot(gam.lr.s,se=T,col="green")
```

7.9 Exercises

Conceptual

1. It was mentioned in the chapter that a cubic regression spline with one knot at ξ can be obtained using a basis of the form x, x^2, x^3, $(x - \xi)_+^3$, where $(x - \xi)_+^3 = (x - \xi)^3$ if $x > \xi$ and equals 0 otherwise. We will now show that a function of the form

$$f(x) = \beta_0 + \beta_1 x + \beta_2 x^2 + \beta_3 x^3 + \beta_4 (x - \xi)_+^3$$

   is indeed a cubic regression spline, regardless of the values of $\beta_0, \beta_1, \beta_2, \beta_3, \beta_4$.

   (a) Find a cubic polynomial

   $$f_1(x) = a_1 + b_1 x + c_1 x^2 + d_1 x^3$$

   such that $f(x) = f_1(x)$ for all $x \le \xi$. Express a_1, b_1, c_1, d_1 in terms of $\beta_0, \beta_1, \beta_2, \beta_3, \beta_4$.

   (b) Find a cubic polynomial

   $$f_2(x) = a_2 + b_2 x + c_2 x^2 + d_2 x^3$$

   such that $f(x) = f_2(x)$ for all $x > \xi$. Express a_2, b_2, c_2, d_2 in terms of $\beta_0, \beta_1, \beta_2, \beta_3, \beta_4$. We have now established that $f(x)$ is a piecewise polynomial.

   (c) Show that $f_1(\xi) = f_2(\xi)$. That is, $f(x)$ is continuous at ξ.

   (d) Show that $f_1'(\xi) = f_2'(\xi)$. That is, $f'(x)$ is continuous at ξ.

(e) Show that $f_1''(\xi) = f_2''(\xi)$. That is, $f''(x)$ is continuous at ξ.

Therefore, $f(x)$ is indeed a cubic spline.

Hint: Parts (d) and (e) of this problem require knowledge of single-variable calculus. As a reminder, given a cubic polynomial

$$f_1(x) = a_1 + b_1 x + c_1 x^2 + d_1 x^3,$$

the first derivative takes the form

$$f_1'(x) = b_1 + 2c_1 x + 3d_1 x^2$$

and the second derivative takes the form

$$f_1''(x) = 2c_1 + 6d_1 x.$$

2. Suppose that a curve $\hat{g}$ is computed to smoothly fit a set of n points using the following formula:

$$\hat{g} = \arg\min_g \left(\sum_{i=1}^n (y_i - g(x_i))^2 + \lambda \int \left[g^{(m)}(x) \right]^2 dx \right),$$

where $g^{(m)}$ represents the mth derivative of g (and $g^{(0)} = g$). Provide example sketches of $\hat{g}$ in each of the following scenarios.

(a) $\lambda = \infty, m = 0$.

(b) $\lambda = \infty, m = 1$.

(c) $\lambda = \infty, m = 2$.

(d) $\lambda = \infty, m = 3$.

(e) $\lambda = 0, m = 3$.

3. Suppose we fit a curve with basis functions $b_1(X) = X$, $b_2(X) = (X-1)^2 I(X \geq 1)$. (Note that $I(X \geq 1)$ equals 1 for $X \geq 1$ and 0 otherwise.) We fit the linear regression model

$$Y = \beta_0 + \beta_1 b_1(X) + \beta_2 b_2(X) + \epsilon,$$

and obtain coefficient estimates $\hat{\beta}_0 = 1, \hat{\beta}_1 = 1, \hat{\beta}_2 = -2$. Sketch the estimated curve between $X = -2$ and $X = 2$. Note the intercepts, slopes, and other relevant information.

4. Suppose we fit a curve with basis functions $b_1(X) = I(0 \leq X \leq 2) - (X-1)I(1 \leq X \leq 2)$, $b_2(X) = (X-3)I(3 \leq X \leq 4) + I(4 < X \leq 5)$. We fit the linear regression model

$$Y = \beta_0 + \beta_1 b_1(X) + \beta_2 b_2(X) + \epsilon,$$

and obtain coefficient estimates $\hat{\beta}_0 = 1, \hat{\beta}_1 = 1, \hat{\beta}_2 = 3$. Sketch the estimated curve between $X = -2$ and $X = 2$. Note the intercepts, slopes, and other relevant information.

5. Consider two curves, $\hat{g}_1$ and $\hat{g}_2$, defined by

$$\hat{g}_1 = \arg\min_g \left(\sum_{i=1}^{n} (y_i - g(x_i))^2 + \lambda \int \left[g^{(3)}(x) \right]^2 dx \right),$$

$$\hat{g}_2 = \arg\min_g \left(\sum_{i=1}^{n} (y_i - g(x_i))^2 + \lambda \int \left[g^{(4)}(x) \right]^2 dx \right),$$

where $g^{(m)}$ represents the mth derivative of g.

(a) As $\lambda \to \infty$, will $\hat{g}_1$ or $\hat{g}_2$ have the smaller training RSS?

(b) As $\lambda \to \infty$, will $\hat{g}_1$ or $\hat{g}_2$ have the smaller test RSS?

(c) For $\lambda = 0$, will $\hat{g}_1$ or $\hat{g}_2$ have the smaller training and test RSS?

Applied

6. In this exercise, you will further analyze the Wage data set considered throughout this chapter.

(a) Perform polynomial regression to predict wage using age. Use cross-validation to select the optimal degree d for the polynomial. What degree was chosen, and how does this compare to the results of hypothesis testing using ANOVA? Make a plot of the resulting polynomial fit to the data.

(b) Fit a step function to predict wage using age, and perform cross-validation to choose the optimal number of cuts. Make a plot of the fit obtained.

7. The Wage data set contains a number of other features not explored in this chapter, such as marital status (maritl), job class (jobclass), and others. Explore the relationships between some of these other predictors and wage, and use non-linear fitting techniques in order to fit flexible models to the data. Create plots of the results obtained, and write a summary of your findings.

8. Fit some of the non-linear models investigated in this chapter to the Auto data set. Is there evidence for non-linear relationships in this data set? Create some informative plots to justify your answer.

9. This question uses the variables dis (the weighted mean of distances to five Boston employment centers) and nox (nitrogen oxides concentration in parts per 10 million) from the Boston data. We will treat dis as the predictor and nox as the response.

(a) Use the poly() function to fit a cubic polynomial regression to predict nox using dis. Report the regression output, and plot the resulting data and polynomial fits.

(b) Plot the polynomial fits for a range of different polynomial degrees (say, from 1 to 10), and report the associated residual sum of squares.

(c) Perform cross-validation or another approach to select the optimal degree for the polynomial, and explain your results.

(d) Use the bs() function to fit a regression spline to predict nox using dis. Report the output for the fit using four degrees of freedom. How did you choose the knots? Plot the resulting fit.

(e) Now fit a regression spline for a range of degrees of freedom, and plot the resulting fits and report the resulting RSS. Describe the results obtained.

(f) Perform cross-validation or another approach in order to select the best degrees of freedom for a regression spline on this data. Describe your results.

10. This question relates to the College data set.

(a) Split the data into a training set and a test set. Using out-of-state tuition as the response and the other variables as the predictors, perform forward stepwise selection on the training set in order to identify a satisfactory model that uses just a subset of the predictors.

(b) Fit a GAM on the training data, using out-of-state tuition as the response and the features selected in the previous step as the predictors. Plot the results, and explain your findings.

(c) Evaluate the model obtained on the test set, and explain the results obtained.

(d) For which variables, if any, is there evidence of a non-linear relationship with the response?

11. In Section 7.7, it was mentioned that GAMs are generally fit using a *backfitting* approach. The idea behind backfitting is actually quite simple. We will now explore backfitting in the context of multiple linear regression.

Suppose that we would like to perform multiple linear regression, but we do not have software to do so. Instead, we only have software to perform simple linear regression. Therefore, we take the following iterative approach: we repeatedly hold all but one coefficient estimate fixed at its current value, and update only that coefficient estimate using a simple linear regression. The process is continued until *convergence*—that is, until the coefficient estimates stop changing.

We now try this out on a toy example.

(a) Generate a response Y and two predictors X_1 and X_2, with $n = 100$.

(b) Initialize $\hat{\beta}_1$ to take on a value of your choice. It does not matter what value you choose.

(c) Keeping $\hat{\beta}_1$ fixed, fit the model

$$Y - \hat{\beta}_1 X_1 = \beta_0 + \beta_2 X_2 + \epsilon.$$

You can do this as follows:

```
> a=y-beta1*x1
> beta2=lm(a~x2)$coef[2]
```

(d) Keeping $\hat{\beta}_2$ fixed, fit the model

$$Y - \hat{\beta}_2 X_2 = \beta_0 + \beta_1 X_1 + \epsilon.$$

You can do this as follows:

```
> a=y-beta2*x2
> beta1=lm(a~x1)$coef[2]
```

(e) Write a for loop to repeat (c) and (d) 1,000 times. Report the estimates of $\hat{\beta}_0$, $\hat{\beta}_1$, and $\hat{\beta}_2$ at each iteration of the for loop. Create a plot in which each of these values is displayed, with $\hat{\beta}_0$, $\hat{\beta}_1$, and $\hat{\beta}_2$ each shown in a different color.

(f) Compare your answer in (e) to the results of simply performing multiple linear regression to predict Y using X_1 and X_2. Use the `abline()` function to overlay those multiple linear regression coefficient estimates on the plot obtained in (e).

(g) On this data set, how many backfitting iterations were required in order to obtain a "good" approximation to the multiple regression coefficient estimates?

12. This problem is a continuation of the previous exercise. In a toy example with $p = 100$, show that one can approximate the multiple linear regression coefficient estimates by repeatedly performing simple linear regression in a backfitting procedure. How many backfitting iterations are required in order to obtain a "good" approximation to the multiple regression coefficient estimates? Create a plot to justify your answer.

8

Tree-Based Methods

In this chapter, we describe *tree-based* methods for regression and classification. These involve *stratifying* or *segmenting* the predictor space into a number of simple regions. In order to make a prediction for a given observation, we typically use the mean or the mode of the training observations in the region to which it belongs. Since the set of splitting rules used to segment the predictor space can be summarized in a tree, these types of approaches are known as *decision tree* methods.

decision tree

Tree-based methods are simple and useful for interpretation. However, they typically are not competitive with the best supervised learning approaches, such as those seen in Chapters 6 and 7, in terms of prediction accuracy. Hence in this chapter we also introduce *bagging*, *random forests*, and *boosting*. Each of these approaches involves producing multiple trees which are then combined to yield a single consensus prediction. We will see that combining a large number of trees can often result in dramatic improvements in prediction accuracy, at the expense of some loss in interpretation.

8.1 The Basics of Decision Trees

Decision trees can be applied to both regression and classification problems. We first consider regression problems, and then move on to classification.

G. James et al., *An Introduction to Statistical Learning: with Applications in R*, 303
Springer Texts in Statistics, DOI 10.1007/978-1-4614-7138-7_8,
© Springer Science+Business Media New York 2013

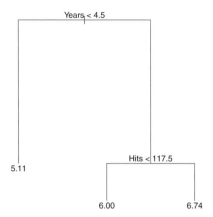

FIGURE 8.1. *For the* Hitters *data, a regression tree for predicting the log salary of a baseball player, based on the number of years that he has played in the major leagues and the number of hits that he made in the previous year. At a given internal node, the label (of the form $X_j < t_k$) indicates the left-hand branch emanating from that split, and the right-hand branch corresponds to $X_j \geq t_k$. For instance, the split at the top of the tree results in two large branches. The left-hand branch corresponds to* Years<4.5*, and the right-hand branch corresponds to* Years>=4.5*. The tree has two internal nodes and three terminal nodes, or leaves. The number in each leaf is the mean of the response for the observations that fall there.*

8.1.1   Regression Trees

In order to motivate *regression trees*, we begin with a simple example.

regression tree

Predicting Baseball Players' Salaries Using Regression Trees

We use the Hitters data set to predict a baseball player's Salary based on Years (the number of years that he has played in the major leagues) and Hits (the number of hits that he made in the previous year). We first remove observations that are missing Salary values, and log-transform Salary so that its distribution has more of a typical bell-shape. (Recall that Salary is measured in thousands of dollars.)

Figure 8.1 shows a regression tree fit to this data. It consists of a series of splitting rules, starting at the top of the tree. The top split assigns observations having Years<4.5 to the left branch.[1] The predicted salary

---

[1] Both Years and Hits are integers in these data; the tree() function in R labels the splits at the midpoint between two adjacent values.

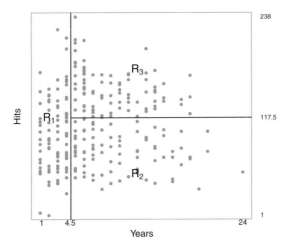

FIGURE 8.2. *The three-region partition for the* Hitters *data set from the regression tree illustrated in Figure 8.1.*

for these players is given by the mean response value for the players in the data set with Years<4.5. For such players, the mean log salary is 5.107, and so we make a prediction of $e^{5.107}$ thousands of dollars, i.e. \$165,174, for these players. Players with Years>=4.5 are assigned to the right branch, and then that group is further subdivided by Hits. Overall, the tree stratifies or segments the players into three regions of predictor space: players who have played for four or fewer years, players who have played for five or more years and who made fewer than 118 hits last year, and players who have played for five or more years and who made at least 118 hits last year. These three regions can be written as $R_1 =\{X \mid \text{Years<4.5}\}$, $R_2 =\{X \mid \text{Years>=4.5},$ Hits<117.5$\}$, and $R_3 =\{X \mid \text{Years>=4.5, Hits>=117.5}\}$. Figure 8.2 illustrates the regions as a function of Years and Hits. The predicted salaries for these three groups are \$1,000$\times e^{5.107}$ =\$165,174, \$1,000$\times e^{5.999}$ =\$402,834, and \$1,000$\times e^{6.740}$ =\$845,346 respectively.

In keeping with the *tree* analogy, the regions R_1, R_2, and R_3 are known as *terminal nodes* or *leaves* of the tree. As is the case for Figure 8.1, decision trees are typically drawn *upside down*, in the sense that the leaves are at the bottom of the tree. The points along the tree where the predictor space is split are referred to as *internal nodes*. In Figure 8.1, the two internal nodes are indicated by the text Years<4.5 and Hits<117.5. We refer to the segments of the trees that connect the nodes as *branches*.

terminal
node
leaf

internal node

branch

We might interpret the regression tree displayed in Figure 8.1 as follows: Years is the most important factor in determining Salary, and players with less experience earn lower salaries than more experienced players. Given that a player is less experienced, the number of hits that he made in the previous year seems to play little role in his salary. But among players who have been in the major leagues for five or more years, the number of hits made in the previous year does affect salary, and players who made more hits last year tend to have higher salaries. The regression tree shown in Figure 8.1 is likely an over-simplification of the true relationship between Hits, Years, and Salary. However, it has advantages over other types of regression models (such as those seen in Chapters 3 and 6): it is easier to interpret, and has a nice graphical representation.

Prediction via Stratification of the Feature Space

We now discuss the process of building a regression tree. Roughly speaking, there are two steps.

1. We divide the predictor space—that is, the set of possible values for $X_1, X_2, \ldots, X_p$—into J distinct and non-overlapping regions, $R_1, R_2, \ldots, R_J$.

2. For every observation that falls into the region R_j, we make the same prediction, which is simply the mean of the response values for the training observations in R_j.

For instance, suppose that in Step 1 we obtain two regions, R_1 and R_2, and that the response mean of the training observations in the first region is 10, while the response mean of the training observations in the second region is 20. Then for a given observation $X = x$, if $x \in R_1$ we will predict a value of 10, and if $x \in R_2$ we will predict a value of 20.

We now elaborate on Step 1 above. How do we construct the regions $R_1, \ldots, R_J$? In theory, the regions could have any shape. However, we choose to divide the predictor space into high-dimensional rectangles, or *boxes*, for simplicity and for ease of interpretation of the resulting predictive model. The goal is to find boxes $R_1, \ldots, R_J$ that minimize the RSS, given by

$$\sum_{j=1}^{J} \sum_{i \in R_j} (y_i - \hat{y}_{R_j})^2, \tag{8.1}$$

where $\hat{y}_{R_j}$ is the mean response for the training observations within the jth box. Unfortunately, it is computationally infeasible to consider every possible partition of the feature space into J boxes. For this reason, we take a *top-down*, *greedy* approach that is known as *recursive binary splitting*. The approach is *top-down* because it begins at the top of the tree (at which point all observations belong to a single region) and then successively splits the predictor space; each split is indicated via two new branches further down on the tree. It is *greedy* because at each step of the tree-building process, the *best* split is made at that particular step, rather than looking ahead and picking a split that will lead to a better tree in some future step.

recursive
binary
splitting

In order to perform recursive binary splitting, we first select the predictor X_j and the cutpoint s such that splitting the predictor space into the regions $\{X|X_j < s\}$ and $\{X|X_j \geq s\}$ leads to the greatest possible reduction in RSS. (The notation $\{X|X_j < s\}$ means *the region of predictor space in which X_j takes on a value less than s.*) That is, we consider all predictors $X_1, \ldots, X_p$, and all possible values of the cutpoint s for each of the predictors, and then choose the predictor and cutpoint such that the resulting tree has the lowest RSS. In greater detail, for any j and s, we define the pair of half-planes

$$R_1(j, s) = \{X|X_j < s\} \text{ and } R_2(j, s) = \{X|X_j \geq s\}, \tag{8.2}$$

and we seek the value of j and s that minimize the equation

$$\sum_{i:\ x_i \in R_1(j,s)} (y_i - \hat{y}_{R_1})^2 + \sum_{i:\ x_i \in R_2(j,s)} (y_i - \hat{y}_{R_2})^2, \tag{8.3}$$

where $\hat{y}_{R_1}$ is the mean response for the training observations in $R_1(j, s)$, and $\hat{y}_{R_2}$ is the mean response for the training observations in $R_2(j, s)$. Finding the values of j and s that minimize (8.3) can be done quite quickly, especially when the number of features p is not too large.

Next, we repeat the process, looking for the best predictor and best cutpoint in order to split the data further so as to minimize the RSS within each of the resulting regions. However, this time, instead of splitting the entire predictor space, we split one of the two previously identified regions. We now have three regions. Again, we look to split one of these three regions further, so as to minimize the RSS. The process continues until a stopping criterion is reached; for instance, we may continue until no region contains more than five observations.

Once the regions $R_1, \ldots, R_J$ have been created, we predict the response for a given test observation using the mean of the training observations in the region to which that test observation belongs.

A five-region example of this approach is shown in Figure 8.3.

Tree Pruning

The process described above may produce good predictions on the training set, but is likely to overfit the data, leading to poor test set performance. This is because the resulting tree might be too complex. A smaller tree with fewer splits (that is, fewer regions $R_1, \ldots, R_J$) might lead to lower variance and better interpretation at the cost of a little bias. One possible alternative to the process described above is to build the tree only so long as the decrease in the RSS due to each split exceeds some (high) threshold. This strategy will result in smaller trees, but is too short-sighted since a seemingly worthless split early on in the tree might be followed by a very good split—that is, a split that leads to a large reduction in RSS later on.

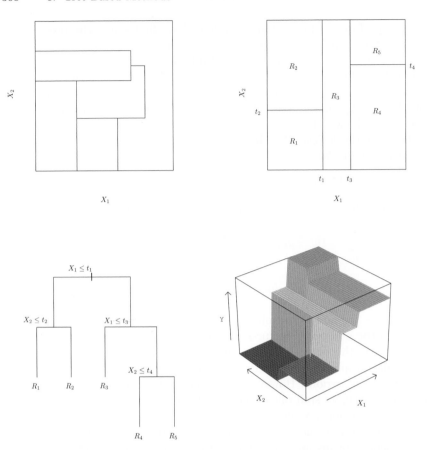

FIGURE 8.3. Top Left: *A partition of two-dimensional feature space that could not result from recursive binary splitting.* Top Right: *The output of recursive binary splitting on a two-dimensional example.* Bottom Left: *A tree corresponding to the partition in the top right panel.* Bottom Right: *A perspective plot of the prediction surface corresponding to that tree.*

Therefore, a better strategy is to grow a very large tree T_0, and then *prune* it back in order to obtain a *subtree*. How do we determine the best   prune
way to prune the tree? Intuitively, our goal is to select a subtree that   subtree
leads to the lowest test error rate. Given a subtree, we can estimate its
test error using cross-validation or the validation set approach. However,
estimating the cross-validation error for every possible subtree would be too
cumbersome, since there is an extremely large number of possible subtrees.
Instead, we need a way to select a small set of subtrees for consideration.

Cost complexity pruning—also known as *weakest link pruning*—gives us   cost
a way to do just this. Rather than considering every possible subtree, we   complexity
consider a sequence of trees indexed by a nonnegative tuning parameter α.   pruning

weakest link
pruning

Algorithm 8.1 *Building a Regression Tree*

1. Use recursive binary splitting to grow a large tree on the training data, stopping only when each terminal node has fewer than some minimum number of observations.

2. Apply cost complexity pruning to the large tree in order to obtain a sequence of best subtrees, as a function of α.

3. Use K-fold cross-validation to choose α. That is, divide the training observations into K folds. For each $k = 1, \ldots, K$:

   (a) Repeat Steps 1 and 2 on all but the kth fold of the training data.

   (b) Evaluate the mean squared prediction error on the data in the left-out kth fold, as a function of α.

   Average the results for each value of α, and pick α to minimize the average error.

4. Return the subtree from Step 2 that corresponds to the chosen value of α.

For each value of α there corresponds a subtree $T \subset T_0$ such that

$$\sum_{m=1}^{|T|} \sum_{i:\ x_i \in R_m} (y_i - \hat{y}_{R_m})^2 + \alpha|T| \tag{8.4}$$

is as small as possible. Here $|T|$ indicates the number of terminal nodes of the tree T, R_m is the rectangle (i.e. the subset of predictor space) corresponding to the mth terminal node, and $\hat{y}_{R_m}$ is the predicted response associated with R_m—that is, the mean of the training observations in R_m. The tuning parameter α controls a trade-off between the subtree's complexity and its fit to the training data. When $\alpha = 0$, then the subtree T will simply equal T_0, because then (8.4) just measures the training error. However, as α increases, there is a price to pay for having a tree with many terminal nodes, and so the quantity (8.4) will tend to be minimized for a smaller subtree. Equation 8.4 is reminiscent of the lasso (6.7) from Chapter 6, in which a similar formulation was used in order to control the complexity of a linear model.

It turns out that as we increase α from zero in (8.4), branches get pruned from the tree in a nested and predictable fashion, so obtaining the whole sequence of subtrees as a function of α is easy. We can select a value of α using a validation set or using cross-validation. We then return to the full data set and obtain the subtree corresponding to α. This process is summarized in Algorithm 8.1.

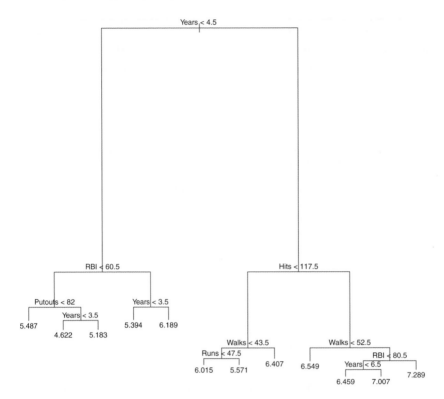

FIGURE 8.4. *Regression tree analysis for the* Hitters *data. The unpruned tree that results from top-down greedy splitting on the training data is shown.*

Figures 8.4 and 8.5 display the results of fitting and pruning a regression tree on the Hitters data, using nine of the features. First, we randomly divided the data set in half, yielding 132 observations in the training set and 131 observations in the test set. We then built a large regression tree on the training data and varied α in (8.4) in order to create subtrees with different numbers of terminal nodes. Finally, we performed six-fold cross-validation in order to estimate the cross-validated MSE of the trees as a function of α. (We chose to perform six-fold cross-validation because 132 is an exact multiple of six.) The unpruned regression tree is shown in Figure 8.4. The green curve in Figure 8.5 shows the CV error as a function of the number of leaves,[2] while the orange curve indicates the test error. Also shown are standard error bars around the estimated errors. For reference, the training error curve is shown in black. The CV error is a reasonable approximation of the test error: the CV error takes on its

---

[2]Although CV error is computed as a function of α, it is convenient to display the result as a function of $|T|$, the number of leaves; this is based on the relationship between α and $|T|$ in the original tree grown to all the training data.

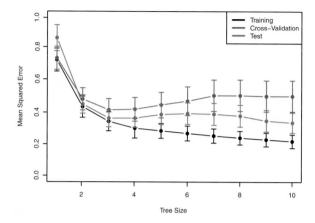

FIGURE 8.5. *Regression tree analysis for the* Hitters *data. The training, cross-validation, and test MSE are shown as a function of the number of terminal nodes in the pruned tree. Standard error bands are displayed. The minimum cross-validation error occurs at a tree size of three.*

minimum for a three-node tree, while the test error also dips down at the three-node tree (though it takes on its lowest value at the ten-node tree). The pruned tree containing three terminal nodes is shown in Figure 8.1.

8.1.2 Classification Trees

A *classification tree* is very similar to a regression tree, except that it is used to predict a qualitative response rather than a quantitative one. Recall that for a regression tree, the predicted response for an observation is given by the mean response of the training observations that belong to the same terminal node. In contrast, for a classification tree, we predict that each observation belongs to the *most commonly occurring class* of training observations in the region to which it belongs. In interpreting the results of a classification tree, we are often interested not only in the class prediction corresponding to a particular terminal node region, but also in the *class proportions* among the training observations that fall into that region.

classification tree

The task of growing a classification tree is quite similar to the task of growing a regression tree. Just as in the regression setting, we use recursive binary splitting to grow a classification tree. However, in the classification setting, RSS cannot be used as a criterion for making the binary splits. A natural alternative to RSS is the *classification error rate*. Since we plan to assign an observation in a given region to the *most commonly occurring class* of training observations in that region, the classification error rate is simply the fraction of the training observations in that region that do not belong to the most common class:

classification error rate

$$E = 1 - \max_k(\hat{p}_{mk}). \qquad (8.5)$$

Here $\hat{p}_{mk}$ represents the proportion of training observations in the mth region that are from the kth class. However, it turns out that classification error is not sufficiently sensitive for tree-growing, and in practice two other measures are preferable.

The *Gini index* is defined by

$$G = \sum_{k=1}^{K} \hat{p}_{mk}(1 - \hat{p}_{mk}), \qquad (8.6)$$

a measure of total variance across the K classes. It is not hard to see that the Gini index takes on a small value if all of the $\hat{p}_{mk}$'s are close to zero or one. For this reason the Gini index is referred to as a measure of node *purity*—a small value indicates that a node contains predominantly observations from a single class.

An alternative to the Gini index is *entropy*, given by

$$D = -\sum_{k=1}^{K} \hat{p}_{mk} \log \hat{p}_{mk}. \qquad (8.7)$$

Since $0 \le \hat{p}_{mk} \le 1$, it follows that $0 \le -\hat{p}_{mk} \log \hat{p}_{mk}$. One can show that the entropy will take on a value near zero if the $\hat{p}_{mk}$'s are all near zero or near one. Therefore, like the Gini index, the entropy will take on a small value if the mth node is pure. In fact, it turns out that the Gini index and the entropy are quite similar numerically.

When building a classification tree, either the Gini index or the entropy are typically used to evaluate the quality of a particular split, since these two approaches are more sensitive to node purity than is the classification error rate. Any of these three approaches might be used when *pruning* the tree, but the classification error rate is preferable if prediction accuracy of the final pruned tree is the goal.

Figure 8.6 shows an example on the Heart data set. These data contain a binary outcome HD for 303 patients who presented with chest pain. An outcome value of Yes indicates the presence of heart disease based on an angiographic test, while No means no heart disease. There are 13 predictors including Age, Sex, Chol (a cholesterol measurement), and other heart and lung function measurements. Cross-validation results in a tree with six terminal nodes.

In our discussion thus far, we have assumed that the predictor variables take on continuous values. However, decision trees can be constructed even in the presence of qualitative predictor variables. For instance, in the Heart data, some of the predictors, such as Sex, Thal (Thallium stress test), and ChestPain, are qualitative. Therefore, a split on one of these variables amounts to assigning some of the qualitative values to one branch and

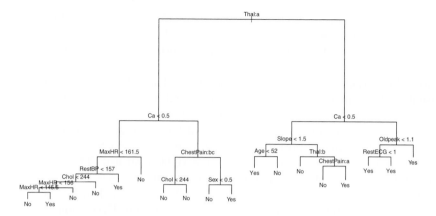

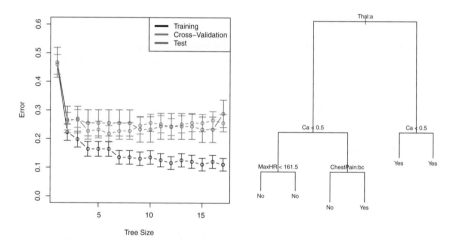

FIGURE 8.6. Heart *data.* Top: *The unpruned tree.* Bottom Left: *Cross-validation error, training, and test error, for different sizes of the pruned tree.* Bottom Right: *The pruned tree corresponding to the minimal cross-validation error.*

assigning the remaining to the other branch. In Figure 8.6, some of the internal nodes correspond to splitting qualitative variables. For instance, the top internal node corresponds to splitting Thal. The text Thal:a indicates that the left-hand branch coming out of that node consists of observations with the first value of the Thal variable (normal), and the right-hand node consists of the remaining observations (fixed or reversible defects). The text ChestPain:bc two splits down the tree on the left indicates that the left-hand branch coming out of that node consists of observations with the second and third values of the ChestPain variable, where the possible values are typical angina, atypical angina, non-anginal pain, and asymptomatic.

Figure 8.6 has a surprising characteristic: some of the splits yield two
terminal nodes that have the *same predicted value*. For instance, consider
the split RestECG<1 near the bottom right of the unpruned tree. Regardless
of the value of RestECG, a response value of Yes is predicted for those observations. Why, then, is the split performed at all? The split is performed
because it leads to increased *node purity*. That is, all 9 of the observations
corresponding to the right-hand leaf have a response value of Yes, whereas
7/11 of those corresponding to the left-hand leaf have a response value of
Yes. Why is node purity important? Suppose that we have a test observation that belongs to the region given by that right-hand leaf. Then we
can be pretty certain that its response value is Yes. In contrast, if a test
observation belongs to the region given by the left-hand leaf, then its response value is probably Yes, but we are much less certain. Even though
the split RestECG<1 does not reduce the classification error, it improves the
Gini index and the entropy, which are more sensitive to node purity.

8.1.3   Trees Versus Linear Models

Regression and classification trees have a very different flavor from the more
classical approaches for regression and classification presented in Chapters 3
and 4. In particular, linear regression assumes a model of the form

$$f(X) = \beta_0 + \sum_{j=1}^{p} X_j \beta_j, \tag{8.8}$$

whereas regression trees assume a model of the form

$$f(X) = \sum_{m=1}^{M} c_m \cdot 1_{(X \in R_m)} \tag{8.9}$$

where $R_1, \ldots, R_M$ represent a partition of feature space, as in Figure 8.3.
Which model is better? It depends on the problem at hand. If the
relationship between the features and the response is well approximated
by a linear model as in (8.8), then an approach such as linear regression
will likely work well, and will outperform a method such as a regression
tree that does not exploit this linear structure. If instead there is a highly
non-linear and complex relationship between the features and the response
as indicated by model (8.9), then decision trees may outperform classical
approaches. An illustrative example is displayed in Figure 8.7. The relative performances of tree-based and classical approaches can be assessed by
estimating the test error, using either cross-validation or the validation set
approach (Chapter 5).

Of course, other considerations beyond simply test error may come into
play in selecting a statistical learning method; for instance, in certain settings, prediction using a tree may be preferred for the sake of interpretability and visualization.

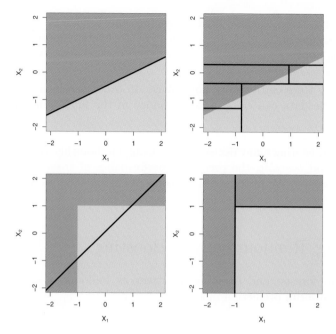

FIGURE 8.7. Top Row: *A two-dimensional classification example in which the true decision boundary is linear, and is indicated by the shaded regions. A classical approach that assumes a linear boundary (left) will outperform a decision tree that performs splits parallel to the axes (right).* Bottom Row: *Here the true decision boundary is non-linear. Here a linear model is unable to capture the true decision boundary (left), whereas a decision tree is successful (right).*

8.1.4 Advantages and Disadvantages of Trees

Decision trees for regression and classification have a number of advantages over the more classical approaches seen in Chapters 3 and 4:

▲ Trees are very easy to explain to people. In fact, they are even easier to explain than linear regression!

▲ Some people believe that decision trees more closely mirror human decision-making than do the regression and classification approaches seen in previous chapters.

▲ Trees can be displayed graphically, and are easily interpreted even by a non-expert (especially if they are small).

▲ Trees can easily handle qualitative predictors without the need to create dummy variables.

▼ Unfortunately, trees generally do not have the same level of predictive accuracy as some of the other regression and classification approaches seen in this book.

▼ Additionally, trees can be very non-robust. In other words, a small change in the data can cause a large change in the final estimated tree.

However, by aggregating many decision trees, using methods like *bagging*, *random forests*, and *boosting*, the predictive performance of trees can be substantially improved. We introduce these concepts in the next section.

8.2   Bagging, Random Forests, Boosting

Bagging, random forests, and boosting use trees as building blocks to construct more powerful prediction models.

8.2.1   Bagging

The bootstrap, introduced in Chapter 5, is an extremely powerful idea. It is used in many situations in which it is hard or even impossible to directly compute the standard deviation of a quantity of interest. We see here that the bootstrap can be used in a completely different context, in order to improve statistical learning methods such as decision trees.

The decision trees discussed in Section 8.1 suffer from *high variance*. This means that if we split the training data into two parts at random, and fit a decision tree to both halves, the results that we get could be quite different. In contrast, a procedure with *low variance* will yield similar results if applied repeatedly to distinct data sets; linear regression tends to have low variance, if the ratio of n to p is moderately large. *Bootstrap aggregation*, or *bagging*, is a general-purpose procedure for reducing the variance of a statistical learning method; we introduce it here because it is particularly useful and frequently used in the context of decision trees.

Recall that given a set of n independent observations $Z_1, \ldots, Z_n$, each with variance σ^2, the variance of the mean $\bar{Z}$ of the observations is given by σ^2/n. In other words, *averaging a set of observations reduces variance*. Hence a natural way to reduce the variance and hence increase the prediction accuracy of a statistical learning method is to take many training sets from the population, build a separate prediction model using each training set, and average the resulting predictions. In other words, we could calculate $\hat{f}^1(x), \hat{f}^2(x), \ldots, \hat{f}^B(x)$ using B separate training sets, and average them in order to obtain a single low-variance statistical learning model,

bagging

given by

$$\hat{f}_{\text{avg}}(x) = \frac{1}{B} \sum_{b=1}^{B} \hat{f}^{b}(x).$$

Of course, this is not practical because we generally do not have access to multiple training sets. Instead, we can bootstrap, by taking repeated samples from the (single) training data set. In this approach we generate B different bootstrapped training data sets. We then train our method on the bth bootstrapped training set in order to get $\hat{f}^{*b}(x)$, and finally average all the predictions, to obtain

$$\hat{f}_{\text{bag}}(x) = \frac{1}{B} \sum_{b=1}^{B} \hat{f}^{*b}(x).$$

This is called bagging.

While bagging can improve predictions for many regression methods, it is particularly useful for decision trees. To apply bagging to regression trees, we simply construct B regression trees using B bootstrapped training sets, and average the resulting predictions. These trees are grown deep, and are not pruned. Hence each individual tree has high variance, but low bias. Averaging these B trees reduces the variance. Bagging has been demonstrated to give impressive improvements in accuracy by combining together hundreds or even thousands of trees into a single procedure.

Thus far, we have described the bagging procedure in the regression context, to predict a quantitative outcome Y. How can bagging be extended to a classification problem where Y is qualitative? In that situation, there are a few possible approaches, but the simplest is as follows. For a given test observation, we can record the class predicted by each of the B trees, and take a *majority vote*: the overall prediction is the most commonly occurring class among the B predictions.

majority
vote

Figure 8.8 shows the results from bagging trees on the Heart data. The test error rate is shown as a function of B, the number of trees constructed using bootstrapped training data sets. We see that the bagging test error rate is slightly lower in this case than the test error rate obtained from a single tree. The number of trees B is not a critical parameter with bagging; using a very large value of B will not lead to overfitting. In practice we use a value of B sufficiently large that the error has settled down. Using $B = 100$ is sufficient to achieve good performance in this example.

Out-of-Bag Error Estimation

It turns out that there is a very straightforward way to estimate the test error of a bagged model, without the need to perform cross-validation or the validation set approach. Recall that the key to bagging is that trees are repeatedly fit to bootstrapped subsets of the observations. One can show

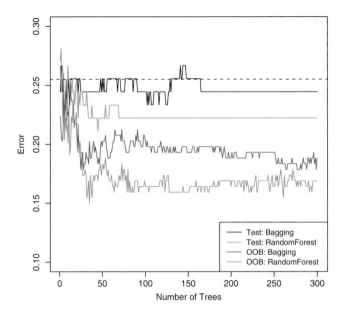

FIGURE 8.8. *Bagging and random forest results for the* Heart *data. The test error (black and orange) is shown as a function of B, the number of bootstrapped training sets used. Random forests were applied with $m = \sqrt{p}$. The dashed line indicates the test error resulting from a single classification tree. The green and blue traces show the OOB error, which in this case is considerably lower.*

that on average, each bagged tree makes use of around two-thirds of the observations.[3] The remaining one-third of the observations not used to fit a given bagged tree are referred to as the *out-of-bag* (OOB) observations. We can predict the response for the ith observation using each of the trees in which that observation was OOB. This will yield around $B/3$ predictions for the ith observation. In order to obtain a single prediction for the ith observation, we can average these predicted responses (if regression is the goal) or can take a majority vote (if classification is the goal). This leads to a single OOB prediction for the ith observation. An OOB prediction can be obtained in this way for each of the n observations, from which the overall OOB MSE (for a regression problem) or classification error (for a classification problem) can be computed. The resulting OOB error is a valid estimate of the test error for the bagged model, since the response for each observation is predicted using only the trees that were not fit using that observation. Figure 8.8 displays the OOB error on the Heart data. It can be shown that with B sufficiently large, OOB error is virtually equivalent to leave-one-out cross-validation error. The OOB approach for estimating

out-of-bag

---

[3]This relates to Exercise 2 of Chapter 5.

the test error is particularly convenient when performing bagging on large data sets for which cross-validation would be computationally onerous.

Variable Importance Measures

As we have discussed, bagging typically results in improved accuracy over prediction using a single tree. Unfortunately, however, it can be difficult to interpret the resulting model. Recall that one of the advantages of decision trees is the attractive and easily interpreted diagram that results, such as the one displayed in Figure 8.1. However, when we bag a large number of trees, it is no longer possible to represent the resulting statistical learning procedure using a single tree, and it is no longer clear which variables are most important to the procedure. Thus, bagging improves prediction accuracy at the expense of interpretability.

Although the collection of bagged trees is much more difficult to interpret than a single tree, one can obtain an overall summary of the importance of each predictor using the RSS (for bagging regression trees) or the Gini index (for bagging classification trees). In the case of bagging regression trees, we can record the total amount that the RSS (8.1) is decreased due to splits over a given predictor, averaged over all B trees. A large value indicates an important predictor. Similarly, in the context of bagging classification trees, we can add up the total amount that the Gini index (8.6) is decreased by splits over a given predictor, averaged over all B trees.

A graphical representation of the *variable importances* in the Heart data is shown in Figure 8.9. We see the mean decrease in Gini index for each variable, relative to the largest. The variables with the largest mean decrease in Gini index are Thal, Ca, and ChestPain.

variable importance

8.2.2 Random Forests

Random forests provide an improvement over bagged trees by way of a small tweak that *decorrelates* the trees. As in bagging, we build a number of decision trees on bootstrapped training samples. But when building these decision trees, each time a split in a tree is considered, *a random sample of m predictors* is chosen as split candidates from the full set of p predictors. The split is allowed to use only one of those m predictors. A fresh sample of m predictors is taken at each split, and typically we choose $m \approx \sqrt{p}$—that is, the number of predictors considered at each split is approximately equal to the square root of the total number of predictors (4 out of the 13 for the Heart data).

random forest

In other words, in building a random forest, at each split in the tree, the algorithm is *not even allowed to consider* a majority of the available predictors. This may sound crazy, but it has a clever rationale. Suppose that there is one very strong predictor in the data set, along with a number of other moderately strong predictors. Then in the collection of bagged

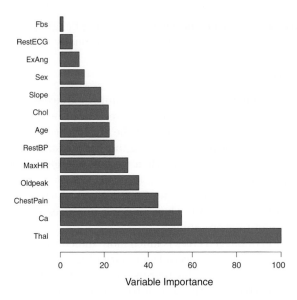

FIGURE 8.9. *A variable importance plot for the* Heart *data. Variable importance is computed using the mean decrease in Gini index, and expressed relative
to the maximum.*

trees, most or all of the trees will use this strong predictor in the top split.
Consequently, all of the bagged trees will look quite similar to each other.
Hence the predictions from the bagged trees will be highly correlated. Unfortunately, averaging many highly correlated quantities does not lead to
as large of a reduction in variance as averaging many uncorrelated quantities. In particular, this means that bagging will not lead to a substantial
reduction in variance over a single tree in this setting.

Random forests overcome this problem by forcing each split to consider
only a subset of the predictors. Therefore, on average $(p - m)/p$ of the
splits will not even consider the strong predictor, and so other predictors
will have more of a chance. We can think of this process as *decorrelating*
the trees, thereby making the average of the resulting trees less variable
and hence more reliable.

The main difference between bagging and random forests is the choice
of predictor subset size m. For instance, if a random forest is built using
$m = p$, then this amounts simply to bagging. On the Heart data, random
forests using $m = \sqrt{p}$ leads to a reduction in both test error and OOB error
over bagging (Figure 8.8).

Using a small value of m in building a random forest will typically be
helpful when we have a large number of correlated predictors. We applied
random forests to a high-dimensional biological data set consisting of expression measurements of 4,718 genes measured on tissue samples from 349
patients. There are around 20,000 genes in humans, and individual genes

have different levels of activity, or expression, in particular cells, tissues, and biological conditions. In this data set, each of the patient samples has a qualitative label with 15 different levels: either normal or 1 of 14 different types of cancer. Our goal was to use random forests to predict cancer type based on the 500 genes that have the largest variance in the training set. We randomly divided the observations into a training and a test set, and applied random forests to the training set for three different values of the number of splitting variables m. The results are shown in Figure 8.10. The error rate of a single tree is 45.7 %, and the null rate is 75.4 %.[4] We see that using 400 trees is sufficient to give good performance, and that the choice $m = \sqrt{p}$ gave a small improvement in test error over bagging $(m = p)$ in this example. As with bagging, random forests will not overfit if we increase B, so in practice we use a value of B sufficiently large for the error rate to have settled down.

8.2.3   Boosting

We now discuss *boosting*, yet another approach for improving the predictions resulting from a decision tree. Like bagging, boosting is a general approach that can be applied to many statistical learning methods for regression or classification. Here we restrict our discussion of boosting to the context of decision trees.

boosting

Recall that bagging involves creating multiple copies of the original training data set using the bootstrap, fitting a separate decision tree to each copy, and then combining all of the trees in order to create a single predictive model. Notably, each tree is built on a bootstrap data set, independent of the other trees. Boosting works in a similar way, except that the trees are grown *sequentially*: each tree is grown using information from previously grown trees. Boosting does not involve bootstrap sampling; instead each tree is fit on a modified version of the original data set.

Consider first the regression setting. Like bagging, boosting involves combining a large number of decision trees, $\hat{f}^1, \ldots, \hat{f}^B$. Boosting is described in Algorithm 8.2.

What is the idea behind this procedure? Unlike fitting a single large decision tree to the data, which amounts to *fitting the data hard* and potentially overfitting, the boosting approach instead *learns slowly*. Given the current model, we fit a decision tree to the residuals from the model. That is, we fit a tree using the current residuals, rather than the outcome Y, as the response. We then add this new decision tree into the fitted function in order to update the residuals. Each of these trees can be rather small, with just a few terminal nodes, determined by the parameter d in the algorithm. By

---

[4]The null rate results from simply classifying each observation to the dominant class overall, which is in this case the normal class.

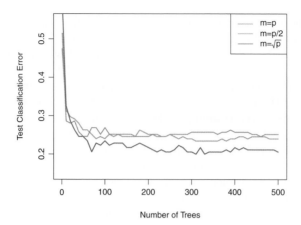

FIGURE 8.10. *Results from random forests for the 15-class gene expression data set with $p = 500$ predictors. The test error is displayed as a function of the number of trees. Each colored line corresponds to a different value of m, the number of predictors available for splitting at each interior tree node. Random forests $(m < p)$ lead to a slight improvement over bagging $(m = p)$. A single classification tree has an error rate of 45.7%.*

fitting small trees to the residuals, we slowly improve $\hat{f}$ in areas where it does not perform well. The shrinkage parameter λ slows the process down even further, allowing more and different shaped trees to attack the residuals. In general, statistical learning approaches that *learn slowly* tend to perform well. Note that in boosting, unlike in bagging, the construction of each tree depends strongly on the trees that have already been grown.

We have just described the process of boosting regression trees. Boosting classification trees proceeds in a similar but slightly more complex way, and the details are omitted here.

Boosting has three tuning parameters:

1. The number of trees B. Unlike bagging and random forests, boosting can overfit if B is too large, although this overfitting tends to occur slowly if at all. We use cross-validation to select B.

2. The shrinkage parameter λ, a small positive number. This controls the rate at which boosting learns. Typical values are 0.01 or 0.001, and the right choice can depend on the problem. Very small λ can require using a very large value of B in order to achieve good performance.

3. The number d of splits in each tree, which controls the complexity of the boosted ensemble. Often $d = 1$ works well, in which case each tree is a *stump*, consisting of a single split. In this case, the boosted ensemble is fitting an additive model, since each term involves only a single variable. More generally d is the *interaction depth*, and controls

stump

interaction
depth

Algorithm 8.2 *Boosting for Regression Trees*

1. Set $\hat{f}(x) = 0$ and $r_i = y_i$ for all i in the training set.

2. For $b = 1, 2, \ldots, B$, repeat:

   (a) Fit a tree $\hat{f}^b$ with d splits ($d+1$ terminal nodes) to the training data (X, r).

   (b) Update $\hat{f}$ by adding in a shrunken version of the new tree:

   $$\hat{f}(x) \leftarrow \hat{f}(x) + \lambda \hat{f}^b(x). \tag{8.10}$$

   (c) Update the residuals,

   $$r_i \leftarrow r_i - \lambda \hat{f}^b(x_i). \tag{8.11}$$

3. Output the boosted model,

$$\hat{f}(x) = \sum_{b=1}^{B} \lambda \hat{f}^b(x). \tag{8.12}$$

---

the interaction order of the boosted model, since d splits can involve at most d variables.

In Figure 8.11, we applied boosting to the 15-class cancer gene expression data set, in order to develop a classifier that can distinguish the normal class from the 14 cancer classes. We display the test error as a function of the total number of trees and the interaction depth d. We see that simple stumps with an interaction depth of one perform well if enough of them are included. This model outperforms the depth-two model, and both outperform a random forest. This highlights one difference between boosting and random forests: in boosting, because the growth of a particular tree takes into account the other trees that have already been grown, smaller trees are typically sufficient. Using smaller trees can aid in interpretability as well; for instance, using stumps leads to an additive model.

8.3 Lab: Decision Trees

8.3.1 Fitting Classification Trees

The tree library is used to construct classification and regression trees.

```
> library(tree)
```

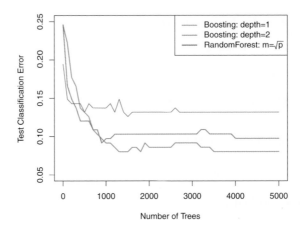

FIGURE 8.11. *Results from performing boosting and random forests on the 15-class gene expression data set in order to predict* cancer *versus* normal. *The test error is displayed as a function of the number of trees. For the two boosted models,* $\lambda = 0.01$. *Depth-1 trees slightly outperform depth-2 trees, and both outperform the random forest, although the standard errors are around 0.02, making none of these differences significant. The test error rate for a single tree is 24%.*

We first use classification trees to analyze the `Carseats` data set. In these data, `Sales` is a continuous variable, and so we begin by recoding it as a binary variable. We use the `ifelse()` function to create a variable, called `High`, which takes on a value of `Yes` if the `Sales` variable exceeds 8, and takes on a value of `No` otherwise.

`ifelse()`

```
> library(ISLR)
> attach(Carseats)
> High=ifelse(Sales<=8,"No","Yes")
```

Finally, we use the `data.frame()` function to merge `High` with the rest of the `Carseats` data.

```
> Carseats=data.frame(Carseats,High)
```

We now use the `tree()` function to fit a classification tree in order to predict `High` using all variables but `Sales`. The syntax of the `tree()` function is quite similar to that of the `lm()` function.

`tree()`

```
> tree.carseats=tree(High~.-Sales,Carseats)
```

The `summary()` function lists the variables that are used as internal nodes in the tree, the number of terminal nodes, and the (training) error rate.

```
> summary(tree.carseats)

Classification tree:
tree(formula = High ~ . - Sales, data = Carseats)
Variables actually used in tree construction:
[1] "ShelveLoc"    "Price"        "Income"       "CompPrice"
```

```
[5] "Population"   "Advertising" "Age"              "US"
Number of terminal nodes:   27
Residual mean deviance:   0.4575 = 170.7 / 373
Misclassification error rate: 0.09 = 36 / 400
```

We see that the training error rate is 9%. For classification trees, the deviance reported in the output of summary() is given by

$$-2\sum_m \sum_k n_{mk}\log\hat{p}_{mk},$$

where n_{mk} is the number of observations in the mth terminal node that belong to the kth class. A small deviance indicates a tree that provides a good fit to the (training) data. The *residual mean deviance* reported is simply the deviance divided by $n-|T_0|$, which in this case is $400-27 = 373$.

One of the most attractive properties of trees is that they can be graphically displayed. We use the plot() function to display the tree structure, and the text() function to display the node labels. The argument pretty=0 instructs R to include the category names for any qualitative predictors, rather than simply displaying a letter for each category.

```
> plot(tree.carseats)
> text(tree.carseats,pretty=0)
```

The most important indicator of Sales appears to be shelving location, since the first branch differentiates Good locations from Bad and Medium locations.

If we just type the name of the tree object, R prints output corresponding to each branch of the tree. R displays the split criterion (e.g. Price<92.5), the number of observations in that branch, the deviance, the overall prediction for the branch (Yes or No), and the fraction of observations in that branch that take on values of Yes and No. Branches that lead to terminal nodes are indicated using asterisks.

```
> tree.carseats
node), split, n, deviance, yval, (yprob)
       * denotes terminal node
 1) root 400 541.5 No ( 0.590 0.410 )
   2) ShelveLoc: Bad,Medium 315 390.6 No ( 0.689 0.311 )
     4) Price < 92.5 46   56.53 Yes ( 0.304 0.696 )
       8) Income < 57 10   12.22 No ( 0.700 0.300 )
```

In order to properly evaluate the performance of a classification tree on these data, we must estimate the test error rather than simply computing the training error. We split the observations into a training set and a test set, build the tree using the training set, and evaluate its performance on the test data. The predict() function can be used for this purpose. In the case of a classification tree, the argument type="class" instructs R to return the actual class prediction. This approach leads to correct predictions for around 71.5% of the locations in the test data set.

```
> set.seed(2)
> train=sample(1:nrow(Carseats), 200)
> Carseats.test=Carseats[-train,]
> High.test=High[-train]
> tree.carseats=tree(High~.-Sales,Carseats,subset=train)
> tree.pred=predict(tree.carseats,Carseats.test,type="class")
> table(tree.pred,High.test)
          High.test
tree.pred No Yes
      No   86  27
      Yes  30  57
> (86+57)/200
[1] 0.715
```

Next, we consider whether pruning the tree might lead to improved results. The function `cv.tree()` performs cross-validation in order to determine the optimal level of tree complexity; cost complexity pruning is used in order to select a sequence of trees for consideration. We use the argument `FUN=prune.misclass` in order to indicate that we want the classification error rate to guide the cross-validation and pruning process, rather than the default for the `cv.tree()` function, which is deviance. The `cv.tree()` function reports the number of terminal nodes of each tree considered (`size`) as well as the corresponding error rate and the value of the cost-complexity parameter used (`k`, which corresponds to α in (8.4)).

`cv.tree()`

```
> set.seed(3)
> cv.carseats=cv.tree(tree.carseats,FUN=prune.misclass)
> names(cv.carseats)
[1] "size"   "dev"    "k"        "method"
> cv.carseats
$size
[1] 19 17 14 13  9  7  3  2  1

$dev
[1] 55 55 53 52 50 56 69 65 80

$k
[1]        -Inf   0.0000000   0.6666667   1.0000000   1.7500000
      2.0000000   4.2500000
[8]   5.0000000  23.0000000

$method
[1] "misclass"

attr(,"class")
[1] "prune"           "tree.sequence"
```

Note that, despite the name, `dev` corresponds to the cross-validation error rate in this instance. The tree with 9 terminal nodes results in the lowest cross-validation error rate, with 50 cross-validation errors. We plot the error rate as a function of both `size` and `k`.

```
> par(mfrow=c(1,2))
```

```
> plot(cv.carseats$size,cv.carseats$dev,type="b")
> plot(cv.carseats$k,cv.carseats$dev,type="b")
```

We now apply the `prune.misclass()` function in order to prune the tree to obtain the nine-node tree.

prune.
misclass()

```
> prune.carseats=prune.misclass(tree.carseats,best=9)
> plot(prune.carseats)
> text(prune.carseats,pretty=0)
```

How well does this pruned tree perform on the test data set? Once again, we apply the `predict()` function.

```
> tree.pred=predict(prune.carseats,Carseats.test,type="class")
> table(tree.pred,High.test)
         High.test
tree.pred No  Yes
      No  94   24
     Yes  22   60
> (94+60)/200
[1] 0.77
```

Now 77 % of the test observations are correctly classified, so not only has the pruning process produced a more interpretable tree, but it has also improved the classification accuracy.

If we increase the value of `best`, we obtain a larger pruned tree with lower classification accuracy:

```
> prune.carseats=prune.misclass(tree.carseats,best=15)
> plot(prune.carseats)
> text(prune.carseats,pretty=0)
> tree.pred=predict(prune.carseats,Carseats.test,type="class")
> table(tree.pred,High.test)
         High.test
tree.pred No  Yes
      No  86   22
     Yes  30   62
> (86+62)/200
[1] 0.74
```

8.3.2 Fitting Regression Trees

Here we fit a regression tree to the `Boston` data set. First, we create a training set, and fit the tree to the training data.

```
> library(MASS)
> set.seed(1)
> train = sample(1:nrow(Boston), nrow(Boston)/2)
> tree.boston=tree(medv~.,Boston,subset=train)
> summary(tree.boston)

Regression tree:
tree(formula = medv ~ ., data = Boston, subset = train)
```

```
Variables actually used in tree construction:
[1] "lstat" "rm"      "dis"
Number of terminal nodes:   8
Residual mean deviance:   12.65 = 3099 / 245
Distribution of residuals:
     Min.   1st Qu.   Median      Mean   3rd Qu.      Max.
 -14.1000   -2.0420   -0.0536    0.0000    1.9600   12.6000
```

Notice that the output of summary() indicates that only three of the variables have been used in constructing the tree. In the context of a regression tree, the deviance is simply the sum of squared errors for the tree. We now plot the tree.

```
> plot(tree.boston)
> text(tree.boston,pretty=0)
```

The variable lstat measures the percentage of individuals with lower socioeconomic status. The tree indicates that lower values of lstat correspond to more expensive houses. The tree predicts a median house price of $46,400$ for larger homes in suburbs in which residents have high socioeconomic status (rm>=7.437 and lstat<9.715).

Now we use the cv.tree() function to see whether pruning the tree will improve performance.

```
> cv.boston=cv.tree(tree.boston)
> plot(cv.boston$size,cv.boston$dev,type='b')
```

In this case, the most complex tree is selected by cross-validation. However, if we wish to prune the tree, we could do so as follows, using the prune.tree() function:

prune.tree()

```
> prune.boston=prune.tree(tree.boston,best=5)
> plot(prune.boston)
> text(prune.boston,pretty=0)
```

In keeping with the cross-validation results, we use the unpruned tree to make predictions on the test set.

```
> yhat=predict(tree.boston,newdata=Boston[-train,])
> boston.test=Boston[-train,"medv"]
> plot(yhat,boston.test)
> abline(0,1)
> mean((yhat-boston.test)^2)
[1] 25.05
```

In other words, the test set MSE associated with the regression tree is 25.05. The square root of the MSE is therefore around 5.005, indicating that this model leads to test predictions that are within around $5,005$ of the true median home value for the suburb.

8.3.3 Bagging and Random Forests

Here we apply bagging and random forests to the Boston data, using the randomForest package in R. The exact results obtained in this section may

depend on the version of R and the version of the randomForest package
installed on your computer. Recall that bagging is simply a special case of
a random forest with $m = p$. Therefore, the randomForest() function can
be used to perform both random forests and bagging. We perform bagging
as follows:

random
Forest()

```
> library (randomForest)
> set.seed(1)
> bag.boston=randomForest(medv~.,data=Boston,subset=train,
    mtry=13,importance=TRUE)
> bag.boston

Call:
 randomForest(formula = medv ~ ., data = Boston, mtry = 13,
      importance = TRUE,          subset = train)
                Type of random forest: regression
                      Number of trees: 500
No. of variables tried at each split: 13

            Mean of squared residuals: 10.77
                    % Var explained: 86.96
```

The argument mtry=13 indicates that all 13 predictors should be considered
for each split of the tree—in other words, that bagging should be done. How
well does this bagged model perform on the test set?

```
> yhat.bag = predict(bag.boston,newdata=Boston[-train,])
> plot(yhat.bag, boston.test)
> abline(0,1)
> mean((yhat.bag-boston.test)^2)
[1] 13.16
```

The test set MSE associated with the bagged regression tree is 13.16, almost
half that obtained using an optimally-pruned single tree. We could change
the number of trees grown by randomForest() using the ntree argument:

```
> bag.boston=randomForest(medv~.,data=Boston,subset=train,
    mtry=13,ntree=25)
> yhat.bag = predict(bag.boston,newdata=Boston[-train,])
> mean((yhat.bag-boston.test)^2)
[1] 13.31
```

Growing a random forest proceeds in exactly the same way, except that
we use a smaller value of the mtry argument. By default, randomForest()
uses $p/3$ variables when building a random forest of regression trees, and
$\sqrt{p}$ variables when building a random forest of classification trees. Here we
use mtry = 6.

```
> set.seed(1)
> rf.boston=randomForest(medv~.,data=Boston,subset=train,
    mtry=6,importance=TRUE)
> yhat.rf = predict(rf.boston,newdata=Boston[-train,])
> mean((yhat.rf-boston.test)^2)
[1] 11.31
```

The test set MSE is 11.31; this indicates that random forests yielded an improvement over bagging in this case.

Using the `importance()` function, we can view the importance of each variable.

importance()

```
> importance (rf.boston)
          %IncMSE IncNodePurity
crim       12.384       1051.54
zn          2.103         50.31
indus       8.390       1017.64
chas        2.294         56.32
nox        12.791       1107.31
rm         30.754       5917.26
age        10.334        552.27
dis        14.641       1223.93
rad         3.583         84.30
tax         8.139        435.71
ptratio    11.274        817.33
black       8.097        367.00
lstat      30.962       7713.63
```

Two measures of variable importance are reported. The former is based upon the mean decrease of accuracy in predictions on the out of bag samples when a given variable is excluded from the model. The latter is a measure of the total decrease in node impurity that results from splits over that variable, averaged over all trees (this was plotted in Figure 8.9). In the case of regression trees, the node impurity is measured by the training RSS, and for classification trees by the deviance. Plots of these importance measures can be produced using the `varImpPlot()` function.

varImpPlot()

```
> varImpPlot (rf.boston)
```

The results indicate that across all of the trees considered in the random forest, the wealth level of the community (`lstat`) and the house size (`rm`) are by far the two most important variables.

8.3.4   Boosting

Here we use the gbm package, and within it the `gbm()` function, to fit boosted regression trees to the Boston data set. We run `gbm()` with the option `distribution="gaussian"` since this is a regression problem; if it were a binary classification problem, we would use `distribution="bernoulli"`. The argument `n.trees=5000` indicates that we want 5000 trees, and the option `interaction.depth=4` limits the depth of each tree.

gbm()

```
> library (gbm)
> set.seed(1)
> boost.boston=gbm(medv~.,data=Boston[train,],distribution=
     "gaussian",n.trees=5000,interaction.depth=4)
```

The `summary()` function produces a relative influence plot and also outputs the relative influence statistics.

```
> summary (boost.boston)
            var    rel.inf
1         lstat    45.96
2            rm    31.22
3           dis     6.81
4          crim     4.07
5           nox     2.56
6       ptratio     2.27
7         black     1.80
8           age     1.64
9           tax     1.36
10        indus     1.27
11         chas     0.80
12          rad     0.20
13           zn     0.015
```

We see that lstat and rm are by far the most important variables. We can also produce *partial dependence plots* for these two variables. These plots illustrate the marginal effect of the selected variables on the response after *integrating* out the other variables. In this case, as we might expect, median house prices are increasing with rm and decreasing with lstat.

partial dependence plot

```
> par(mfrow=c(1,2))
> plot(boost.boston,i="rm")
> plot(boost.boston,i="lstat")
```

We now use the boosted model to predict medv on the test set:

```
> yhat.boost=predict(boost.boston,newdata=Boston[-train,],
            n.trees=5000)
> mean((yhat.boost-boston.test)^2)
[1] 11.8
```

The test MSE obtained is 11.8; similar to the test MSE for random forests and superior to that for bagging. If we want to, we can perform boosting with a different value of the shrinkage parameter λ in (8.10). The default value is 0.001, but this is easily modified. Here we take $\lambda = 0.2$.

```
> boost.boston=gbm(medv~.,data=Boston[train,],distribution=
    "gaussian",n.trees=5000,interaction.depth=4,shrinkage=0.2,
    verbose=F)
> yhat.boost=predict(boost.boston,newdata=Boston[-train,],
            n.trees=5000)
> mean((yhat.boost-boston.test)^2)
[1] 11.5
```

In this case, using $\lambda = 0.2$ leads to a slightly lower test MSE than $\lambda = 0.001$.

8.4   Exercises

Conceptual

1. Draw an example (of your own invention) of a partition of two-dimensional feature space that could result from recursive binary splitting. Your example should contain at least six regions. Draw a decision tree corresponding to this partition. Be sure to label all aspects of your figures, including the regions $R_1, R_2, \ldots$, the cutpoints $t_1, t_2, \ldots$, and so forth.

   Hint: Your result should look something like Figures 8.1 and 8.2.

2. It is mentioned in Section 8.2.3 that boosting using depth-one trees (or *stumps*) leads to an *additive* model: that is, a model of the form

$$f(X) = \sum_{j=1}^{p} f_j(X_j).$$

   Explain why this is the case. You can begin with (8.12) in Algorithm 8.2.

3. Consider the Gini index, classification error, and entropy in a simple classification setting with two classes. Create a single plot that displays each of these quantities as a function of $\hat{p}_{m1}$. The x-axis should display $\hat{p}_{m1}$, ranging from 0 to 1, and the y-axis should display the value of the Gini index, classification error, and entropy.

   Hint: In a setting with two classes, $\hat{p}_{m1} = 1 - \hat{p}_{m2}$. You could make this plot by hand, but it will be much easier to make in R.

4. This question relates to the plots in Figure 8.12.

   (a) Sketch the tree corresponding to the partition of the predictor space illustrated in the left-hand panel of Figure 8.12. The numbers inside the boxes indicate the mean of Y within each region.

   (b) Create a diagram similar to the left-hand panel of Figure 8.12, using the tree illustrated in the right-hand panel of the same figure. You should divide up the predictor space into the correct regions, and indicate the mean for each region.

5. Suppose we produce ten bootstrapped samples from a data set containing red and green classes. We then apply a classification tree to each bootstrapped sample and, for a specific value of X, produce 10 estimates of $P(\text{Class is Red}|X)$:

$$0.1, 0.15, 0.2, 0.2, 0.55, 0.6, 0.6, 0.65, 0.7, \text{ and } 0.75.$$

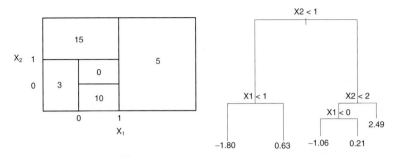

FIGURE 8.12. Left: *A partition of the predictor space corresponding to Exercise 4a.* Right: *A tree corresponding to Exercise 4b.*

There are two common ways to combine these results together into a single class prediction. One is the majority vote approach discussed in this chapter. The second approach is to classify based on the average probability. In this example, what is the final classification under each of these two approaches?

6. Provide a detailed explanation of the algorithm that is used to fit a regression tree.

Applied

7. In the lab, we applied random forests to the Boston data using mtry=6 and using ntree=25 and ntree=500. Create a plot displaying the test error resulting from random forests on this data set for a more comprehensive range of values for mtry and ntree. You can model your plot after Figure 8.10. Describe the results obtained.

8. In the lab, a classification tree was applied to the Carseats data set after converting Sales into a qualitative response variable. Now we will seek to predict Sales using regression trees and related approaches, treating the response as a quantitative variable.

   (a) Split the data set into a training set and a test set.

   (b) Fit a regression tree to the training set. Plot the tree, and interpret the results. What test MSE do you obtain?

   (c) Use cross-validation in order to determine the optimal level of tree complexity. Does pruning the tree improve the test MSE?

   (d) Use the bagging approach in order to analyze this data. What test MSE do you obtain? Use the importance() function to determine which variables are most important.

(e) Use random forests to analyze this data. What test MSE do you obtain? Use the `importance()` function to determine which variables are most important. Describe the effect of m, the number of variables considered at each split, on the error rate obtained.

9. This problem involves the `OJ` data set which is part of the `ISLR` package.

   (a) Create a training set containing a random sample of 800 observations, and a test set containing the remaining observations.

   (b) Fit a tree to the training data, with `Purchase` as the response and the other variables as predictors. Use the `summary()` function to produce summary statistics about the tree, and describe the results obtained. What is the training error rate? How many terminal nodes does the tree have?

   (c) Type in the name of the tree object in order to get a detailed text output. Pick one of the terminal nodes, and interpret the information displayed.

   (d) Create a plot of the tree, and interpret the results.

   (e) Predict the response on the test data, and produce a confusion matrix comparing the test labels to the predicted test labels. What is the test error rate?

   (f) Apply the `cv.tree()` function to the training set in order to determine the optimal tree size.

   (g) Produce a plot with tree size on the x-axis and cross-validated classification error rate on the y-axis.

   (h) Which tree size corresponds to the lowest cross-validated classification error rate?

   (i) Produce a pruned tree corresponding to the optimal tree size obtained using cross-validation. If cross-validation does not lead to selection of a pruned tree, then create a pruned tree with five terminal nodes.

   (j) Compare the training error rates between the pruned and unpruned trees. Which is higher?

   (k) Compare the test error rates between the pruned and unpruned trees. Which is higher?

10. We now use boosting to predict `Salary` in the `Hitters` data set.

   (a) Remove the observations for whom the salary information is unknown, and then log-transform the salaries.

(b) Create a training set consisting of the first 200 observations, and a test set consisting of the remaining observations.

(c) Perform boosting on the training set with 1,000 trees for a range of values of the shrinkage parameter λ. Produce a plot with different shrinkage values on the x-axis and the corresponding training set MSE on the y-axis.

(d) Produce a plot with different shrinkage values on the x-axis and the corresponding test set MSE on the y-axis.

(e) Compare the test MSE of boosting to the test MSE that results from applying two of the regression approaches seen in Chapters 3 and 6.

(f) Which variables appear to be the most important predictors in the boosted model?

(g) Now apply bagging to the training set. What is the test set MSE for this approach?

11. This question uses the Caravan data set.

(a) Create a training set consisting of the first 1,000 observations, and a test set consisting of the remaining observations.

(b) Fit a boosting model to the training set with Purchase as the response and the other variables as predictors. Use 1,000 trees, and a shrinkage value of 0.01. Which predictors appear to be the most important?

(c) Use the boosting model to predict the response on the test data. Predict that a person will make a purchase if the estimated probability of purchase is greater than 20 %. Form a confusion matrix. What fraction of the people predicted to make a purchase do in fact make one? How does this compare with the results obtained from applying KNN or logistic regression to this data set?

12. Apply boosting, bagging, and random forests to a data set of your choice. Be sure to fit the models on a training set and to evaluate their performance on a test set. How accurate are the results compared to simple methods like linear or logistic regression? Which of these approaches yields the best performance?

9
Support Vector Machines

In this chapter, we discuss the *support vector machine* (SVM), an approach for classification that was developed in the computer science community in the 1990s and that has grown in popularity since then. SVMs have been shown to perform well in a variety of settings, and are often considered one of the best "out of the box" classifiers.

The support vector machine is a generalization of a simple and intuitive classifier called the *maximal margin classifier*, which we introduce in Section 9.1. Though it is elegant and simple, we will see that this classifier unfortunately cannot be applied to most data sets, since it requires that the classes be separable by a linear boundary. In Section 9.2, we introduce the *support vector classifier*, an extension of the maximal margin classifier that can be applied in a broader range of cases. Section 9.3 introduces the *support vector machine*, which is a further extension of the support vector classifier in order to accommodate non-linear class boundaries. Support vector machines are intended for the binary classification setting in which there are two classes; in Section 9.4 we discuss extensions of support vector machines to the case of more than two classes. In Section 9.5 we discuss the close connections between support vector machines and other statistical methods such as logistic regression.

People often loosely refer to the maximal margin classifier, the support vector classifier, and the support vector machine as "support vector machines". To avoid confusion, we will carefully distinguish between these three notions in this chapter.

G. James et al., *An Introduction to Statistical Learning: with Applications in R*, 337
Springer Texts in Statistics, DOI 10.1007/978-1-4614-7138-7_9,
© Springer Science+Business Media New York 2013

9.1 Maximal Margin Classifier

In this section, we define a hyperplane and introduce the concept of an optimal separating hyperplane.

9.1.1 What Is a Hyperplane?

In a p-dimensional space, a *hyperplane* is a flat affine subspace of dimension $p - 1$.[1] For instance, in two dimensions, a hyperplane is a flat one-dimensional subspace—in other words, a line. In three dimensions, a hyperplane is a flat two-dimensional subspace—that is, a plane. In $p > 3$ dimensions, it can be hard to visualize a hyperplane, but the notion of a $(p - 1)$-dimensional flat subspace still applies.

hyperplane

The mathematical definition of a hyperplane is quite simple. In two dimensions, a hyperplane is defined by the equation

$$\beta_0 + \beta_1 X_1 + \beta_2 X_2 = 0 \tag{9.1}$$

for parameters β_0, β_1, and β_2. When we say that (9.1) "defines" the hyperplane, we mean that any $X = (X_1, X_2)^T$ for which (9.1) holds is a point on the hyperplane. Note that (9.1) is simply the equation of a line, since indeed in two dimensions a hyperplane is a line.

Equation 9.1 can be easily extended to the p-dimensional setting:

$$\beta_0 + \beta_1 X_1 + \beta_2 X_2 + \ldots + \beta_p X_p = 0 \tag{9.2}$$

defines a p-dimensional hyperplane, again in the sense that if a point $X = (X_1, X_2, \ldots, X_p)^T$ in p-dimensional space (i.e. a vector of length p) satisfies (9.2), then X lies on the hyperplane.

Now, suppose that X does not satisfy (9.2); rather,

$$\beta_0 + \beta_1 X_1 + \beta_2 X_2 + \ldots + \beta_p X_p > 0. \tag{9.3}$$

Then this tells us that X lies to one side of the hyperplane. On the other hand, if

$$\beta_0 + \beta_1 X_1 + \beta_2 X_2 + \ldots + \beta_p X_p < 0, \tag{9.4}$$

then X lies on the other side of the hyperplane. So we can think of the hyperplane as dividing p-dimensional space into two halves. One can easily determine on which side of the hyperplane a point lies by simply calculating the sign of the left hand side of (9.2). A hyperplane in two-dimensional space is shown in Figure 9.1.

---

[1] The word *affine* indicates that the subspace need not pass through the origin.

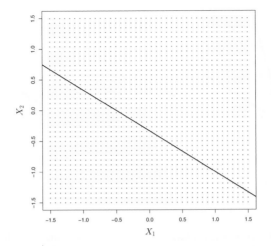

FIGURE 9.1. *The hyperplane* $1 + 2X_1 + 3X_2 = 0$ *is shown. The blue region is the set of points for which* $1 + 2X_1 + 3X_2 > 0$, *and the purple region is the set of points for which* $1 + 2X_1 + 3X_2 < 0$.

9.1.2 Classification Using a Separating Hyperplane

Now suppose that we have a $n \times p$ data matrix $\mathbf{X}$ that consists of n training observations in p-dimensional space,

$$x_1 = \begin{pmatrix} x_{11} \\ \vdots \\ x_{1p} \end{pmatrix}, \ldots, x_n = \begin{pmatrix} x_{n1} \\ \vdots \\ x_{np} \end{pmatrix}, \tag{9.5}$$

and that these observations fall into two classes—that is, $y_1, \ldots, y_n \in \{-1, 1\}$ where -1 represents one class and 1 the other class. We also have a test observation, a p-vector of observed features $x^* = \begin{pmatrix} x_1^* & \cdots & x_p^* \end{pmatrix}^T$. Our goal is to develop a classifier based on the training data that will correctly classify the test observation using its feature measurements. We have seen a number of approaches for this task, such as linear discriminant analysis and logistic regression in Chapter 4, and classification trees, bagging, and boosting in Chapter 8. We will now see a new approach that is based upon the concept of a *separating hyperplane*.

separating hyperplane

Suppose that it is possible to construct a hyperplane that separates the training observations perfectly according to their class labels. Examples of three such *separating hyperplanes* are shown in the left-hand panel of Figure 9.2. We can label the observations from the blue class as $y_i = 1$ and

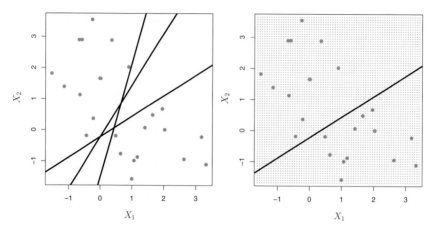

FIGURE 9.2. Left: *There are two classes of observations, shown in blue and in purple, each of which has measurements on two variables. Three separating hyperplanes, out of many possible, are shown in black.* Right: *A separating hyperplane is shown in black. The blue and purple grid indicates the decision rule made by a classifier based on this separating hyperplane: a test observation that falls in the blue portion of the grid will be assigned to the blue class, and a test observation that falls into the purple portion of the grid will be assigned to the purple class.*

those from the purple class as $y_i = -1$. Then a separating hyperplane has the property that

$$\beta_0 + \beta_1 x_{i1} + \beta_2 x_{i2} + \ldots + \beta_p x_{ip} > 0 \text{ if } y_i = 1, \tag{9.6}$$

and

$$\beta_0 + \beta_1 x_{i1} + \beta_2 x_{i2} + \ldots + \beta_p x_{ip} < 0 \text{ if } y_i = -1. \tag{9.7}$$

Equivalently, a separating hyperplane has the property that

$$y_i(\beta_0 + \beta_1 x_{i1} + \beta_2 x_{i2} + \ldots + \beta_p x_{ip}) > 0 \tag{9.8}$$

for all $i = 1, \ldots, n$.

If a separating hyperplane exists, we can use it to construct a very natural classifier: a test observation is assigned a class depending on which side of the hyperplane it is located. The right-hand panel of Figure 9.2 shows an example of such a classifier. That is, we classify the test observation x^* based on the sign of $f(x^*) = \beta_0 + \beta_1 x_1^* + \beta_2 x_2^* + \ldots + \beta_p x_p^*$. If $f(x^*)$ is positive, then we assign the test observation to class 1, and if $f(x^*)$ is negative, then we assign it to class -1. We can also make use of the *magnitude* of $f(x^*)$. If $f(x^*)$ is far from zero, then this means that x^* lies far from the hyperplane, and so we can be confident about our class assignment for x^*. On the other hand, if $f(x^*)$ is close to zero, then x^* is located near the hyperplane, and so we are less certain about the class assignment for x^*. Not surprisingly, and as we see in Figure 9.2, a classifier that is based on a separating hyperplane leads to a linear decision boundary.

9.1.3 The Maximal Margin Classifier

In general, if our data can be perfectly separated using a hyperplane, then there will in fact exist an infinite number of such hyperplanes. This is because a given separating hyperplane can usually be shifted a tiny bit up or down, or rotated, without coming into contact with any of the observations. Three possible separating hyperplanes are shown in the left-hand panel of Figure 9.2. In order to construct a classifier based upon a separating hyperplane, we must have a reasonable way to decide which of the infinite possible separating hyperplanes to use.

A natural choice is the *maximal margin hyperplane* (also known as the *optimal separating hyperplane*), which is the separating hyperplane that is farthest from the training observations. That is, we can compute the (perpendicular) distance from each training observation to a given separating hyperplane; the smallest such distance is the minimal distance from the observations to the hyperplane, and is known as the *margin*. The maximal margin hyperplane is the separating hyperplane for which the margin is largest—that is, it is the hyperplane that has the farthest minimum distance to the training observations. We can then classify a test observation based on which side of the maximal margin hyperplane it lies. This is known as the *maximal margin classifier*. We hope that a classifier that has a large margin on the training data will also have a large margin on the test data, and hence will classify the test observations correctly. Although the maximal margin classifier is often successful, it can also lead to overfitting when p is large.

maximal margin hyperplane

optimal separating hyperplane

margin

maximal margin classifier

If $\beta_0, \beta_1, \ldots, \beta_p$ are the coefficients of the maximal margin hyperplane, then the maximal margin classifier classifies the test observation x^* based on the sign of $f(x^*) = \beta_0 + \beta_1 x_1^* + \beta_2 x_2^* + \ldots + \beta_p x_p^*$.

Figure 9.3 shows the maximal margin hyperplane on the data set of Figure 9.2. Comparing the right-hand panel of Figure 9.2 to Figure 9.3, we see that the maximal margin hyperplane shown in Figure 9.3 does indeed result in a greater minimal distance between the observations and the separating hyperplane—that is, a larger margin. In a sense, the maximal margin hyperplane represents the mid-line of the widest "slab" that we can insert between the two classes.

Examining Figure 9.3, we see that three training observations are equidistant from the maximal margin hyperplane and lie along the dashed lines indicating the width of the margin. These three observations are known as

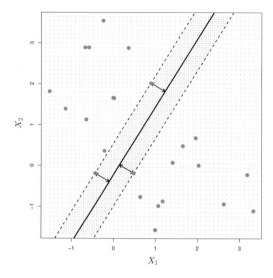

X_1

FIGURE 9.3. *There are two classes of observations, shown in blue and in purple. The maximal margin hyperplane is shown as a solid line. The margin is the distance from the solid line to either of the dashed lines. The two blue points and the purple point that lie on the dashed lines are the support vectors, and the distance from those points to the hyperplane is indicated by arrows. The purple and blue grid indicates the decision rule made by a classifier based on this separating hyperplane.*

support vectors, since they are vectors in p-dimensional space (in Figure 9.3, $p = 2$) and they "support" the maximal margin hyperplane in the sense that if these points were moved slightly then the maximal margin hyperplane would move as well. Interestingly, the maximal margin hyperplane depends directly on the support vectors, but not on the other observations: a movement to any of the other observations would not affect the separating hyperplane, provided that the observation's movement does not cause it to cross the boundary set by the margin. The fact that the maximal margin hyperplane depends directly on only a small subset of the observations is an important property that will arise later in this chapter when we discuss the support vector classifier and support vector machines.

support vector

9.1.4 *Construction of the Maximal Margin Classifier*

We now consider the task of constructing the maximal margin hyperplane based on a set of n training observations $x_1, \ldots, x_n \in \mathbb{R}^p$ and associated class labels $y_1, \ldots, y_n \in \{-1, 1\}$. Briefly, the maximal margin hyperplane is the solution to the optimization problem

$$\underset{\beta_0,\beta_1,\ldots,\beta_p,M}{\text{maximize}} \ M \qquad\qquad\qquad (9.9)$$

$$\text{subject to } \sum_{j=1}^{p} \beta_j^2 = 1, \qquad\qquad\qquad (9.10)$$

$$y_i(\beta_0 + \beta_1 x_{i1} + \beta_2 x_{i2} + \ldots + \beta_p x_{ip}) \geq M \ \ \forall \, i = 1,\ldots,n. \ (9.11)$$

This optimization problem (9.9)–(9.11) is actually simpler than it looks. First of all, the constraint in (9.11) that

$$y_i(\beta_0 + \beta_1 x_{i1} + \beta_2 x_{i2} + \ldots + \beta_p x_{ip}) \geq M \ \ \forall \, i = 1,\ldots,n$$

guarantees that each observation will be on the correct side of the hyperplane, provided that M is positive. (Actually, for each observation to be on the correct side of the hyperplane we would simply need $y_i(\beta_0 + \beta_1 x_{i1} + \beta_2 x_{i2} + \ldots + \beta_p x_{ip}) > 0$, so the constraint in (9.11) in fact requires that each observation be on the correct side of the hyperplane, with some cushion, provided that M is positive.)

Second, note that (9.10) is not really a constraint on the hyperplane, since if $\beta_0 + \beta_1 x_{i1} + \beta_2 x_{i2} + \ldots + \beta_p x_{ip} = 0$ defines a hyperplane, then so does $k(\beta_0 + \beta_1 x_{i1} + \beta_2 x_{i2} + \ldots + \beta_p x_{ip}) = 0$ for any $k \neq 0$. However, (9.10) adds meaning to (9.11); one can show that with this constraint the perpendicular distance from the ith observation to the hyperplane is given by

$$y_i(\beta_0 + \beta_1 x_{i1} + \beta_2 x_{i2} + \ldots + \beta_p x_{ip}).$$

Therefore, the constraints (9.10) and (9.11) ensure that each observation is on the correct side of the hyperplane and at least a distance M from the hyperplane. Hence, M represents the margin of our hyperplane, and the optimization problem chooses $\beta_0, \beta_1, \ldots, \beta_p$ to maximize M. This is exactly the definition of the maximal margin hyperplane! The problem (9.9)–(9.11) can be solved efficiently, but details of this optimization are outside of the scope of this book.

9.1.5 *The Non-separable Case*

The maximal margin classifier is a very natural way to perform classification, *if a separating hyperplane exists*. However, as we have hinted, in many cases no separating hyperplane exists, and so there is no maximal margin classifier. In this case, the optimization problem (9.9)–(9.11) has no solution with $M > 0$. An example is shown in Figure 9.4. In this case, we cannot *exactly* separate the two classes. However, as we will see in the next section, we can extend the concept of a separating hyperplane in order to develop a hyperplane that *almost* separates the classes, using a so-called *soft margin*. The generalization of the maximal margin classifier to the non-separable case is known as the *support vector classifier*.

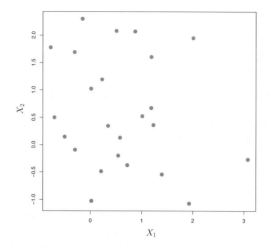

FIGURE 9.4. *There are two classes of observations, shown in blue and in purple. In this case, the two classes are not separable by a hyperplane, and so the maximal margin classifier cannot be used.*

9.2   Support Vector Classifiers

9.2.1   Overview of the Support Vector Classifier

In Figure 9.4, we see that observations that belong to two classes are not necessarily separable by a hyperplane. In fact, even if a separating hyperplane does exist, then there are instances in which a classifier based on a separating hyperplane might not be desirable. A classifier based on a separating hyperplane will necessarily perfectly classify all of the training observations; this can lead to sensitivity to individual observations. An example is shown in Figure 9.5. The addition of a single observation in the right-hand panel of Figure 9.5 leads to a dramatic change in the maximal margin hyperplane. The resulting maximal margin hyperplane is not satisfactory—for one thing, it has only a tiny margin. This is problematic because as discussed previously, the distance of an observation from the hyperplane can be seen as a measure of our confidence that the observation was correctly classified. Moreover, the fact that the maximal margin hyperplane is extremely sensitive to a change in a single observation suggests that it may have overfit the training data.

In this case, we might be willing to consider a classifier based on a hyperplane that does *not* perfectly separate the two classes, in the interest of

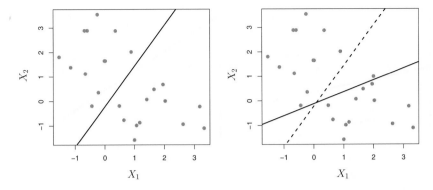

FIGURE 9.5. Left: *Two classes of observations are shown in blue and in purple, along with the maximal margin hyperplane.* Right: *An additional blue observation has been added, leading to a dramatic shift in the maximal margin hyperplane shown as a solid line. The dashed line indicates the maximal margin hyperplane that was obtained in the absence of this additional point.*

- Greater robustness to individual observations, and

- Better classification of *most* of the training observations.

That is, it could be worthwhile to misclassify a few training observations in order to do a better job in classifying the remaining observations.

The *support vector classifier*, sometimes called a *soft margin classifier*, does exactly this. Rather than seeking the largest possible margin so that every observation is not only on the correct side of the hyperplane but also on the correct side of the margin, we instead allow some observations to be on the incorrect side of the margin, or even the incorrect side of the hyperplane. (The margin is *soft* because it can be violated by some of the training observations.) An example is shown in the left-hand panel of Figure 9.6. Most of the observations are on the correct side of the margin. However, a small subset of the observations are on the wrong side of the margin.

support vector classifier
soft margin classifier

An observation can be not only on the wrong side of the margin, but also on the wrong side of the hyperplane. In fact, when there is no separating hyperplane, such a situation is inevitable. Observations on the wrong side of the hyperplane correspond to training observations that are misclassified by the support vector classifier. The right-hand panel of Figure 9.6 illustrates such a scenario.

9.2.2   Details of the Support Vector Classifier

The support vector classifier classifies a test observation depending on which side of a hyperplane it lies. The hyperplane is chosen to correctly

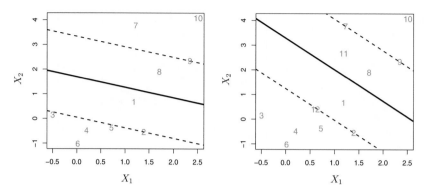

FIGURE 9.6. *Left: A support vector classifier was fit to a small data set. The hyperplane is shown as a solid line and the margins are shown as dashed lines. Purple observations: Observations 3, 4, 5, and 6 are on the correct side of the margin, observation 2 is on the margin, and observation 1 is on the wrong side of the margin. Blue observations: Observations 7 and 10 are on the correct side of the margin, observation 9 is on the margin, and observation 8 is on the wrong side of the margin. No observations are on the wrong side of the hyperplane. Right: Same as left panel with two additional points, 11 and 12. These two observations are on the wrong side of the hyperplane and the wrong side of the margin.*

separate most of the training observations into the two classes, but may misclassify a few observations. It is the solution to the optimization problem

$$\underset{\beta_0, \beta_1, \ldots, \beta_p, \epsilon_1, \ldots, \epsilon_n, M}{\text{maximize}} \quad M \tag{9.12}$$

$$\text{subject to} \ \sum_{j=1}^{p} \beta_j^2 = 1, \tag{9.13}$$

$$y_i(\beta_0 + \beta_1 x_{i1} + \beta_2 x_{i2} + \ldots + \beta_p x_{ip}) \geq M(1 - \epsilon_i), \tag{9.14}$$

$$\epsilon_i \geq 0, \ \sum_{i=1}^{n} \epsilon_i \leq C, \tag{9.15}$$

where C is a nonnegative tuning parameter. As in (9.11), M is the width of the margin; we seek to make this quantity as large as possible. In (9.14), $\epsilon_1, \ldots, \epsilon_n$ are *slack variables* that allow individual observations to be on the wrong side of the margin or the hyperplane; we will explain them in greater detail momentarily. Once we have solved (9.12)–(9.15), we classify a test observation x^* as before, by simply determining on which side of the hyperplane it lies. That is, we classify the test observation based on the sign of $f(x^*) = \beta_0 + \beta_1 x_1^* + \ldots + \beta_p x_p^*$.

slack variable

The problem (9.12)–(9.15) seems complex, but insight into its behavior can be made through a series of simple observations presented below. First of all, the slack variable ϵ_i tells us where the ith observation is located, relative to the hyperplane and relative to the margin. If $\epsilon_i = 0$ then the ith observation is on the correct side of the margin, as we saw in Section 9.1.4. If $\epsilon_i > 0$ then the ith observation is on the wrong side of the margin, and we say that the ith observation has *violated* the margin. If $\epsilon_i > 1$ then it is on the wrong side of the hyperplane.

We now consider the role of the tuning parameter C. In (9.15), C bounds the sum of the ϵ_i's, and so it determines the number and severity of the violations to the margin (and to the hyperplane) that we will tolerate. We can think of C as a *budget* for the amount that the margin can be violated by the n observations. If $C = 0$ then there is no budget for violations to the margin, and it must be the case that $\epsilon_1 = \ldots = \epsilon_n = 0$, in which case (9.12)–(9.15) simply amounts to the maximal margin hyperplane optimization problem (9.9)–(9.11). (Of course, a maximal margin hyperplane exists only if the two classes are separable.) For $C > 0$ no more than C observations can be on the wrong side of the hyperplane, because if an observation is on the wrong side of the hyperplane then $\epsilon_i > 1$, and (9.15) requires that $\sum_{i=1}^{n} \epsilon_i \leq C$. As the budget C increases, we become more tolerant of violations to the margin, and so the margin will widen. Conversely, as C decreases, we become less tolerant of violations to the margin and so the margin narrows. An example in shown in Figure 9.7.

In practice, C is treated as a tuning parameter that is generally chosen via cross-validation. As with the tuning parameters that we have seen throughout this book, C controls the bias-variance trade-off of the statistical learning technique. When C is small, we seek narrow margins that are rarely violated; this amounts to a classifier that is highly fit to the data, which may have low bias but high variance. On the other hand, when C is larger, the margin is wider and we allow more violations to it; this amounts to fitting the data less hard and obtaining a classifier that is potentially more biased but may have lower variance.

The optimization problem (9.12)–(9.15) has a very interesting property: it turns out that only observations that either lie on the margin or that violate the margin will affect the hyperplane, and hence the classifier obtained. In other words, an observation that lies strictly on the correct side of the margin does not affect the support vector classifier! Changing the position of that observation would not change the classifier at all, provided that its position remains on the correct side of the margin. Observations that lie directly on the margin, or on the wrong side of the margin for their class, are known as *support vectors*. These observations do affect the support vector classifier.

The fact that only support vectors affect the classifier is in line with our previous assertion that C controls the bias-variance trade-off of the support vector classifier. When the tuning parameter C is large, then the margin is wide, many observations violate the margin, and so there are many support vectors. In this case, many observations are involved in determining the hyperplane. The top left panel in Figure 9.7 illustrates this setting: this classifier has low variance (since many observations are support vectors)

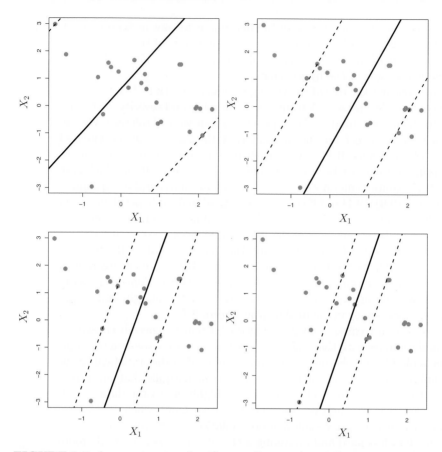

FIGURE 9.7. *A support vector classifier was fit using four different values of the tuning parameter C in (9.12)–(9.15). The largest value of C was used in the top left panel, and smaller values were used in the top right, bottom left, and bottom right panels. When C is large, then there is a high tolerance for observations being on the wrong side of the margin, and so the margin will be large. As C decreases, the tolerance for observations being on the wrong side of the margin decreases, and the margin narrows.*

but potentially high bias. In contrast, if C is small, then there will be fewer support vectors and hence the resulting classifier will have low bias but high variance. The bottom right panel in Figure 9.7 illustrates this setting, with only eight support vectors.

The fact that the support vector classifier's decision rule is based only on a potentially small subset of the training observations (the support vectors) means that it is quite robust to the behavior of observations that are far away from the hyperplane. This property is distinct from some of the other classification methods that we have seen in preceding chapters, such as linear discriminant analysis. Recall that the LDA classification rule

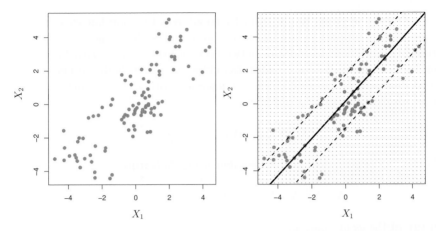

FIGURE 9.8. Left: *The observations fall into two classes, with a non-linear boundary between them.* Right: *The support vector classifier seeks a linear boundary, and consequently performs very poorly.*

depends on the mean of *all* of the observations within each class, as well as the within-class covariance matrix computed using *all* of the observations. In contrast, logistic regression, unlike LDA, has very low sensitivity to observations far from the decision boundary. In fact we will see in Section 9.5 that the support vector classifier and logistic regression are closely related.

9.3   Support Vector Machines

We first discuss a general mechanism for converting a linear classifier into one that produces non-linear decision boundaries. We then introduce the support vector machine, which does this in an automatic way.

9.3.1   Classification with Non-linear Decision Boundaries

The support vector classifier is a natural approach for classification in the two-class setting, if the boundary between the two classes is linear. However, in practice we are sometimes faced with non-linear class boundaries. For instance, consider the data in the left-hand panel of Figure 9.8. It is clear that a support vector classifier or any linear classifier will perform poorly here. Indeed, the support vector classifier shown in the right-hand panel of Figure 9.8 is useless here.

In Chapter 7, we are faced with an analogous situation. We see there that the performance of linear regression can suffer when there is a non-linear relationship between the predictors and the outcome. In that case, we consider enlarging the feature space using functions of the predictors, such as quadratic and cubic terms, in order to address this non-linearity. In the case of the support vector classifier, we could address the problem of possibly non-linear boundaries between classes in a similar way, by enlarging the feature space using quadratic, cubic, and even higher-order polynomial functions of the predictors. For instance, rather than fitting a support vector classifier using p features

$$X_1, X_2, \ldots, X_p,$$

we could instead fit a support vector classifier using $2p$ features

$$X_1, X_1^2, X_2, X_2^2, \ldots, X_p, X_p^2.$$

Then (9.12)–(9.15) would become

$$\underset{\beta_0, \beta_{11}, \beta_{12} \ldots, \beta_{p1}, \beta_{p2}, \epsilon_1, \ldots, \epsilon_n, M}{\text{maximize}} M \tag{9.16}$$

$$\text{subject to } y_i \left(\beta_0 + \sum_{j=1}^{p} \beta_{j1} x_{ij} + \sum_{j=1}^{p} \beta_{j2} x_{ij}^2 \right) \geq M(1 - \epsilon_i),$$

$$\sum_{i=1}^{n} \epsilon_i \leq C, \quad \epsilon_i \geq 0, \quad \sum_{j=1}^{p} \sum_{k=1}^{2} \beta_{jk}^2 = 1.$$

Why does this lead to a non-linear decision boundary? In the enlarged feature space, the decision boundary that results from (9.16) is in fact linear. But in the original feature space, the decision boundary is of the form $q(x) = 0$, where q is a quadratic polynomial, and its solutions are generally non-linear. One might additionally want to enlarge the feature space with higher-order polynomial terms, or with interaction terms of the form $X_j X_{j'}$ for $j \neq j'$. Alternatively, other functions of the predictors could be considered rather than polynomials. It is not hard to see that there are many possible ways to enlarge the feature space, and that unless we are careful, we could end up with a huge number of features. Then computations would become unmanageable. The support vector machine, which we present next, allows us to enlarge the feature space used by the support vector classifier in a way that leads to efficient computations.

9.3.2 The Support Vector Machine

The *support vector machine* (SVM) is an extension of the support vector classifier that results from enlarging the feature space in a specific way, using *kernels*. We will now discuss this extension, the details of which are somewhat complex and beyond the scope of this book. However, the main idea is described in Section 9.3.1: we may want to enlarge our feature space

support
vector
machine
kernel

in order to accommodate a non-linear boundary between the classes. The kernel approach that we describe here is simply an efficient computational approach for enacting this idea.

We have not discussed exactly how the support vector classifier is computed because the details become somewhat technical. However, it turns out that the solution to the support vector classifier problem (9.12)–(9.15) involves only the *inner products* of the observations (as opposed to the observations themselves). The inner product of two r-vectors a and b is defined as $\langle a, b \rangle = \sum_{i=1}^{r} a_i b_i$. Thus the inner product of two observations x_i, $x_{i'}$ is given by

$$\langle x_i, x_{i'} \rangle = \sum_{j=1}^{p} x_{ij} x_{i'j}. \tag{9.17}$$

It can be shown that

- The linear support vector classifier can be represented as

$$f(x) = \beta_0 + \sum_{i=1}^{n} \alpha_i \langle x, x_i \rangle, \tag{9.18}$$

  where there are n parameters α_i, $i = 1, \ldots, n$, one per training observation.

- To estimate the parameters $\alpha_1, \ldots, \alpha_n$ and β_0, all we need are the $\binom{n}{2}$ inner products $\langle x_i, x_{i'} \rangle$ between all pairs of training observations. (The notation $\binom{n}{2}$ means $n(n-1)/2$, and gives the number of pairs among a set of n items.)

Notice that in (9.18), in order to evaluate the function $f(x)$, we need to compute the inner product between the new point x and each of the training points x_i. However, it turns out that α_i is nonzero only for the support vectors in the solution—that is, if a training observation is not a support vector, then its α_i equals zero. So if $\mathcal{S}$ is the collection of indices of these support points, we can rewrite any solution function of the form (9.18) as

$$f(x) = \beta_0 + \sum_{i \in \mathcal{S}} \alpha_i \langle x, x_i \rangle, \tag{9.19}$$

which typically involves far fewer terms than in (9.18).[2]

To summarize, in representing the linear classifier $f(x)$, and in computing its coefficients, all we need are inner products.

Now suppose that every time the inner product (9.17) appears in the representation (9.18), or in a calculation of the solution for the support

---

[2]By expanding each of the inner products in (9.19), it is easy to see that $f(x)$ is a linear function of the coordinates of x. Doing so also establishes the correspondence between the α_i and the original parameters β_j.

vector classifier, we replace it with a *generalization* of the inner product of the form

$$K(x_i, x_{i'}),\tag{9.20}$$

where K is some function that we will refer to as a *kernel*. A kernel is a function that quantifies the similarity of two observations. For instance, we could simply take

kernel

$$K(x_i, x_{i'}) = \sum_{j=1}^{p} x_{ij} x_{i'j},\tag{9.21}$$

which would just give us back the support vector classifier. Equation 9.21 is known as a *linear* kernel because the support vector classifier is linear in the features; the linear kernel essentially quantifies the similarity of a pair of observations using Pearson (standard) correlation. But one could instead choose another form for (9.20). For instance, one could replace every instance of $\sum_{j=1}^{p} x_{ij} x_{i'j}$ with the quantity

$$K(x_i, x_{i'}) = (1 + \sum_{j=1}^{p} x_{ij} x_{i'j})^d.\tag{9.22}$$

This is known as a *polynomial kernel* of degree d, where d is a positive integer. Using such a kernel with $d > 1$, instead of the standard linear kernel (9.21), in the support vector classifier algorithm leads to a much more flexible decision boundary. It essentially amounts to fitting a support vector classifier in a higher-dimensional space involving polynomials of degree d, rather than in the original feature space. When the support vector classifier is combined with a non-linear kernel such as (9.22), the resulting classifier is known as a support vector machine. Note that in this case the (non-linear) function has the form

polynomial kernel

$$f(x) = \beta_0 + \sum_{i \in \mathcal{S}} \alpha_i K(x, x_i).\tag{9.23}$$

The left-hand panel of Figure 9.9 shows an example of an SVM with a polynomial kernel applied to the non-linear data from Figure 9.8. The fit is a substantial improvement over the linear support vector classifier. When $d = 1$, then the SVM reduces to the support vector classifier seen earlier in this chapter.

The polynomial kernel shown in (9.22) is one example of a possible non-linear kernel, but alternatives abound. Another popular choice is the *radial kernel*, which takes the form

radial kernel

$$K(x_i, x_{i'}) = \exp(-\gamma \sum_{j=1}^{p} (x_{ij} - x_{i'j})^2).\tag{9.24}$$

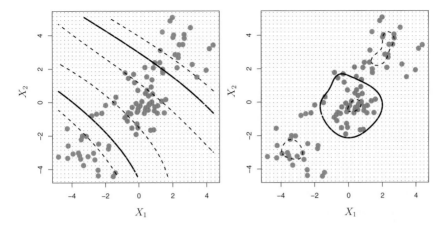

FIGURE 9.9. Left: *An SVM with a polynomial kernel of degree 3 is applied to the non-linear data from Figure 9.8, resulting in a far more appropriate decision rule.* Right: *An SVM with a radial kernel is applied. In this example, either kernel is capable of capturing the decision boundary.*

In (9.24), γ is a positive constant. The right-hand panel of Figure 9.9 shows an example of an SVM with a radial kernel on this non-linear data; it also does a good job in separating the two classes.

How does the radial kernel (9.24) actually work? If a given test observation $x^* = (x_1^* \ldots x_p^*)^T$ is far from a training observation x_i in terms of Euclidean distance, then $\sum_{j=1}^{p}(x_j^* - x_{ij})^2$ will be large, and so $K(x_i, x_{i'}) = \exp(-\gamma \sum_{j=1}^{p}(x_j^* - x_{ij})^2)$ will be very tiny. This means that in (9.23), x_i will play virtually no role in $f(x^*)$. Recall that the predicted class label for the test observation x^* is based on the sign of $f(x^*)$. In other words, training observations that are far from x^* will play essentially no role in the predicted class label for x^*. This means that the radial kernel has very *local* behavior, in the sense that only nearby training observations have an effect on the class label of a test observation.

What is the advantage of using a kernel rather than simply enlarging the feature space using functions of the original features, as in (9.16)? One advantage is computational, and it amounts to the fact that using kernels, one need only compute $K(x_i, x_{i'})$ for all $\binom{n}{2}$ distinct pairs i, i'. This can be done without explicitly working in the enlarged feature space. This is important because in many applications of SVMs, the enlarged feature space is so large that computations are intractable. For some kernels, such as the radial kernel (9.24), the feature space is *implicit* and infinite-dimensional, so we could never do the computations there anyway!

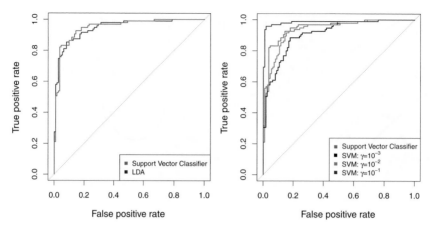

FIGURE 9.10. *ROC curves for the* Heart *data training set.* Left: *The support vector classifier and LDA are compared.* Right: *The support vector classifier is compared to an SVM using a radial basis kernel with* $\gamma = 10^{-3}$, 10^{-2}, *and* 10^{-1}.

9.3.3   An Application to the Heart Disease Data

In Chapter 8 we apply decision trees and related methods to the Heart data. The aim is to use 13 predictors such as Age, Sex, and Chol in order to predict whether an individual has heart disease. We now investigate how an SVM compares to LDA on this data. After removing 6 missing observations, the data consist of 297 subjects, which we randomly split into 207 training and 90 test observations.

We first fit LDA and the support vector classifier to the training data. Note that the support vector classifier is equivalent to a SVM using a polynomial kernel of degree $d = 1$. The left-hand panel of Figure 9.10 displays ROC curves (described in Section 4.4.3) for the training set predictions for both LDA and the support vector classifier. Both classifiers compute scores of the form $\hat{f}(X) = \hat{\beta}_0 + \hat{\beta}_1 X_1 + \hat{\beta}_2 X_2 + \ldots + \hat{\beta}_p X_p$ for each observation. For any given cutoff t, we classify observations into the *heart disease* or *no heart disease* categories depending on whether $\hat{f}(X) < t$ or $\hat{f}(X) \geq t$. The ROC curve is obtained by forming these predictions and computing the false positive and true positive rates for a range of values of t. An optimal classifier will hug the top left corner of the ROC plot. In this instance LDA and the support vector classifier both perform well, though there is a suggestion that the support vector classifier may be slightly superior.

The right-hand panel of Figure 9.10 displays ROC curves for SVMs using a radial kernel, with various values of γ. As γ increases and the fit becomes more non-linear, the ROC curves improve. Using $\gamma = 10^{-1}$ appears to give an almost perfect ROC curve. However, these curves represent training error rates, which can be misleading in terms of performance on new test data. Figure 9.11 displays ROC curves computed on the 90 test observa-

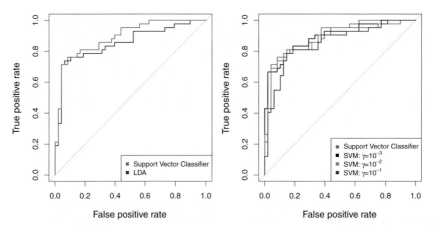

FIGURE 9.11. *ROC curves for the test set of the* Heart *data. Left: The support vector classifier and LDA are compared. Right: The support vector classifier is compared to an SVM using a radial basis kernel with* $\gamma = 10^{-3}$, 10^{-2}, *and* 10^{-1}.

tions. We observe some differences from the training ROC curves. In the left-hand panel of Figure 9.11, the support vector classifier appears to have a small advantage over LDA (although these differences are not statistically significant). In the right-hand panel, the SVM using $\gamma = 10^{-1}$, which showed the best results on the training data, produces the worst estimates on the test data. This is once again evidence that while a more flexible method will often produce lower training error rates, this does not necessarily lead to improved performance on test data. The SVMs with $\gamma = 10^{-2}$ and $\gamma = 10^{-3}$ perform comparably to the support vector classifier, and all three outperform the SVM with $\gamma = 10^{-1}$.

9.4   SVMs with More than Two Classes

So far, our discussion has been limited to the case of binary classification: that is, classification in the two-class setting. How can we extend SVMs to the more general case where we have some arbitrary number of classes? It turns out that the concept of separating hyperplanes upon which SVMs are based does not lend itself naturally to more than two classes. Though a number of proposals for extending SVMs to the K-class case have been made, the two most popular are the *one-versus-one* and *one-versus-all* approaches. We briefly discuss those two approaches here.

9.4.1   One-Versus-One Classification

Suppose that we would like to perform classification using SVMs, and there are $K > 2$ classes. A *one-versus-one* or *all-pairs* approach constructs $\binom{K}{2}$ one-versus-one

SVMs, each of which compares a pair of classes. For example, one such SVM might compare the kth class, coded as $+1$, to the k'th class, coded as -1. We classify a test observation using each of the $\binom{K}{2}$ classifiers, and we tally the number of times that the test observation is assigned to each of the K classes. The final classification is performed by assigning the test observation to the class to which it was most frequently assigned in these $\binom{K}{2}$ pairwise classifications.

9.4.2 One-Versus-All Classification

The *one-versus-all* approach is an alternative procedure for applying SVMs in the case of $K > 2$ classes. We fit K SVMs, each time comparing one of the K classes to the remaining $K - 1$ classes. Let $\beta_{0k}, \beta_{1k}, \ldots, \beta_{pk}$ denote the parameters that result from fitting an SVM comparing the kth class (coded as $+1$) to the others (coded as -1). Let x^* denote a test observation. We assign the observation to the class for which $\beta_{0k} + \beta_{1k}x_1^* + \beta_{2k}x_2^* + \ldots + \beta_{pk}x_p^*$ is largest, as this amounts to a high level of confidence that the test observation belongs to the kth class rather than to any of the other classes.

one-versus-all

9.5 Relationship to Logistic Regression

When SVMs were first introduced in the mid-1990s, they made quite a splash in the statistical and machine learning communities. This was due in part to their good performance, good marketing, and also to the fact that the underlying approach seemed both novel and mysterious. The idea of finding a hyperplane that separates the data as well as possible, while allowing some violations to this separation, seemed distinctly different from classical approaches for classification, such as logistic regression and linear discriminant analysis. Moreover, the idea of using a kernel to expand the feature space in order to accommodate non-linear class boundaries appeared to be a unique and valuable characteristic.

However, since that time, deep connections between SVMs and other more classical statistical methods have emerged. It turns out that one can rewrite the criterion (9.12)–(9.15) for fitting the support vector classifier $f(X) = \beta_0 + \beta_1 X_1 + \ldots + \beta_p X_p$ as

$$\underset{\beta_0, \beta_1, \ldots, \beta_p}{\text{minimize}} \left\{ \sum_{i=1}^{n} \max\left[0, 1 - y_i f(x_i)\right] + \lambda \sum_{j=1}^{p} \beta_j^2 \right\}, \tag{9.25}$$

where λ is a nonnegative tuning parameter. When λ is large then $\beta_1, \ldots, \beta_p$ are small, more violations to the margin are tolerated, and a low-variance but high-bias classifier will result. When λ is small then few violations to the margin will occur; this amounts to a high-variance but low-bias classifier. Thus, a small value of λ in (9.25) amounts to a small value of C in (9.15). Note that the $\lambda \sum_{j=1}^{p} \beta_j^2$ term in (9.25) is the ridge penalty term from Section 6.2.1, and plays a similar role in controlling the bias-variance trade-off for the support vector classifier.

Now (9.25) takes the "Loss + Penalty" form that we have seen repeatedly throughout this book:

$$\underset{\beta_0, \beta_1, \ldots, \beta_p}{\text{minimize}} \left\{ L(\mathbf{X}, \mathbf{y}, \beta) + \lambda P(\beta) \right\}. \tag{9.26}$$

In (9.26), $L(\mathbf{X}, \mathbf{y}, \beta)$ is some loss function quantifying the extent to which the model, parametrized by β, fits the data $(\mathbf{X}, \mathbf{y})$, and $P(\beta)$ is a penalty function on the parameter vector β whose effect is controlled by a nonnegative tuning parameter λ. For instance, ridge regression and the lasso both take this form with

$$L(\mathbf{X}, \mathbf{y}, \beta) = \sum_{i=1}^{n} \left(y_i - \beta_0 - \sum_{j=1}^{p} x_{ij} \beta_j \right)^2$$

and with $P(\beta) = \sum_{j=1}^{p} \beta_j^2$ for ridge regression and $P(\beta) = \sum_{j=1}^{p} |\beta_j|$ for the lasso. In the case of (9.25) the loss function instead takes the form

$$L(\mathbf{X}, \mathbf{y}, \beta) = \sum_{i=1}^{n} \max \left[0, 1 - y_i(\beta_0 + \beta_1 x_{i1} + \ldots + \beta_p x_{ip}) \right].$$

This is known as *hinge loss*, and is depicted in Figure 9.12. However, it turns out that the hinge loss function is closely related to the loss function used in logistic regression, also shown in Figure 9.12.

hinge loss

An interesting characteristic of the support vector classifier is that only support vectors play a role in the classifier obtained; observations on the correct side of the margin do not affect it. This is due to the fact that the loss function shown in Figure 9.12 is exactly zero for observations for which $y_i(\beta_0 + \beta_1 x_{i1} + \ldots + \beta_p x_{ip}) \geq 1$; these correspond to observations that are on the correct side of the margin.[3] In contrast, the loss function for logistic regression shown in Figure 9.12 is not exactly zero anywhere. But it is very small for observations that are far from the decision boundary. Due to the similarities between their loss functions, logistic regression and the support vector classifier often give very similar results. When the classes are well separated, SVMs tend to behave better than logistic regression; in more overlapping regimes, logistic regression is often preferred.

---

[3] With this hinge-loss + penalty representation, the margin corresponds to the value one, and the width of the margin is determined by $\sum \beta_j^2$.

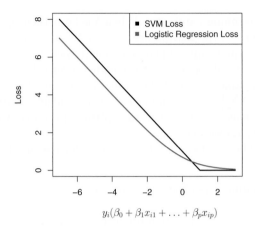

$$y_i(\beta_0 + \beta_1 x_{i1} + \ldots + \beta_p x_{ip})$$

FIGURE 9.12. *The SVM and logistic regression loss functions are compared, as a function of $y_i(\beta_0 + \beta_1 x_{i1} + \ldots + \beta_p x_{ip})$. When $y_i(\beta_0 + \beta_1 x_{i1} + \ldots + \beta_p x_{ip})$ is greater than 1, then the SVM loss is zero, since this corresponds to an observation that is on the correct side of the margin. Overall, the two loss functions have quite similar behavior.*

When the support vector classifier and SVM were first introduced, it was thought that the tuning parameter C in (9.15) was an unimportant "nuisance" parameter that could be set to some default value, like 1. However, the "Loss + Penalty" formulation (9.25) for the support vector classifier indicates that this is not the case. The choice of tuning parameter is very important and determines the extent to which the model underfits or overfits the data, as illustrated, for example, in Figure 9.7.

We have established that the support vector classifier is closely related to logistic regression and other preexisting statistical methods. Is the SVM unique in its use of kernels to enlarge the feature space to accommodate non-linear class boundaries? The answer to this question is "no". We could just as well perform logistic regression or many of the other classification methods seen in this book using non-linear kernels; this is closely related to some of the non-linear approaches seen in Chapter 7. However, for historical reasons, the use of non-linear kernels is much more widespread in the context of SVMs than in the context of logistic regression or other methods.

Though we have not addressed it here, there is in fact an extension of the SVM for regression (i.e. for a quantitative rather than a qualitative response), called *support vector regression*. In Chapter 3, we saw that least squares regression seeks coefficients $\beta_0, \beta_1, \ldots, \beta_p$ such that the sum of squared residuals is as small as possible. (Recall from Chapter 3 that residuals are defined as $y_i - \beta_0 - \beta_1 x_{i1} - \cdots - \beta_p x_{ip}$.) Support vector regression instead seeks coefficients that minimize a different type of loss, where only residuals larger in absolute value than some positive constant

support vector regression

contribute to the loss function. This is an extension of the margin used in support vector classifiers to the regression setting.

9.6 Lab: Support Vector Machines

We use the e1071 library in R to demonstrate the support vector classifier and the SVM. Another option is the LiblineaR library, which is useful for very large linear problems.

9.6.1 Support Vector Classifier

The e1071 library contains implementations for a number of statistical learning methods. In particular, the svm() function can be used to fit a support vector classifier when the argument kernel="linear" is used. This function uses a slightly different formulation from (9.14) and (9.25) for the support vector classifier. A cost argument allows us to specify the cost of a violation to the margin. When the cost argument is small, then the margins will be wide and many support vectors will be on the margin or will violate the margin. When the cost argument is large, then the margins will be narrow and there will be few support vectors on the margin or violating the margin.

svm()

We now use the svm() function to fit the support vector classifier for a given value of the cost parameter. Here we demonstrate the use of this function on a two-dimensional example so that we can plot the resulting decision boundary. We begin by generating the observations, which belong to two classes, and checking whether the classes are linearly separable.

```
> set.seed(1)
> x=matrix(rnorm(20*2), ncol=2)
> y=c(rep(-1,10), rep(1,10))
> x[y==1,]=x[y==1,] + 1
> plot(x, col=(3-y))
```

They are not. Next, we fit the support vector classifier. Note that in order for the svm() function to perform classification (as opposed to SVM-based regression), we must encode the response as a factor variable. We now create a data frame with the response coded as a factor.

```
> dat=data.frame(x=x, y=as.factor(y))
> library(e1071)
> svmfit=svm(y~., data=dat, kernel="linear", cost=10,
    scale=FALSE)
```

The argument `scale=FALSE` tells the `svm()` function not to scale each feature to have mean zero or standard deviation one; depending on the application, one might prefer to use `scale=TRUE`.

We can now plot the support vector classifier obtained:

```
> plot(svmfit, dat)
```

Note that the two arguments to the `plot.svm()` function are the output of the call to `svm()`, as well as the data used in the call to `svm()`. The region of feature space that will be assigned to the −1 class is shown in light blue, and the region that will be assigned to the +1 class is shown in purple. The decision boundary between the two classes is linear (because we used the argument `kernel="linear"`), though due to the way in which the plotting function is implemented in this library the decision boundary looks somewhat jagged in the plot. (Note that here the second feature is plotted on the x-axis and the first feature is plotted on the y-axis, in contrast to the behavior of the usual `plot()` function in R.) The support vectors are plotted as crosses and the remaining observations are plotted as circles; we see here that there are seven support vectors. We can determine their identities as follows:

```
> svmfit$index
[1]   1   2   5   7  14  16  17
```

We can obtain some basic information about the support vector classifier fit using the `summary()` command:

```
> summary(svmfit)
Call:
svm(formula = y ~ ., data = dat, kernel = "linear", cost = 10,
    scale = FALSE)
Parameters:
   SVM-Type:  C-classification
 SVM-Kernel:  linear
       cost:  10
      gamma:  0.5
Number of Support Vectors:   7
 ( 4 3 )
Number of Classes:   2
Levels:
 -1 1
```

This tells us, for instance, that a linear kernel was used with `cost=10`, and that there were seven support vectors, four in one class and three in the other.

What if we instead used a smaller value of the cost parameter?

```
> svmfit=svm(y~., data=dat, kernel="linear", cost=0.1,
    scale=FALSE)
> plot(svmfit, dat)
> svmfit$index
[1]   1   2   3   4   5   7   9  10  12  13  14  15  16  17  18  20
```

Now that a smaller value of the cost parameter is being used, we obtain a larger number of support vectors, because the margin is now wider. Unfortunately, the svm() function does not explicitly output the coefficients of the linear decision boundary obtained when the support vector classifier is fit, nor does it output the width of the margin.

The e1071 library includes a built-in function, tune(), to perform cross-validation. By default, tune() performs ten-fold cross-validation on a set of models of interest. In order to use this function, we pass in relevant information about the set of models that are under consideration. The following command indicates that we want to compare SVMs with a linear kernel, using a range of values of the cost parameter.

tune()

```
> set.seed(1)
> tune.out=tune(svm,y~.,data=dat,kernel="linear",
    ranges=list(cost=c(0.001, 0.01, 0.1, 1,5,10,100)))
```

We can easily access the cross-validation errors for each of these models using the summary() command:

```
> summary(tune.out)
Parameter tuning of 'svm':
- sampling method: 10-fold cross validation
- best parameters:
  cost
  0.1
- best performance: 0.1
- Detailed performance results:
   cost error dispersion
1 1e-03  0.70      0.422
2 1e-02  0.70      0.422
3 1e-01  0.10      0.211
4 1e+00  0.15      0.242
5 5e+00  0.15      0.242
6 1e+01  0.15      0.242
7 1e+02  0.15      0.242
```

We see that cost=0.1 results in the lowest cross-validation error rate. The tune() function stores the best model obtained, which can be accessed as follows:

```
> bestmod=tune.out$best.model
> summary(bestmod)
```

The predict() function can be used to predict the class label on a set of test observations, at any given value of the cost parameter. We begin by generating a test data set.

```
> xtest=matrix(rnorm(20*2), ncol=2)
> ytest=sample(c(-1,1), 20, rep=TRUE)
> xtest[ytest==1,]=xtest[ytest==1,] + 1
> testdat=data.frame(x=xtest, y=as.factor(ytest))
```

Now we predict the class labels of these test observations. Here we use the best model obtained through cross-validation in order to make predictions.

```
> ypred=predict(bestmod,testdat)
> table(predict=ypred, truth=testdat$y)
        truth
predict -1  1
     -1 11  1
      1  0  8
```

Thus, with this value of cost, 19 of the test observations are correctly classified. What if we had instead used cost=0.01?

```
> svmfit=svm(y~., data=dat, kernel="linear", cost=.01,
    scale=FALSE)
> ypred=predict(svmfit,testdat)
> table(predict=ypred, truth=testdat$y)
        truth
predict -1  1
     -1 11  2
      1  0  7
```

In this case one additional observation is misclassified.

Now consider a situation in which the two classes are linearly separable. Then we can find a separating hyperplane using the svm() function. We first further separate the two classes in our simulated data so that they are linearly separable:

```
> x[y==1,]=x[y==1,]+0.5
> plot(x, col=(y+5)/2, pch=19)
```

Now the observations are just barely linearly separable. We fit the support vector classifier and plot the resulting hyperplane, using a very large value of cost so that no observations are misclassified.

```
> dat=data.frame(x=x,y=as.factor(y))
> svmfit=svm(y~., data=dat, kernel="linear", cost=1e5)
> summary(svmfit)
Call:
svm(formula = y ~ ., data = dat, kernel = "linear", cost = 1e
    +05)
Parameters:
   SVM-Type:  C-classification
 SVM-Kernel:  linear
       cost:  1e+05
      gamma:  0.5
Number of Support Vectors:  3
 ( 1 2 )
Number of Classes:  2
Levels:
 -1 1
> plot(svmfit, dat)
```

No training errors were made and only three support vectors were used. However, we can see from the figure that the margin is very narrow (because the observations that are not support vectors, indicated as circles, are very close to the decision boundary). It seems likely that this model will perform poorly on test data. We now try a smaller value of `cost`:

```
> svmfit=svm(y~., data=dat, kernel="linear", cost=1)
> summary(svmfit)
> plot(svmfit,dat)
```

Using `cost=1`, we misclassify a training observation, but we also obtain a much wider margin and make use of seven support vectors. It seems likely that this model will perform better on test data than the model with `cost=1e5`.

9.6.2 Support Vector Machine

In order to fit an SVM using a non-linear kernel, we once again use the `svm()` function. However, now we use a different value of the parameter `kernel`. To fit an SVM with a polynomial kernel we use `kernel="polynomial"`, and to fit an SVM with a radial kernel we use `kernel="radial"`. In the former case we also use the `degree` argument to specify a degree for the polynomial kernel (this is d in (9.22)), and in the latter case we use `gamma` to specify a value of γ for the radial basis kernel (9.24).

We first generate some data with a non-linear class boundary, as follows:

```
> set.seed(1)
> x=matrix(rnorm(200*2), ncol=2)
> x[1:100,]=x[1:100,]+2
> x[101:150,]=x[101:150,]-2
> y=c(rep(1,150),rep(2,50))
> dat=data.frame(x=x,y=as.factor(y))
```

Plotting the data makes it clear that the class boundary is indeed non-linear:

```
> plot(x, col=y)
```

The data is randomly split into training and testing groups. We then fit the training data using the `svm()` function with a radial kernel and $\gamma = 1$:

```
> train=sample(200,100)
> svmfit=svm(y~., data=dat[train,], kernel="radial", gamma=1,
      cost=1)
> plot(svmfit, dat[train,])
```

The plot shows that the resulting SVM has a decidedly non-linear boundary. The `summary()` function can be used to obtain some information about the SVM fit:

```
> summary(svmfit)
Call:
svm(formula = y ~ ., data = dat, kernel = "radial",
    gamma = 1, cost = 1)
Parameters:
   SVM-Type:  C-classification
```

```
SVM-Kernel:   radial
       cost:   1
      gamma:   1
Number of Support Vectors:   37
 ( 17 20 )
Number of Classes:   2
Levels:
 1 2
```

We can see from the figure that there are a fair number of training errors in this SVM fit. If we increase the value of cost, we can reduce the number of training errors. However, this comes at the price of a more irregular decision boundary that seems to be at risk of overfitting the data.

```
> svmfit=svm(y~., data=dat[train,], kernel="radial",gamma=1,
    cost=1e5)

> plot(svmfit,dat[train,])
```

We can perform cross-validation using tune() to select the best choice of γ and cost for an SVM with a radial kernel:

```
> set.seed(1)
> tune.out=tune(svm, y~., data=dat[train,], kernel="radial",
    ranges=list(cost=c(0.1,1,10,100,1000),
    gamma=c(0.5,1,2,3,4)))
> summary(tune.out)
Parameter tuning of 'svm':
- sampling method: 10-fold cross validation
- best parameters:
 cost gamma
    1     2
- best performance: 0.12
- Detailed performance results:
    cost gamma error dispersion
1  1e-01   0.5  0.27    0.1160
2  1e+00   0.5  0.13    0.0823
3  1e+01   0.5  0.15    0.0707
4  1e+02   0.5  0.17    0.0823
5  1e+03   0.5  0.21    0.0994
6  1e-01   1.0  0.25    0.1354
7  1e+00   1.0  0.13    0.0823
. . .
```

Therefore, the best choice of parameters involves cost=1 and gamma=2. We can view the test set predictions for this model by applying the predict() function to the data. Notice that to do this we subset the dataframe dat using -train as an index set.

```
> table(true=dat[-train,"y"], pred=predict(tune.out$best.model,
    newdata=dat[-train,]))
```

10% of test observations are misclassified by this SVM.

9.6.3 ROC Curves

The ROCR package can be used to produce ROC curves such as those in Figures 9.10 and 9.11. We first write a short function to plot an ROC curve given a vector containing a numerical score for each observation, pred, and a vector containing the class label for each observation, truth.

```
> library(ROCR)
> rocplot=function(pred, truth, ...){
+     predob = prediction(pred, truth)
+     perf = performance(predob, "tpr", "fpr")
+     plot(perf,...)}
```

SVMs and support vector classifiers output class labels for each observation. However, it is also possible to obtain *fitted values* for each observation, which are the numerical scores used to obtain the class labels. For instance, in the case of a support vector classifier, the fitted value for an observation $X = (X_1, X_2, \ldots, X_p)^T$ takes the form $\hat{\beta}_0 + \hat{\beta}_1 X_1 + \hat{\beta}_2 X_2 + \ldots + \hat{\beta}_p X_p$. For an SVM with a non-linear kernel, the equation that yields the fitted value is given in (9.23). In essence, the sign of the fitted value determines on which side of the decision boundary the observation lies. Therefore, the relationship between the fitted value and the class prediction for a given observation is simple: if the fitted value exceeds zero then the observation is assigned to one class, and if it is less than zero then it is assigned to the other. In order to obtain the fitted values for a given SVM model fit, we use decision.values=TRUE when fitting svm(). Then the predict() function will output the fitted values.

```
> svmfit.opt=svm(y~., data=dat[train,], kernel="radial",
            gamma=2, cost=1,decision.values=T)
> fitted=attributes(predict(svmfit.opt,dat[train,],decision.
     values=TRUE))$decision.values
```

Now we can produce the ROC plot.

```
> par(mfrow=c(1,2))
> rocplot(fitted,dat[train,"y"],main="Training Data")
```

SVM appears to be producing accurate predictions. By increasing γ we can produce a more flexible fit and generate further improvements in accuracy.

```
> svmfit.flex=svm(y~., data=dat[train,], kernel="radial",
      gamma=50, cost=1, decision.values=T)
> fitted=attributes(predict(svmfit.flex,dat[train,],decision.
     values=T))$decision.values
> rocplot(fitted,dat[train,"y"],add=T,col="red")
```

However, these ROC curves are all on the training data. We are really more interested in the level of prediction accuracy on the test data. When we compute the ROC curves on the test data, the model with $\gamma = 2$ appears to provide the most accurate results.

```
> fitted=attributes(predict(svmfit.opt,dat[-train,],decision.
     values=T))$decision.values
> rocplot(fitted,dat[-train,"y"],main="Test Data")
> fitted=attributes(predict(svmfit.flex,dat[-train,],decision.
     values=T))$decision.values
> rocplot(fitted,dat[-train,"y"],add=T,col="red")
```

9.6.4   SVM with Multiple Classes

If the response is a factor containing more than two levels, then the svm()
function will perform multi-class classification using the one-versus-one approach. We explore that setting here by generating a third class of observations.

```
> set.seed(1)
> x=rbind(x, matrix(rnorm(50*2), ncol=2))
> y=c(y, rep(0,50))
> x[y==0,2]=x[y==0,2]+2
> dat=data.frame(x=x, y=as.factor(y))
> par(mfrow=c(1,1))
> plot(x,col=(y+1))
```

We now fit an SVM to the data:

```
> svmfit=svm(y~., data=dat, kernel="radial", cost=10, gamma=1)
> plot(svmfit, dat)
```

The e1071 library can also be used to perform support vector regression,
if the response vector that is passed in to svm() is numerical rather than a
factor.

9.6.5   Application to Gene Expression Data

We now examine the Khan data set, which consists of a number of tissue
samples corresponding to four distinct types of small round blue cell tu-
mors. For each tissue sample, gene expression measurements are available.
The data set consists of training data, xtrain and ytrain, and testing data,
xtest and ytest.
    We examine the dimension of the data:

```
> library(ISLR)
> names(Khan)
[1]   "xtrain"   "xtest"   "ytrain"   "ytest"
> dim(Khan$xtrain)
[1]   63 2308
> dim(Khan$xtest)
[1]   20 2308
> length(Khan$ytrain)
[1]  63
> length(Khan$ytest)
[1]  20
```

This data set consists of expression measurements for 2,308 genes.
The training and test sets consist of 63 and 20 observations respectively.

```
> table(Khan$ytrain)
 1  2  3  4
 8 23 12 20
> table(Khan$ytest)
1 2 3 4
3 6 6 5
```

We will use a support vector approach to predict cancer subtype using gene
expression measurements. In this data set, there are a very large number
of features relative to the number of observations. This suggests that we
should use a linear kernel, because the additional flexibility that will result
from using a polynomial or radial kernel is unnecessary.

```
> dat=data.frame(x=Khan$xtrain, y=as.factor(Khan$ytrain))
> out=svm(y~., data=dat, kernel="linear",cost=10)
> summary(out)
Call:
svm(formula = y ~ ., data = dat, kernel = "linear",
      cost = 10)
Parameters:
   SVM-Type:  C-classification
 SVM-Kernel:  linear
       cost:  10
      gamma:  0.000433
Number of Support Vectors:  58
 ( 20 20 11 7 )
Number of Classes:  4
Levels:
 1 2 3 4
> table(out$fitted, dat$y)

     1  2  3  4
  1  8  0  0  0
  2  0 23  0  0
  3  0  0 12  0
  4  0  0  0 20
```

We see that there are *no* training errors. In fact, this is not surprising,
because the large number of variables relative to the number of observations
implies that it is easy to find hyperplanes that fully separate the classes. We
are most interested not in the support vector classifier's performance on the
training observations, but rather its performance on the test observations.

```
> dat.te=data.frame(x=Khan$xtest, y=as.factor(Khan$ytest))
> pred.te=predict(out, newdata=dat.te)
> table(pred.te, dat.te$y)

pred.te 1 2 3 4
      1 3 0 0 0
      2 0 6 2 0
      3 0 0 4 0
      4 0 0 0 5
```

We see that using `cost=10` yields two test set errors on this data.

9.7 Exercises

Conceptual

1. This problem involves hyperplanes in two dimensions.

    (a) Sketch the hyperplane $1 + 3X_1 - X_2 = 0$. Indicate the set of points for which $1 + 3X_1 - X_2 > 0$, as well as the set of points for which $1 + 3X_1 - X_2 < 0$.

    (b) On the same plot, sketch the hyperplane $-2 + X_1 + 2X_2 = 0$. Indicate the set of points for which $-2 + X_1 + 2X_2 > 0$, as well as the set of points for which $-2 + X_1 + 2X_2 < 0$.

2. We have seen that in $p = 2$ dimensions, a linear decision boundary takes the form $\beta_0 + \beta_1 X_1 + \beta_2 X_2 = 0$. We now investigate a non-linear decision boundary.

    (a) Sketch the curve

    $$(1 + X_1)^2 + (2 - X_2)^2 = 4.$$

    (b) On your sketch, indicate the set of points for which

    $$(1 + X_1)^2 + (2 - X_2)^2 > 4,$$

    as well as the set of points for which

    $$(1 + X_1)^2 + (2 - X_2)^2 \leq 4.$$

    (c) Suppose that a classifier assigns an observation to the blue class if

    $$(1 + X_1)^2 + (2 - X_2)^2 > 4,$$

    and to the red class otherwise. To what class is the observation $(0, 0)$ classified? $(-1, 1)$? $(2, 2)$? $(3, 8)$?

    (d) Argue that while the decision boundary in (c) is not linear in terms of X_1 and X_2, it is linear in terms of X_1, X_1^2, X_2, and X_2^2.

3. Here we explore the maximal margin classifier on a toy data set.

    (a) We are given $n = 7$ observations in $p = 2$ dimensions. For each observation, there is an associated class label.

Obs.	X_1	X_2	Y
1	3	4	Red
2	2	2	Red
3	4	4	Red
4	1	4	Red
5	2	1	Blue
6	4	3	Blue
7	4	1	Blue

Sketch the observations.

(b) Sketch the optimal separating hyperplane, and provide the equation for this hyperplane (of the form (9.1)).

(c) Describe the classification rule for the maximal margin classifier. It should be something along the lines of "Classify to Red if $\beta_0 + \beta_1 X_1 + \beta_2 X_2 > 0$, and classify to Blue otherwise." Provide the values for β_0, β_1, and β_2.

(d) On your sketch, indicate the margin for the maximal margin hyperplane.

(e) Indicate the support vectors for the maximal margin classifier.

(f) Argue that a slight movement of the seventh observation would not affect the maximal margin hyperplane.

(g) Sketch a hyperplane that is *not* the optimal separating hyperplane, and provide the equation for this hyperplane.

(h) Draw an additional observation on the plot so that the two classes are no longer separable by a hyperplane.

Applied

4. Generate a simulated two-class data set with 100 observations and two features in which there is a visible but non-linear separation between the two classes. Show that in this setting, a support vector machine with a polynomial kernel (with degree greater than 1) or a radial kernel will outperform a support vector classifier on the training data. Which technique performs best on the test data? Make plots and report training and test error rates in order to back up your assertions.

5. We have seen that we can fit an SVM with a non-linear kernel in order to perform classification using a non-linear decision boundary. We will now see that we can also obtain a non-linear decision boundary by performing logistic regression using non-linear transformations of the features.

(a) Generate a data set with $n = 500$ and $p = 2$, such that the observations belong to two classes with a quadratic decision boundary between them. For instance, you can do this as follows:

```
> x1=runif(500)-0.5
> x2=runif(500)-0.5
> y=1*(x1^2-x2^2 > 0)
```

(b) Plot the observations, colored according to their class labels. Your plot should display X_1 on the x-axis, and X_2 on the y-axis.

(c) Fit a logistic regression model to the data, using X_1 and X_2 as predictors.

(d) Apply this model to the *training data* in order to obtain a predicted class label for each training observation. Plot the observations, colored according to the *predicted* class labels. The decision boundary should be linear.

(e) Now fit a logistic regression model to the data using non-linear functions of X_1 and X_2 as predictors (e.g. X_1^2, $X_1 \times X_2$, $\log(X_2)$, and so forth).

(f) Apply this model to the *training data* in order to obtain a predicted class label for each training observation. Plot the observations, colored according to the *predicted* class labels. The decision boundary should be obviously non-linear. If it is not, then repeat (a)-(e) until you come up with an example in which the predicted class labels are obviously non-linear.

(g) Fit a support vector classifier to the data with X_1 and X_2 as predictors. Obtain a class prediction for each training observation. Plot the observations, colored according to the *predicted class labels*.

(h) Fit a SVM using a non-linear kernel to the data. Obtain a class prediction for each training observation. Plot the observations, colored according to the *predicted class labels*.

(i) Comment on your results.

6. At the end of Section 9.6.1, it is claimed that in the case of data that is just barely linearly separable, a support vector classifier with a small value of cost that misclassifies a couple of training observations may perform better on test data than one with a huge value of cost that does not misclassify any training observations. You will now investigate this claim.

(a) Generate two-class data with $p = 2$ in such a way that the classes are just barely linearly separable.

(b) Compute the cross-validation error rates for support vector classifiers with a range of cost values. How many training errors are misclassified for each value of cost considered, and how does this relate to the cross-validation errors obtained?

(c) Generate an appropriate test data set, and compute the test errors corresponding to each of the values of cost considered. Which value of cost leads to the fewest test errors, and how does this compare to the values of cost that yield the fewest training errors and the fewest cross-validation errors?

(d) Discuss your results.

7. In this problem, you will use support vector approaches in order to predict whether a given car gets high or low gas mileage based on the Auto data set.

(a) Create a binary variable that takes on a 1 for cars with gas mileage above the median, and a 0 for cars with gas mileage below the median.

(b) Fit a support vector classifier to the data with various values of cost, in order to predict whether a car gets high or low gas mileage. Report the cross-validation errors associated with different values of this parameter. Comment on your results.

(c) Now repeat (b), this time using SVMs with radial and polynomial basis kernels, with different values of gamma and degree and cost. Comment on your results.

(d) Make some plots to back up your assertions in (b) and (c).

   Hint: In the lab, we used the plot() *function for* svm *objects only in cases with p = 2. When p > 2, you can use the* plot() *function to create plots displaying pairs of variables at a time. Essentially, instead of typing*

   ```
   > plot(svmfit, dat)
   ```

   where svmfit *contains your fitted model and* dat *is a data frame containing your data, you can type*

   ```
   > plot(svmfit, dat, x1~x4)
   ```

   in order to plot just the first and fourth variables. However, you must replace x1 *and* x4 *with the correct variable names. To find out more, type* ?plot.svm.

8. This problem involves the OJ data set which is part of the ISLR package.

(a) Create a training set containing a random sample of 800 observations, and a test set containing the remaining observations.

(b) Fit a support vector classifier to the training data using `cost=0.01`, with `Purchase` as the response and the other variables as predictors. Use the `summary()` function to produce summary statistics, and describe the results obtained.

(c) What are the training and test error rates?

(d) Use the `tune()` function to select an optimal `cost`. Consider values in the range 0.01 to 10.

(e) Compute the training and test error rates using this new value for `cost`.

(f) Repeat parts (b) through (e) using a support vector machine with a radial kernel. Use the default value for `gamma`.

(g) Repeat parts (b) through (e) using a support vector machine with a polynomial kernel. Set `degree=2`.

(h) Overall, which approach seems to give the best results on this data?

10
Unsupervised Learning

Most of this book concerns *supervised learning* methods such as regression and classification. In the supervised learning setting, we typically have access to a set of p features $X_1, X_2, \ldots, X_p$, measured on n observations, and a response Y also measured on those same n observations. The goal is then to predict Y using $X_1, X_2, \ldots, X_p$.

This chapter will instead focus on *unsupervised learning*, a set of statistical tools intended for the setting in which we have only a set of features $X_1, X_2, \ldots, X_p$ measured on n observations. We are not interested in prediction, because we do not have an associated response variable Y. Rather, the goal is to discover interesting things about the measurements on $X_1, X_2, \ldots, X_p$. Is there an informative way to visualize the data? Can we discover subgroups among the variables or among the observations? Unsupervised learning refers to a diverse set of techniques for answering questions such as these. In this chapter, we will focus on two particular types of unsupervised learning: *principal components analysis*, a tool used for data visualization or data pre-processing before supervised techniques are applied, and *clustering*, a broad class of methods for discovering unknown subgroups in data.

10.1 The Challenge of Unsupervised Learning

Supervised learning is a well-understood area. In fact, if you have read the preceding chapters in this book, then you should by now have a good

G. James et al., *An Introduction to Statistical Learning: with Applications in R*,
Springer Texts in Statistics, DOI 10.1007/978-1-4614-7138-7_10,
© Springer Science+Business Media New York 2013

grasp of supervised learning. For instance, if you are asked to predict a binary outcome from a data set, you have a very well developed set of tools at your disposal (such as logistic regression, linear discriminant analysis, classification trees, support vector machines, and more) as well as a clear understanding of how to assess the quality of the results obtained (using cross-validation, validation on an independent test set, and so forth).

In contrast, unsupervised learning is often much more challenging. The exercise tends to be more subjective, and there is no simple goal for the analysis, such as prediction of a response. Unsupervised learning is often performed as part of an *exploratory data analysis*. Furthermore, it can be hard to assess the results obtained from unsupervised learning methods, since there is no universally accepted mechanism for performing cross-validation or validating results on an independent data set. The reason for this difference is simple. If we fit a predictive model using a supervised learning technique, then it is possible to *check our work* by seeing how well our model predicts the response Y on observations not used in fitting the model. However, in unsupervised learning, there is no way to check our work because we don't know the true answer—the problem is unsupervised.

Techniques for unsupervised learning are of growing importance in a number of fields. A cancer researcher might assay gene expression levels in 100 patients with breast cancer. He or she might then look for subgroups among the breast cancer samples, or among the genes, in order to obtain a better understanding of the disease. An online shopping site might try to identify groups of shoppers with similar browsing and purchase histories, as well as items that are of particular interest to the shoppers within each group. Then an individual shopper can be preferentially shown the items in which he or she is particularly likely to be interested, based on the purchase histories of similar shoppers. A search engine might choose what search results to display to a particular individual based on the click histories of other individuals with similar search patterns. These statistical learning tasks, and many more, can be performed via unsupervised learning techniques.

10.2 Principal Components Analysis

Principal components are discussed in Section 6.3.1 in the context of principal components regression. When faced with a large set of correlated variables, principal components allow us to summarize this set with a smaller number of representative variables that collectively explain most of the variability in the original set. The principal component directions are presented in Section 6.3.1 as directions in feature space along which the original data are *highly variable*. These directions also define lines and subspaces that are *as close as possible* to the data cloud. To perform principal components regression, we simply use principal components as predictors in a regression model in place of the original larger set of variables.

Principal component analysis (PCA) refers to the process by which principal components are computed, and the subsequent use of these components in understanding the data. PCA is an unsupervised approach, since it involves only a set of features $X_1, X_2, \ldots, X_p$, and no associated response Y. Apart from producing derived variables for use in supervised learning problems, PCA also serves as a tool for data visualization (visualization of the observations or visualization of the variables). We now discuss PCA in greater detail, focusing on the use of PCA as a tool for unsupervised data exploration, in keeping with the topic of this chapter.

principal
component
analysis

10.2.1 *What Are Principal Components?*

Suppose that we wish to visualize n observations with measurements on a set of p features, $X_1, X_2, \ldots, X_p$, as part of an exploratory data analysis. We could do this by examining two-dimensional scatterplots of the data, each of which contains the n observations' measurements on two of the features. However, there are $\binom{p}{2} = p(p-1)/2$ such scatterplots; for example, with $p = 10$ there are 45 plots! If p is large, then it will certainly not be possible to look at all of them; moreover, most likely none of them will be informative since they each contain just a small fraction of the total information present in the data set. Clearly, a better method is required to visualize the n observations when p is large. In particular, we would like to find a low-dimensional representation of the data that captures as much of the information as possible. For instance, if we can obtain a two-dimensional representation of the data that captures most of the information, then we can plot the observations in this low-dimensional space.

PCA provides a tool to do just this. It finds a low-dimensional representation of a data set that contains as much as possible of the variation. The idea is that each of the n observations lives in p-dimensional space, but not all of these dimensions are equally interesting. PCA seeks a small number of dimensions that are as interesting as possible, where the concept of *interesting* is measured by the amount that the observations vary along each dimension. Each of the dimensions found by PCA is a linear combination of the p features. We now explain the manner in which these dimensions, or *principal components*, are found.

The *first principal component* of a set of features $X_1, X_2, \ldots, X_p$ is the normalized linear combination of the features

$$Z_1 = \phi_{11}X_1 + \phi_{21}X_2 + \ldots + \phi_{p1}X_p \tag{10.1}$$

that has the largest variance. By *normalized*, we mean that $\sum_{j=1}^{p} \phi_{j1}^2 = 1$. We refer to the elements $\phi_{11}, \ldots, \phi_{p1}$ as the *loadings* of the first principal

loading

component; together, the loadings make up the principal component loading vector, $\phi_1 = (\phi_{11}\ \phi_{21}\ \ldots\ \phi_{p1})^T$. We constrain the loadings so that their sum of squares is equal to one, since otherwise setting these elements to be arbitrarily large in absolute value could result in an arbitrarily large variance.

Given a $n \times p$ data set $\mathbf{X}$, how do we compute the first principal component? Since we are only interested in variance, we assume that each of the variables in $\mathbf{X}$ has been centered to have mean zero (that is, the column means of $\mathbf{X}$ are zero). We then look for the linear combination of the sample feature values of the form

$$z_{i1} = \phi_{11}x_{i1} + \phi_{21}x_{i2} + \ldots + \phi_{p1}x_{ip} \tag{10.2}$$

that has largest sample variance, subject to the constraint that $\sum_{j=1}^{p} \phi_{j1}^2 = 1$. In other words, the first principal component loading vector solves the optimization problem

$$\underset{\phi_{11},\ldots,\phi_{p1}}{\text{maximize}} \left\{ \frac{1}{n} \sum_{i=1}^{n} \left(\sum_{j=1}^{p} \phi_{j1}x_{ij} \right)^2 \right\} \text{ subject to } \sum_{j=1}^{p} \phi_{j1}^2 = 1. \tag{10.3}$$

From (10.2) we can write the objective in (10.3) as $\frac{1}{n}\sum_{i=1}^{n} z_{i1}^2$. Since $\frac{1}{n}\sum_{i=1}^{n} x_{ij} = 0$, the average of the $z_{11}, \ldots, z_{n1}$ will be zero as well. Hence the objective that we are maximizing in (10.3) is just the sample variance of the n values of z_{i1}. We refer to $z_{11}, \ldots, z_{n1}$ as the *scores* of the first princi- score
pal component. Problem (10.3) can be solved via an eigen decomposition, a standard technique in linear algebra, but details are outside of the scope of this book.

There is a nice geometric interpretation for the first principal component. The loading vector ϕ_1 with elements $\phi_{11}, \phi_{21}, \ldots, \phi_{p1}$ defines a direction in feature space along which the data vary the most. If we project the n data points $x_1, \ldots, x_n$ onto this direction, the projected values are the principal component scores $z_{11}, \ldots, z_{n1}$ themselves. For instance, Figure 6.14 on page 230 displays the first principal component loading vector (green solid line) on an advertising data set. In these data, there are only two features, and so the observations as well as the first principal component loading vector can be easily displayed. As can be seen from (6.19), in that data set $\phi_{11} = 0.839$ and $\phi_{21} = 0.544$.

After the first principal component Z_1 of the features has been determined, we can find the second principal component Z_2. The second principal component is the linear combination of $X_1, \ldots, X_p$ that has maximal variance out of all linear combinations that are *uncorrelated* with Z_1. The second principal component scores $z_{12}, z_{22}, \ldots, z_{n2}$ take the form

$$z_{i2} = \phi_{12}x_{i1} + \phi_{22}x_{i2} + \ldots + \phi_{p2}x_{ip}, \tag{10.4}$$

	PC1	PC2
Murder	0.5358995	−0.4181809
Assault	0.5831836	−0.1879856
UrbanPop	0.2781909	0.8728062
Rape	0.5434321	0.1673186

TABLE 10.1. *The principal component loading vectors, ϕ_1 and ϕ_2, for the* USArrests *data. These are also displayed in Figure 10.1.*

where ϕ_2 is the second principal component loading vector, with elements $\phi_{12}, \phi_{22}, \dots, \phi_{p2}$. It turns out that constraining Z_2 to be uncorrelated with Z_1 is equivalent to constraining the direction ϕ_2 to be orthogonal (perpendicular) to the direction ϕ_1. In the example in Figure 6.14, the observations lie in two-dimensional space (since $p = 2$), and so once we have found ϕ_1, there is only one possibility for ϕ_2, which is shown as a blue dashed line. (From Section 6.3.1, we know that $\phi_{12} = 0.544$ and $\phi_{22} = -0.839$.) But in a larger data set with $p > 2$ variables, there are multiple distinct principal components, and they are defined in a similar manner. To find ϕ_2, we solve a problem similar to (10.3) with ϕ_2 replacing ϕ_1, and with the additional constraint that ϕ_2 is orthogonal to ϕ_1.[1]

Once we have computed the principal components, we can plot them against each other in order to produce low-dimensional views of the data. For instance, we can plot the score vector Z_1 against Z_2, Z_1 against Z_3, Z_2 against Z_3, and so forth. Geometrically, this amounts to projecting the original data down onto the subspace spanned by ϕ_1, ϕ_2, and ϕ_3, and plotting the projected points.

We illustrate the use of PCA on the USArrests data set. For each of the 50 states in the United States, the data set contains the number of arrests per 100,000 residents for each of three crimes: Assault, Murder, and Rape. We also record UrbanPop (the percent of the population in each state living in urban areas). The principal component score vectors have length $n = 50$, and the principal component loading vectors have length $p = 4$. PCA was performed after standardizing each variable to have mean zero and standard deviation one. Figure 10.1 plots the first two principal components of these data. The figure represents both the principal component scores and the loading vectors in a single *biplot* display. The loadings are also given in Table 10.1.

In Figure 10.1, we see that the first loading vector places approximately equal weight on Assault, Murder, and Rape, with much less weight on

biplot

---

[1] On a technical note, the principal component directions ϕ_1, ϕ_2, ϕ_3, ... are the ordered sequence of eigenvectors of the matrix $\mathbf{X}^T\mathbf{X}$, and the variances of the components are the eigenvalues. There are at most $\min(n - 1, p)$ principal components.

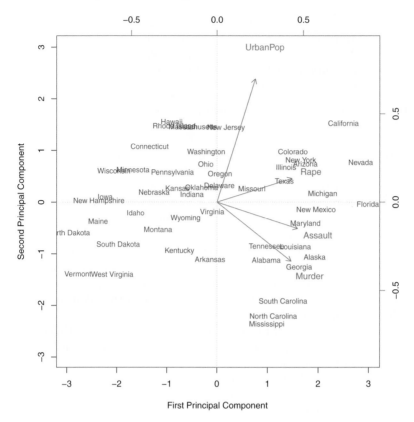

FIGURE 10.1. *The first two principal components for the* USArrests *data. The blue state names represent the scores for the first two principal components. The orange arrows indicate the first two principal component loading vectors (with axes on the top and right). For example, the loading for* Rape *on the first component is* 0.54, *and its loading on the second principal component* 0.17 *(the word* Rape *is centered at the point* (0.54, 0.17)). *This figure is known as a* biplot, *because it displays both the principal component scores and the principal component loadings.*

UrbanPop. Hence this component roughly corresponds to a measure of overall rates of serious crimes. The second loading vector places most of its weight on UrbanPop and much less weight on the other three features. Hence, this component roughly corresponds to the level of urbanization of the state. Overall, we see that the crime-related variables (Murder, Assault, and Rape) are located close to each other, and that the UrbanPop variable is far from the other three. This indicates that the crime-related variables are correlated with each other—states with high murder rates tend to have high assault and rape rates—and that the UrbanPop variable is less correlated with the other three.

We can examine differences between the states via the two principal component score vectors shown in Figure 10.1. Our discussion of the loading vectors suggests that states with large positive scores on the first component, such as California, Nevada and Florida, have high crime rates, while states like North Dakota, with negative scores on the first component, have low crime rates. California also has a high score on the second component, indicating a high level of urbanization, while the opposite is true for states like Mississippi. States close to zero on both components, such as Indiana, have approximately average levels of both crime and urbanization.

10.2.2 *Another Interpretation of Principal Components*

The first two principal component loading vectors in a simulated three-dimensional data set are shown in the left-hand panel of Figure 10.2; these two loading vectors span a plane along which the observations have the highest variance.

In the previous section, we describe the principal component loading vectors as the directions in feature space along which the data vary the most, and the principal component scores as projections along these directions. However, an alternative interpretation for principal components can also be useful: principal components provide low-dimensional linear surfaces that are *closest* to the observations. We expand upon that interpretation here.

The first principal component loading vector has a very special property: it is the line in p-dimensional space that is *closest* to the n observations (using average squared Euclidean distance as a measure of closeness). This interpretation can be seen in the left-hand panel of Figure 6.15; the dashed lines indicate the distance between each observation and the first principal component loading vector. The appeal of this interpretation is clear: we seek a single dimension of the data that lies as close as possible to all of the data points, since such a line will likely provide a good summary of the data.

The notion of principal components as the dimensions that are closest to the n observations extends beyond just the first principal component. For instance, the first two principal components of a data set span the plane that is closest to the n observations, in terms of average squared Euclidean distance. An example is shown in the left-hand panel of Figure 10.2. The first three principal components of a data set span the three-dimensional hyperplane that is closest to the n observations, and so forth.

Using this interpretation, together the first M principal component score vectors and the first M principal component loading vectors provide the best M-dimensional approximation (in terms of Euclidean distance) to the ith observation x_{ij}. This representation can be written

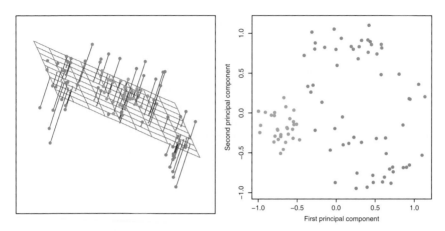

FIGURE 10.2. *Ninety observations simulated in three dimensions. Left: the first two principal component directions span the plane that best fits the data. It minimizes the sum of squared distances from each point to the plane. Right: the first two principal component score vectors give the coordinates of the projection of the 90 observations onto the plane. The variance in the plane is maximized.*

$$x_{ij} \approx \sum_{m=1}^{M} z_{im}\phi_{jm} \tag{10.5}$$

(assuming the original data matrix $\mathbf{X}$ is column-centered). In other words, together the M principal component score vectors and M principal component loading vectors can give a good approximation to the data when M is sufficiently large. When $M = \min(n-1, p)$, then the representation is exact: $x_{ij} = \sum_{m=1}^{M} z_{im}\phi_{jm}$.

10.2.3 More on PCA

Scaling the Variables

We have already mentioned that before PCA is performed, the variables should be centered to have mean zero. Furthermore, *the results obtained when we perform PCA will also depend on whether the variables have been individually scaled* (each multiplied by a different constant). This is in contrast to some other supervised and unsupervised learning techniques, such as linear regression, in which scaling the variables has no effect. (In linear regression, multiplying a variable by a factor of c will simply lead to multiplication of the corresponding coefficient estimate by a factor of $1/c$, and thus will have no substantive effect on the model obtained.)

For instance, Figure 10.1 was obtained after scaling each of the variables to have standard deviation one. This is reproduced in the left-hand plot in Figure 10.3. Why does it matter that we scaled the variables? In these data,

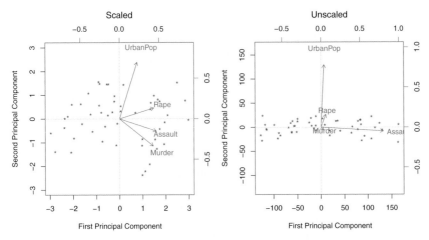

FIGURE 10.3. *Two principal component biplots for the* USArrests *data.* Left: *the same as Figure 10.1, with the variables scaled to have unit standard deviations.* Right: *principal components using unscaled data.* Assault *has by far the largest loading on the first principal component because it has the highest variance among the four variables. In general, scaling the variables to have standard deviation one is recommended.*

the variables are measured in different units; Murder, Rape, and Assault are reported as the number of occurrences per 100,000 people, and UrbanPop is the percentage of the state's population that lives in an urban area. These four variables have variance 18.97, 87.73, 6945.16, and 209.5, respectively. Consequently, if we perform PCA on the unscaled variables, then the first principal component loading vector will have a very large loading for Assault, since that variable has by far the highest variance. The right-hand plot in Figure 10.3 displays the first two principal components for the USArrests data set, without scaling the variables to have standard deviation one. As predicted, the first principal component loading vector places almost all of its weight on Assault, while the second principal component loading vector places almost all of its weight on UrpanPop. Comparing this to the left-hand plot, we see that scaling does indeed have a substantial effect on the results obtained.

However, this result is simply a consequence of the scales on which the variables were measured. For instance, if Assault were measured in units of the number of occurrences per 100 people (rather than number of occurrences per 100,000 people), then this would amount to dividing all of the elements of that variable by 1,000. Then the variance of the variable would be tiny, and so the first principal component loading vector would have a very small value for that variable. Because it is undesirable for the principal components obtained to depend on an arbitrary choice of scaling, we typically scale each variable to have standard deviation one before we perform PCA.

In certain settings, however, the variables may be measured in the same units. In this case, we might not wish to scale the variables to have standard deviation one before performing PCA. For instance, suppose that the variables in a given data set correspond to expression levels for p genes. Then since expression is measured in the same "units" for each gene, we might choose not to scale the genes to each have standard deviation one.

Uniqueness of the Principal Components

Each principal component loading vector is unique, up to a sign flip. This means that two different software packages will yield the same principal component loading vectors, although the signs of those loading vectors may differ. The signs may differ because each principal component loading vector specifies a direction in p-dimensional space: flipping the sign has no effect as the direction does not change. (Consider Figure 6.14—the principal component loading vector is a line that extends in either direction, and flipping its sign would have no effect.) Similarly, the score vectors are unique up to a sign flip, since the variance of Z is the same as the variance of $-Z$. It is worth noting that when we use (10.5) to approximate x_{ij} we multiply z_{im} by ϕ_{jm}. Hence, if the sign is flipped on both the loading and score vectors, the final product of the two quantities is unchanged.

The Proportion of Variance Explained

In Figure 10.2, we performed PCA on a three-dimensional data set (left-hand panel) and projected the data onto the first two principal component loading vectors in order to obtain a two-dimensional view of the data (i.e. the principal component score vectors; right-hand panel). We see that this two-dimensional representation of the three-dimensional data does successfully capture the major pattern in the data: the orange, green, and cyan observations that are near each other in three-dimensional space remain nearby in the two-dimensional representation. Similarly, we have seen on the USArrests data set that we can summarize the 50 observations and 4 variables using just the first two principal component score vectors and the first two principal component loading vectors.

We can now ask a natural question: how much of the information in a given data set is lost by projecting the observations onto the first few principal components? That is, how much of the variance in the data is *not* contained in the first few principal components? More generally, we are interested in knowing the *proportion of variance explained* (PVE) by each principal component. The *total variance* present in a data set (assuming that the variables have been centered to have mean zero) is defined as

$$\sum_{j=1}^{p} \operatorname{Var}(X_j) = \sum_{j=1}^{p} \frac{1}{n} \sum_{i=1}^{n} x_{ij}^2, \qquad (10.6)$$

proportion of variance explained

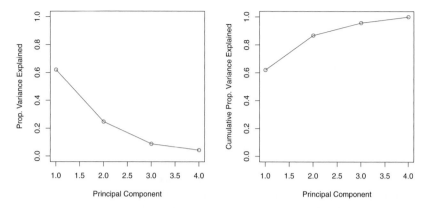

FIGURE 10.4. *Left: a scree plot depicting the proportion of variance explained by each of the four principal components in the* USArrests *data. Right: the cumulative proportion of variance explained by the four principal components in the* USArrests *data.*

and the variance explained by the mth principal component is

$$\frac{1}{n}\sum_{i=1}^{n} z_{im}^2 = \frac{1}{n}\sum_{i=1}^{n}\left(\sum_{j=1}^{p}\phi_{jm}x_{ij}\right)^2. \qquad (10.7)$$

Therefore, the PVE of the mth principal component is given by

$$\frac{\sum_{i=1}^{n}\left(\sum_{j=1}^{p}\phi_{jm}x_{ij}\right)^2}{\sum_{j=1}^{p}\sum_{i=1}^{n}x_{ij}^2}. \qquad (10.8)$$

The PVE of each principal component is a positive quantity. In order to compute the cumulative PVE of the first M principal components, we can simply sum (10.8) over each of the first M PVEs. In total, there are $\min(n-1,p)$ principal components, and their PVEs sum to one.

In the USArrests data, the first principal component explains 62.0 % of the variance in the data, and the next principal component explains 24.7 % of the variance. Together, the first two principal components explain almost 87 % of the variance in the data, and the last two principal components explain only 13 % of the variance. This means that Figure 10.1 provides a pretty accurate summary of the data using just two dimensions. The PVE of each principal component, as well as the cumulative PVE, is shown in Figure 10.4. The left-hand panel is known as a *scree plot*, and will be discussed next.

scree plot

Deciding How Many Principal Components to Use

In general, a $n \times p$ data matrix $\mathbf{X}$ has $\min(n-1,p)$ distinct principal components. However, we usually are not interested in all of them; rather, we would like to use just the first few principal components in order to visualize or interpret the data. In fact, we would like to use the smallest number of principal components required to get a *good* understanding of the data. How many principal components are needed? Unfortunately, there is no single (or simple!) answer to this question.

We typically decide on the number of principal components required to visualize the data by examining a *scree plot*, such as the one shown in the left-hand panel of Figure 10.4. We choose the smallest number of principal components that are required in order to explain a sizable amount of the variation in the data. This is done by eyeballing the scree plot, and looking for a point at which the proportion of variance explained by each subsequent principal component drops off. This is often referred to as an *elbow* in the scree plot. For instance, by inspection of Figure 10.4, one might conclude that a fair amount of variance is explained by the first two principal components, and that there is an elbow after the second component. After all, the third principal component explains less than ten percent of the variance in the data, and the fourth principal component explains less than half that and so is essentially worthless.

However, this type of visual analysis is inherently *ad hoc*. Unfortunately, there is no well-accepted objective way to decide how many principal components are *enough*. In fact, the question of how many principal components are enough is inherently ill-defined, and will depend on the specific area of application and the specific data set. In practice, we tend to look at the first few principal components in order to find interesting patterns in the data. If no interesting patterns are found in the first few principal components, then further principal components are unlikely to be of interest. Conversely, if the first few principal components are interesting, then we typically continue to look at subsequent principal components until no further interesting patterns are found. This is admittedly a subjective approach, and is reflective of the fact that PCA is generally used as a tool for exploratory data analysis.

On the other hand, if we compute principal components for use in a supervised analysis, such as the principal components regression presented in Section 6.3.1, then there is a simple and objective way to determine how many principal components to use: we can treat the number of principal component score vectors to be used in the regression as a tuning parameter to be selected via cross-validation or a related approach. The comparative simplicity of selecting the number of principal components for a supervised analysis is one manifestation of the fact that supervised analyses tend to be more clearly defined and more objectively evaluated than unsupervised analyses.

10.2.4   Other Uses for Principal Components

We saw in Section 6.3.1 that we can perform regression using the principal component score vectors as features. In fact, many statistical techniques, such as regression, classification, and clustering, can be easily adapted to use the $n \times M$ matrix whose columns are the first $M \ll p$ principal component score vectors, rather than using the full $n \times p$ data matrix. This can lead to *less noisy* results, since it is often the case that the signal (as opposed to the noise) in a data set is concentrated in its first few principal components.

10.3   Clustering Methods

Clustering refers to a very broad set of techniques for finding *subgroups*, or *clusters*, in a data set. When we cluster the observations of a data set, we seek to partition them into distinct groups so that the observations within each group are quite similar to each other, while observations in different groups are quite different from each other. Of course, to make this concrete, we must define what it means for two or more observations to be *similar* or *different*. Indeed, this is often a domain-specific consideration that must be made based on knowledge of the data being studied.

clustering

For instance, suppose that we have a set of n observations, each with p features. The n observations could correspond to tissue samples for patients with breast cancer, and the p features could correspond to measurements collected for each tissue sample; these could be clinical measurements, such as tumor stage or grade, or they could be gene expression measurements. We may have a reason to believe that there is some heterogeneity among the n tissue samples; for instance, perhaps there are a few different *unknown* subtypes of breast cancer. Clustering could be used to find these subgroups. This is an unsupervised problem because we are trying to discover structure—in this case, distinct clusters—on the basis of a data set. The goal in supervised problems, on the other hand, is to try to predict some outcome vector such as survival time or response to drug treatment.

Both clustering and PCA seek to simplify the data via a small number of summaries, but their mechanisms are different:

- PCA looks to find a low-dimensional representation of the observations that explain a good fraction of the variance;

- Clustering looks to find homogeneous subgroups among the observations.

Another application of clustering arises in marketing. We may have access to a large number of measurements (e.g. median household income, occupation, distance from nearest urban area, and so forth) for a large number of people. Our goal is to perform *market segmentation* by identifying subgroups of people who might be more receptive to a particular form of advertising, or more likely to purchase a particular product. The task of performing market segmentation amounts to clustering the people in the data set.

Since clustering is popular in many fields, there exist a great number of clustering methods. In this section we focus on perhaps the two best-known clustering approaches: *K-means clustering* and *hierarchical clustering*. In *K*-means clustering, we seek to partition the observations into a pre-specified number of clusters. On the other hand, in hierarchical clustering, we do not know in advance how many clusters we want; in fact, we end up with a tree-like visual representation of the observations, called a *dendrogram*, that allows us to view at once the clusterings obtained for each possible number of clusters, from 1 to n. There are advantages and disadvantages to each of these clustering approaches, which we highlight in this chapter.

In general, we can cluster observations on the basis of the features in order to identify subgroups among the observations, or we can cluster features on the basis of the observations in order to discover subgroups among the features. In what follows, for simplicity we will discuss clustering observations on the basis of the features, though the converse can be performed by simply transposing the data matrix.

K-means clustering

hierarchical clustering

dendrogram

10.3.1   K-Means Clustering

K-means clustering is a simple and elegant approach for partitioning a data set into K distinct, non-overlapping clusters. To perform K-means clustering, we must first specify the desired number of clusters K; then the K-means algorithm will assign each observation to exactly one of the K clusters. Figure 10.5 shows the results obtained from performing K-means clustering on a simulated example consisting of 150 observations in two dimensions, using three different values of K.

The K-means clustering procedure results from a simple and intuitive mathematical problem. We begin by defining some notation. Let $C_1, \ldots, C_K$ denote sets containing the indices of the observations in each cluster. These sets satisfy two properties:

1. $C_1 \cup C_2 \cup \ldots \cup C_K = \{1, \ldots, n\}$. In other words, each observation belongs to at least one of the K clusters.

2. $C_k \cap C_{k'} = \emptyset$ for all $k \neq k'$. In other words, the clusters are non-overlapping: no observation belongs to more than one cluster.

For instance, if the ith observation is in the kth cluster, then $i \in C_k$. The idea behind K-means clustering is that a *good* clustering is one for which the *within-cluster variation* is as small as possible. The within-cluster variation

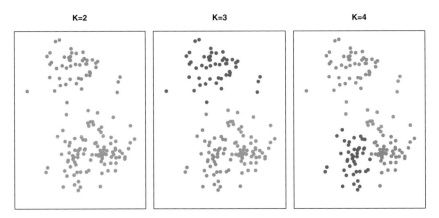

FIGURE 10.5. *A simulated data set with 150 observations in two-dimensional space. Panels show the results of applying K-means clustering with different values of K, the number of clusters. The color of each observation indicates the cluster to which it was assigned using the K-means clustering algorithm. Note that there is no ordering of the clusters, so the cluster coloring is arbitrary. These cluster labels were not used in clustering; instead, they are the outputs of the clustering procedure.*

for cluster C_k is a measure $W(C_k)$ of the amount by which the observations within a cluster differ from each other. Hence we want to solve the problem

$$\underset{C_1,\ldots,C_K}{\text{minimize}}\left\{\sum_{k=1}^{K}W(C_k)\right\}. \tag{10.9}$$

In words, this formula says that we want to partition the observations into K clusters such that the total within-cluster variation, summed over all K clusters, is as small as possible.

Solving (10.9) seems like a reasonable idea, but in order to make it actionable we need to define the within-cluster variation. There are many possible ways to define this concept, but by far the most common choice involves *squared Euclidean distance*. That is, we define

$$W(C_k) = \frac{1}{|C_k|}\sum_{i,i'\in C_k}\sum_{j=1}^{p}(x_{ij}-x_{i'j})^2, \tag{10.10}$$

where $|C_k|$ denotes the number of observations in the kth cluster. In other words, the within-cluster variation for the kth cluster is the sum of all of the pairwise squared Euclidean distances between the observations in the kth cluster, divided by the total number of observations in the kth cluster. Combining (10.9) and (10.10) gives the optimization problem that defines K-means clustering,

$$\underset{C_1,\ldots,C_K}{\text{minimize}}\left\{\sum_{k=1}^{K}\frac{1}{|C_k|}\sum_{i,i'\in C_k}\sum_{j=1}^{p}(x_{ij}-x_{i'j})^2\right\}. \tag{10.11}$$

Now, we would like to find an algorithm to solve (10.11)—that is, a method to partition the observations into K clusters such that the objective of (10.11) is minimized. This is in fact a very difficult problem to solve precisely, since there are almost K^n ways to partition n observations into K clusters. This is a huge number unless K and n are tiny! Fortunately, a very simple algorithm can be shown to provide a local optimum—a *pretty good solution*—to the K-means optimization problem (10.11). This approach is laid out in Algorithm 10.1.

Algorithm 10.1 *K-Means Clustering*

---

1. Randomly assign a number, from 1 to K, to each of the observations. These serve as initial cluster assignments for the observations.

2. Iterate until the cluster assignments stop changing:

    (a) For each of the K clusters, compute the cluster *centroid*. The kth cluster centroid is the vector of the p feature means for the observations in the kth cluster.

    (b) Assign each observation to the cluster whose centroid is closest (where *closest* is defined using Euclidean distance).

Algorithm 10.1 is guaranteed to decrease the value of the objective (10.11) at each step. To understand why, the following identity is illuminating:

$$\frac{1}{|C_k|} \sum_{i,i' \in C_k} \sum_{j=1}^{p} (x_{ij} - x_{i'j})^2 = 2 \sum_{i \in C_k} \sum_{j=1}^{p} (x_{ij} - \bar{x}_{kj})^2, \qquad (10.12)$$

where $\bar{x}_{kj} = \frac{1}{|C_k|} \sum_{i \in C_k} x_{ij}$ is the mean for feature j in cluster C_k. In Step 2(a) the cluster means for each feature are the constants that minimize the sum-of-squared deviations, and in Step 2(b), reallocating the observations can only improve (10.12). This means that as the algorithm is run, the clustering obtained will continually improve until the result no longer changes; the objective of (10.11) will never increase. When the result no longer changes, a *local optimum* has been reached. Figure 10.6 shows the progression of the algorithm on the toy example from Figure 10.5. K-means clustering derives its name from the fact that in Step 2(a), the cluster centroids are computed as the mean of the observations assigned to each cluster.

Because the K-means algorithm finds a local rather than a global optimum, the results obtained will depend on the initial (random) cluster assignment of each observation in Step 1 of Algorithm 10.1. For this reason, it is important to run the algorithm multiple times from different random

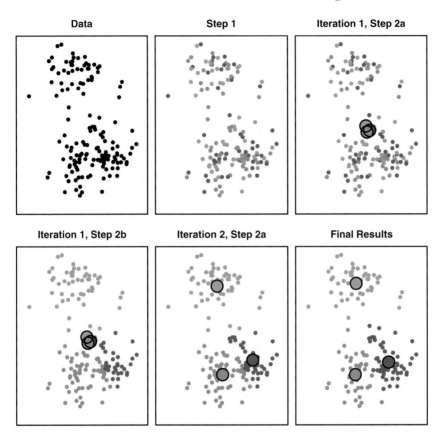

FIGURE 10.6. *The progress of the K-means algorithm on the example of Figure 10.5 with K=3. Top left: the observations are shown.* Top center: *in Step 1 of the algorithm, each observation is randomly assigned to a cluster.* Top right: *in Step 2(a), the cluster centroids are computed. These are shown as large colored disks. Initially the centroids are almost completely overlapping because the initial cluster assignments were chosen at random.* Bottom left: *in Step 2(b), each observation is assigned to the nearest centroid.* Bottom center: *Step 2(a) is once again performed, leading to new cluster centroids.* Bottom right: *the results obtained after ten iterations.*

initial configurations. Then one selects the *best* solution, i.e. that for which the objective (10.11) is smallest. Figure 10.7 shows the local optima obtained by running K-means clustering six times using six different initial cluster assignments, using the toy data from Figure 10.5. In this case, the best clustering is the one with an objective value of 235.8.

As we have seen, to perform K-means clustering, we must decide how many clusters we expect in the data. The problem of selecting K is far from simple. This issue, along with other practical considerations that arise in performing K-means clustering, is addressed in Section 10.3.3.

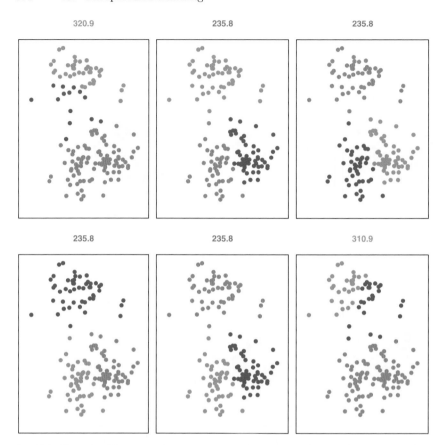

FIGURE 10.7. *K-means clustering performed six times on the data from Figure 10.5 with K = 3, each time with a different random assignment of the observations in Step 1 of the K-means algorithm. Above each plot is the value of the objective (10.11). Three different local optima were obtained, one of which resulted in a smaller value of the objective and provides better separation between the clusters. Those labeled in red all achieved the same best solution, with an objective value of 235.8.*

10.3.2 Hierarchical Clustering

One potential disadvantage of K-means clustering is that it requires us to pre-specify the number of clusters K. *Hierarchical clustering* is an alternative approach which does not require that we commit to a particular choice of K. Hierarchical clustering has an added advantage over K-means clustering in that it results in an attractive tree-based representation of the observations, called a *dendrogram*.

In this section, we describe *bottom-up* or *agglomerative*   clustering. This is the most common type of hierarchical clustering, and refers to the fact that a dendrogram (generally depicted as an upside-down tree; see

bottom-up
agglomerative

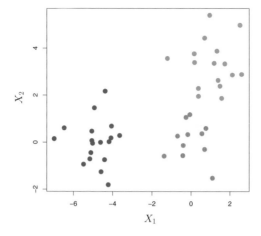

FIGURE 10.8. *Forty-five observations generated in two-dimensional space. In reality there are three distinct classes, shown in separate colors. However, we will treat these class labels as unknown and will seek to cluster the observations in order to discover the classes from the data.*

Figure 10.9) is built starting from the leaves and combining clusters up to the trunk. We will begin with a discussion of how to interpret a dendrogram and then discuss how hierarchical clustering is actually performed—that is, how the dendrogram is built.

Interpreting a Dendrogram

We begin with the simulated data set shown in Figure 10.8, consisting of 45 observations in two-dimensional space. The data were generated from a three-class model; the true class labels for each observation are shown in distinct colors. However, suppose that the data were observed without the class labels, and that we wanted to perform hierarchical clustering of the data. Hierarchical clustering (with complete linkage, to be discussed later) yields the result shown in the left-hand panel of Figure 10.9. How can we interpret this dendrogram?

In the left-hand panel of Figure 10.9, each *leaf* of the dendrogram represents one of the 45 observations in Figure 10.8. However, as we move up the tree, some leaves begin to *fuse* into branches. These correspond to observations that are similar to each other. As we move higher up the tree, branches themselves fuse, either with leaves or other branches. The earlier (lower in the tree) fusions occur, the more similar the groups of observations are to each other. On the other hand, observations that fuse later (near the top of the tree) can be quite different. In fact, this statement can be made precise: for any two observations, we can look for the point in the tree where branches containing those two observations are first fused. The height of this fusion, as measured on the vertical axis, indicates how

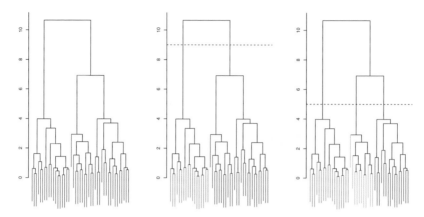

FIGURE 10.9. Left: *dendrogram obtained from hierarchically clustering the data from Figure 10.8 with complete linkage and Euclidean distance.* Center: *the dendrogram from the left-hand panel, cut at a height of nine (indicated by the dashed line). This cut results in two distinct clusters, shown in different colors.* Right: *the dendrogram from the left-hand panel, now cut at a height of five. This cut results in three distinct clusters, shown in different colors. Note that the colors were not used in clustering, but are simply used for display purposes in this figure.*

different the two observations are. Thus, observations that fuse at the very bottom of the tree are quite similar to each other, whereas observations that fuse close to the top of the tree will tend to be quite different.

This highlights a very important point in interpreting dendrograms that is often misunderstood. Consider the left-hand panel of Figure 10.10, which shows a simple dendrogram obtained from hierarchically clustering nine observations. One can see that observations 5 and 7 are quite similar to each other, since they fuse at the lowest point on the dendrogram. Observations 1 and 6 are also quite similar to each other. However, it is tempting but incorrect to conclude from the figure that observations 9 and 2 are quite similar to each other on the basis that they are located near each other on the dendrogram. In fact, based on the information contained in the dendrogram, observation 9 is no more similar to observation 2 than it is to observations 8, 5, and 7. (This can be seen from the right-hand panel of Figure 10.10, in which the raw data are displayed.) To put it mathematically, there are 2^{n-1} possible reorderings of the dendrogram, where n is the number of leaves. This is because at each of the $n - 1$ points where fusions occur, the positions of the two fused branches could be swapped without affecting the meaning of the dendrogram. Therefore, we cannot draw conclusions about the similarity of two observations based on their proximity along the *horizontal axis*. Rather, we draw conclusions about the similarity of two observations based on the location on the *vertical axis* where branches containing those two observations first are fused.

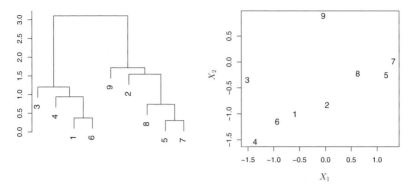

FIGURE 10.10. *An illustration of how to properly interpret a dendrogram with nine observations in two-dimensional space. Left: a dendrogram generated using Euclidean distance and complete linkage. Observations 5 and 7 are quite similar to each other, as are observations 1 and 6. However, observation 9 is no more similar to observation 2 than it is to observations 8, 5, and 7, even though observations 9 and 2 are close together in terms of horizontal distance. This is because observations 2, 8, 5, and 7 all fuse with observation 9 at the same height, approximately 1.8. Right: the raw data used to generate the dendrogram can be used to confirm that indeed, observation 9 is no more similar to observation 2 than it is to observations 8, 5, and 7.*

Now that we understand how to interpret the left-hand panel of Figure 10.9, we can move on to the issue of identifying clusters on the basis of a dendrogram. In order to do this, we make a horizontal cut across the dendrogram, as shown in the center and right-hand panels of Figure 10.9. The distinct sets of observations beneath the cut can be interpreted as clusters. In the center panel of Figure 10.9, cutting the dendrogram at a height of nine results in two clusters, shown in distinct colors. In the right-hand panel, cutting the dendrogram at a height of five results in three clusters. Further cuts can be made as one descends the dendrogram in order to obtain any number of clusters, between 1 (corresponding to no cut) and n (corresponding to a cut at height 0, so that each observation is in its own cluster). In other words, the height of the cut to the dendrogram serves the same role as the K in K-means clustering: it controls the number of clusters obtained.

Figure 10.9 therefore highlights a very attractive aspect of hierarchical clustering: one single dendrogram can be used to obtain any number of clusters. In practice, people often look at the dendrogram and select by eye a sensible number of clusters, based on the heights of the fusion and the number of clusters desired. In the case of Figure 10.9, one might choose to select either two or three clusters. However, often the choice of where to cut the dendrogram is not so clear.

The term *hierarchical* refers to the fact that clusters obtained by cutting the dendrogram at a given height are necessarily nested within the clusters obtained by cutting the dendrogram at any greater height. However, on an arbitrary data set, this assumption of hierarchical structure might be unrealistic. For instance, suppose that our observations correspond to a group of people with a 50–50 split of males and females, evenly split among Americans, Japanese, and French. We can imagine a scenario in which the best division into two groups might split these people by gender, and the best division into three groups might split them by nationality. In this case, the true clusters are not nested, in the sense that the best division into three groups does not result from taking the best division into two groups and splitting up one of those groups. Consequently, this situation could not be well-represented by hierarchical clustering. Due to situations such as this one, hierarchical clustering can sometimes yield *worse* (i.e. less accurate) results than K-means clustering for a given number of clusters.

The Hierarchical Clustering Algorithm

The hierarchical clustering dendrogram is obtained via an extremely simple algorithm. We begin by defining some sort of *dissimilarity* measure between each pair of observations. Most often, Euclidean distance is used; we will discuss the choice of dissimilarity measure later in this chapter. The algorithm proceeds iteratively. Starting out at the bottom of the dendrogram, each of the n observations is treated as its own cluster. The two clusters that are most similar to each other are then *fused* so that there now are $n-1$ clusters. Next the two clusters that are most similar to each other are fused again, so that there now are $n-2$ clusters. The algorithm proceeds in this fashion until all of the observations belong to one single cluster, and the dendrogram is complete. Figure 10.11 depicts the first few steps of the algorithm, for the data from Figure 10.9. To summarize, the hierarchical clustering algorithm is given in Algorithm 10.2.

This algorithm seems simple enough, but one issue has not been addressed. Consider the bottom right panel in Figure 10.11. How did we determine that the cluster $\{5, 7\}$ should be fused with the cluster $\{8\}$? We have a concept of the dissimilarity between pairs of observations, but how do we define the dissimilarity between two clusters if one or both of the clusters contains multiple observations? The concept of dissimilarity between a pair of observations needs to be extended to a pair of *groups of observations*. This extension is achieved by developing the notion of *linkage*, which defines the dissimilarity between two groups of observations. The four most common types of linkage—*complete, average, single,* and *centroid*—are briefly described in Table 10.2. Average, complete, and single linkage are most popular among statisticians. Average and complete

linkage

Algorithm 10.2 *Hierarchical Clustering*

---

1. Begin with n observations and a measure (such as Euclidean distance) of all the $\binom{n}{2} = n(n-1)/2$ pairwise dissimilarities. Treat each observation as its own cluster.

2. For $i = n, n-1, \ldots, 2$:

   (a) Examine all pairwise inter-cluster dissimilarities among the i clusters and identify the pair of clusters that are least dissimilar (that is, most similar). Fuse these two clusters. The dissimilarity between these two clusters indicates the height in the dendrogram at which the fusion should be placed.

   (b) Compute the new pairwise inter-cluster dissimilarities among the $i-1$ remaining clusters.

Linkage	*Description*
Complete	Maximal intercluster dissimilarity. Compute all pairwise dissimilarities between the observations in cluster A and the observations in cluster B, and record the *largest* of these dissimilarities.
Single	Minimal intercluster dissimilarity. Compute all pairwise dissimilarities between the observations in cluster A and the observations in cluster B, and record the *smallest* of these dissimilarities. Single linkage can result in extended, trailing clusters in which single observations are fused one-at-a-time.
Average	Mean intercluster dissimilarity. Compute all pairwise dissimilarities between the observations in cluster A and the observations in cluster B, and record the *average* of these dissimilarities.
Centroid	Dissimilarity between the centroid for cluster A (a mean vector of length p) and the centroid for cluster B. Centroid linkage can result in undesirable *inversions*.

TABLE 10.2. *A summary of the four most commonly-used types of linkage in hierarchical clustering.*

linkage are generally preferred over single linkage, as they tend to yield more balanced dendrograms. Centroid linkage is often used in genomics, but suffers from a major drawback in that an *inversion* can occur, whereby two clusters are fused at a height *below* either of the individual clusters in the dendrogram. This can lead to difficulties in visualization as well as in interpretation of the dendrogram. The dissimilarities computed in Step 2(b) of the hierarchical clustering algorithm will depend on the type of linkage used, as well as on the choice of dissimilarity measure. Hence, the resulting

inversion

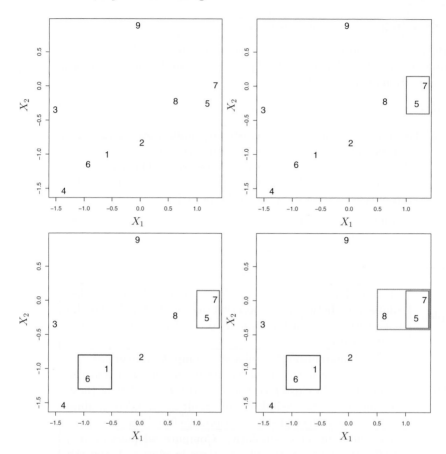

FIGURE 10.11. *An illustration of the first few steps of the hierarchical clustering algorithm, using the data from Figure 10.10, with complete linkage and Euclidean distance.* Top Left: *initially, there are nine distinct clusters,* $\{1\}, \{2\}, \ldots, \{9\}$. Top Right: *the two clusters that are closest together,* $\{5\}$ *and* $\{7\}$, *are fused into a single cluster.* Bottom Left: *the two clusters that are closest together,* $\{6\}$ *and* $\{1\}$, *are fused into a single cluster.* Bottom Right: *the two clusters that are closest together using* complete linkage, $\{8\}$ *and the cluster* $\{5, 7\}$, *are fused into a single cluster.*

dendrogram typically depends quite strongly on the type of linkage used, as is shown in Figure 10.12.

Choice of Dissimilarity Measure

Thus far, the examples in this chapter have used Euclidean distance as the dissimilarity measure. But sometimes other dissimilarity measures might be preferred. For example, *correlation-based distance* considers two observations to be similar if their features are highly correlated, even though the observed values may be far apart in terms of Euclidean distance. This is

Average Linkage   Complete Linkage   Single Linkage

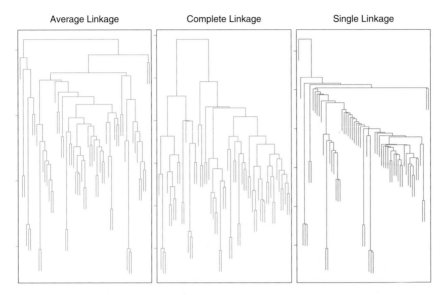

FIGURE 10.12. *Average, complete, and single linkage applied to an example data set. Average and complete linkage tend to yield more balanced clusters.*

an unusual use of correlation, which is normally computed between variables; here it is computed between the observation profiles for each pair of observations. Figure 10.13 illustrates the difference between Euclidean and correlation-based distance. Correlation-based distance focuses on the shapes of observation profiles rather than their magnitudes.

The choice of dissimilarity measure is very important, as it has a strong effect on the resulting dendrogram. In general, careful attention should be paid to the type of data being clustered and the scientific question at hand. These considerations should determine what type of dissimilarity measure is used for hierarchical clustering.

For instance, consider an online retailer interested in clustering shoppers based on their past shopping histories. The goal is to identify subgroups of *similar* shoppers, so that shoppers within each subgroup can be shown items and advertisements that are particularly likely to interest them. Suppose the data takes the form of a matrix where the rows are the shoppers and the columns are the items available for purchase; the elements of the data matrix indicate the number of times a given shopper has purchased a given item (i.e. a 0 if the shopper has never purchased this item, a 1 if the shopper has purchased it once, etc.) What type of dissimilarity measure should be used to cluster the shoppers? If Euclidean distance is used, then shoppers who have bought very few items overall (i.e. infrequent users of the online shopping site) will be clustered together. This may not be desirable. On the other hand, if correlation-based distance is used, then shoppers with similar preferences (e.g. shoppers who have bought items A and B but

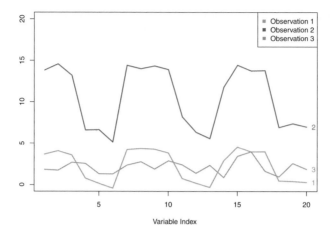

FIGURE 10.13. *Three observations with measurements on 20 variables are shown. Observations 1 and 3 have similar values for each variable and so there is a small Euclidean distance between them. But they are very weakly correlated, so they have a large correlation-based distance. On the other hand, observations 1 and 2 have quite different values for each variable, and so there is a large Euclidean distance between them. But they are highly correlated, so there is a small correlation-based distance between them.*

never items C or D) will be clustered together, even if some shoppers with these preferences are higher-volume shoppers than others. Therefore, for this application, correlation-based distance may be a better choice.

In addition to carefully selecting the dissimilarity measure used, one must also consider whether or not the variables should be scaled to have standard deviation one before the dissimilarity between the observations is computed. To illustrate this point, we continue with the online shopping example just described. Some items may be purchased more frequently than others; for instance, a shopper might buy ten pairs of socks a year, but a computer very rarely. High-frequency purchases like socks therefore tend to have a much larger effect on the inter-shopper dissimilarities, and hence on the clustering ultimately obtained, than rare purchases like computers. This may not be desirable. If the variables are scaled to have standard deviation one before the inter-observation dissimilarities are computed, then each variable will in effect be given equal importance in the hierarchical clustering performed. We might also want to scale the variables to have standard deviation one if they are measured on different scales; otherwise, the choice of units (e.g. centimeters versus kilometers) for a particular variable will greatly affect the dissimilarity measure obtained. It should come as no surprise that whether or not it is a good decision to scale the variables before computing the dissimilarity measure depends on the application at hand. An example is shown in Figure 10.14. We note that the issue of whether or not to scale the variables before performing clustering applies to K-means clustering as well.

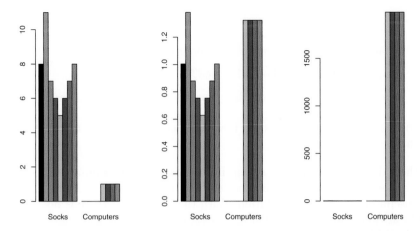

FIGURE 10.14. *An eclectic online retailer sells two items: socks and computers. Left: the number of pairs of socks, and computers, purchased by eight online shoppers is displayed. Each shopper is shown in a different color. If inter-observation dissimilarities are computed using Euclidean distance on the raw variables, then the number of socks purchased by an individual will drive the dissimilarities obtained, and the number of computers purchased will have little effect. This might be undesirable, since (1) computers are more expensive than socks and so the online retailer may be more interested in encouraging shoppers to buy computers than socks, and (2) a large difference in the number of socks purchased by two shoppers may be less informative about the shoppers' overall shopping preferences than a small difference in the number of computers purchased. Center: the same data is shown, after scaling each variable by its standard deviation. Now the number of computers purchased will have a much greater effect on the inter-observation dissimilarities obtained. Right: the same data are displayed, but now the y-axis represents the number of dollars spent by each online shopper on socks and on computers. Since computers are much more expensive than socks, now computer purchase history will drive the inter-observation dissimilarities obtained.*

10.3.3 Practical Issues in Clustering

Clustering can be a very useful tool for data analysis in the unsupervised setting. However, there are a number of issues that arise in performing clustering. We describe some of these issues here.

Small Decisions with Big Consequences

In order to perform clustering, some decisions must be made.

- Should the observations or features first be standardized in some way? For instance, maybe the variables should be centered to have mean zero and scaled to have standard deviation one.

- In the case of hierarchical clustering,

  - What dissimilarity measure should be used?
  - What type of linkage should be used?
  - Where should we cut the dendrogram in order to obtain clusters?

- In the case of K-means clustering, how many clusters should we look for in the data?

Each of these decisions can have a strong impact on the results obtained. In practice, we try several different choices, and look for the one with the most useful or interpretable solution. With these methods, there is no single right answer—any solution that exposes some interesting aspects of the data should be considered.

Validating the Clusters Obtained

Any time clustering is performed on a data set we will find clusters. But we really want to know whether the clusters that have been found represent true subgroups in the data, or whether they are simply a result of *clustering the noise*. For instance, if we were to obtain an independent set of observations, then would those observations also display the same set of clusters? This is a hard question to answer. There exist a number of techniques for assigning a p-value to a cluster in order to assess whether there is more evidence for the cluster than one would expect due to chance. However, there has been no consensus on a single best approach. More details can be found in Hastie et al. (2009).

Other Considerations in Clustering

Both K-means and hierarchical clustering will assign each observation to a cluster. However, sometimes this might not be appropriate. For instance, suppose that most of the observations truly belong to a small number of (unknown) subgroups, and a small subset of the observations are quite different from each other and from all other observations. Then since K-means and hierarchical clustering force *every* observation into a cluster, the clusters found may be heavily distorted due to the presence of outliers that do not belong to any cluster. Mixture models are an attractive approach for accommodating the presence of such outliers. These amount to a *soft* version of K-means clustering, and are described in Hastie et al. (2009).

In addition, clustering methods generally are not very robust to perturbations to the data. For instance, suppose that we cluster n observations, and then cluster the observations again after removing a subset of the n observations at random. One would hope that the two sets of clusters obtained would be quite similar, but often this is not the case!

A Tempered Approach to Interpreting the Results of Clustering

We have described some of the issues associated with clustering. However, clustering can be a very useful and valid statistical tool if used properly. We mentioned that small decisions in how clustering is performed, such as how the data are standardized and what type of linkage is used, can have a large effect on the results. Therefore, we recommend performing clustering with different choices of these parameters, and looking at the full set of results in order to see what patterns consistently emerge. Since clustering can be non-robust, we recommend clustering subsets of the data in order to get a sense of the robustness of the clusters obtained. Most importantly, we must be careful about how the results of a clustering analysis are reported. These results should not be taken as the absolute truth about a data set. Rather, they should constitute a starting point for the development of a scientific hypothesis and further study, preferably on an independent data set.

10.4 Lab 1: Principal Components Analysis

In this lab, we perform PCA on the `USArrests` data set, which is part of the base R package. The rows of the data set contain the 50 states, in alphabetical order.

```
> states=row.names(USArrests)
> states
```

The columns of the data set contain the four variables.

```
> names(USArrests)
[1] "Murder"   "Assault"  "UrbanPop" "Rape"
```

We first briefly examine the data. We notice that the variables have vastly different means.

```
> apply(USArrests, 2, mean)
  Murder   Assault UrbanPop     Rape
    7.79    170.76    65.54    21.23
```

Note that the `apply()` function allows us to apply a function—in this case, the `mean()` function—to each row or column of the data set. The second input here denotes whether we wish to compute the mean of the rows, 1, or the columns, 2. We see that there are on average three times as many rapes as murders, and more than eight times as many assaults as rapes. We can also examine the variances of the four variables using the `apply()` function.

```
> apply(USArrests, 2, var)
  Murder   Assault UrbanPop     Rape
    19.0    6945.2    209.5    87.7
```

Not surprisingly, the variables also have vastly different variances: the
UrbanPop variable measures the percentage of the population in each state
living in an urban area, which is not a comparable number to the number of rapes in each state per 100,000 individuals. If we failed to scale the
variables before performing PCA, then most of the principal components
that we observed would be driven by the Assault variable, since it has by
far the largest mean and variance. Thus, it is important to standardize the
variables to have mean zero and standard deviation one before performing
PCA.

We now perform principal components analysis using the prcomp() function, which is one of several functions in R that perform PCA.

prcomp()

```
> pr.out=prcomp(USArrests, scale=TRUE)
```

By default, the prcomp() function centers the variables to have mean zero.
By using the option scale=TRUE, we scale the variables to have standard
deviation one. The output from prcomp() contains a number of useful quantities.

```
> names(pr.out)
[1] "sdev"      "rotation" "center"   "scale"    "x"
```

The center and scale components correspond to the means and standard
deviations of the variables that were used for scaling prior to implementing
PCA.

```
> pr.out$center
  Murder  Assault UrbanPop     Rape
    7.79   170.76    65.54    21.23
> pr.out$scale
  Murder  Assault UrbanPop     Rape
    4.36    83.34    14.47     9.37
```

The rotation matrix provides the principal component loadings; each column of pr.out$rotation contains the corresponding principal component
loading vector.[2]

```
> pr.out$rotation
              PC1     PC2     PC3     PC4
Murder     -0.536   0.418  -0.341   0.649
Assault    -0.583   0.188  -0.268  -0.743
UrbanPop   -0.278  -0.873  -0.378   0.134
Rape       -0.543  -0.167   0.818   0.089
```

We see that there are four distinct principal components. This is to be
expected because there are in general $\min(n-1, p)$ informative principal
components in a data set with n observations and p variables.

---

[2]This function names it the rotation matrix, because when we matrix-multiply the
X matrix by pr.out$rotation, it gives us the coordinates of the data in the rotated
coordinate system. These coordinates are the principal component scores.

Using the `prcomp()` function, we do not need to explicitly multiply the data by the principal component loading vectors in order to obtain the principal component score vectors. Rather the 50×4 matrix x has as its columns the principal component score vectors. That is, the kth column is the kth principal component score vector.

```
> dim(pr.out$x)
[1] 50   4
```

We can plot the first two principal components as follows:

```
> biplot(pr.out, scale=0)
```

The `scale=0` argument to `biplot()` ensures that the arrows are scaled to represent the loadings; other values for scale give slightly different biplots with different interpretations.

`biplot()`

Notice that this figure is a mirror image of Figure 10.1. Recall that the principal components are only unique up to a sign change, so we can reproduce Figure 10.1 by making a few small changes:

```
> pr.out$rotation=-pr.out$rotation
> pr.out$x=-pr.out$x
> biplot(pr.out, scale=0)
```

The `prcomp()` function also outputs the standard deviation of each principal component. For instance, on the `USArrests` data set, we can access these standard deviations as follows:

```
> pr.out$sdev
[1] 1.575 0.995 0.597 0.416
```

The variance explained by each principal component is obtained by squaring these:

```
> pr.var=pr.out$sdev^2
> pr.var
[1] 2.480 0.990 0.357 0.173
```

To compute the proportion of variance explained by each principal component, we simply divide the variance explained by each principal component by the total variance explained by all four principal components:

```
> pve=pr.var/sum(pr.var)
> pve
[1] 0.6201 0.2474 0.0891 0.0434
```

We see that the first principal component explains 62.0 % of the variance in the data, the next principal component explains 24.7 % of the variance, and so forth. We can plot the PVE explained by each component, as well as the cumulative PVE, as follows:

```
> plot(pve, xlab="Principal Component", ylab="Proportion of
    Variance Explained", ylim=c(0,1),type='b')
> plot(cumsum(pve), xlab="Principal Component", ylab="
    Cumulative Proportion of Variance Explained", ylim=c(0,1),
    type='b')
```

The result is shown in Figure 10.4. Note that the function `cumsum()` computes the cumulative sum of the elements of a numeric vector. For instance:

`cumsum()`

```
> a=c(1,2,8,-3)
> cumsum(a)
[1]   1   3  11   8
```

10.5   Lab 2: Clustering

10.5.1   K-Means Clustering

The function `kmeans()` performs K-means clustering in R. We begin with a simple simulated example in which there truly are two clusters in the data: the first 25 observations have a mean shift relative to the next 25 observations.

`kmeans()`

```
> set.seed(2)
> x=matrix(rnorm(50*2), ncol=2)
> x[1:25,1]=x[1:25,1]+3
> x[1:25,2]=x[1:25,2]-4
```

We now perform K-means clustering with $K = 2$.

```
> km.out=kmeans(x,2,nstart=20)
```

The cluster assignments of the 50 observations are contained in `km.out$cluster`.

```
> km.out$cluster
 [1] 2 2 2 2 2 2 2 2 2 2 2 2 2 2 2 2 2 2 2 2 2 2 2 1 1 1 1
[30] 1 1 1 1 1 1 1 1 1 1 1 1 1 1 1 1 1 1 1 1 1
```

The K-means clustering perfectly separated the observations into two clusters even though we did not supply any group information to `kmeans()`. We can plot the data, with each observation colored according to its cluster assignment.

```
> plot(x, col=(km.out$cluster+1), main="K-Means Clustering
     Results with K=2", xlab="", ylab="", pch=20, cex=2)
```

Here the observations can be easily plotted because they are two-dimensional. If there were more than two variables then we could instead perform PCA and plot the first two principal components score vectors.

In this example, we knew that there really were two clusters because we generated the data. However, for real data, in general we do not know the true number of clusters. We could instead have performed K-means clustering on this example with $K = 3$.

```
> set.seed(4)
> km.out=kmeans(x,3,nstart=20)
> km.out
K-means clustering with 3 clusters of sizes 10, 23, 17
```

```
Cluster means:
        [,1]             [,2]
1   2.3001545  -2.69622023
2  -0.3820397  -0.08740753
3   3.7789567  -4.56200798


Clustering vector:
 [1] 3 1 3 1 3 3 3 1 3 1 3 1 3 1 3 1 3 3 3 3 3 1 3 3 3 2 2 2 2
     2 2 2 2 2 2 2 2 2 2 2 2 2 1 2 1 2 2 2 2


Within cluster sum of squares by cluster:
[1] 19.56137 52.67700 25.74089
 (between_SS / total_SS =  79.3 %)


Available components:

[1] "cluster"        "centers"        "totss"          "withinss"
    "tot.withinss"  "betweenss"      "size"
> plot(x, col=(km.out$cluster+1), main="K-Means Clustering
    Results with K=3", xlab="", ylab="", pch=20, cex=2)
```

When $K = 3$, K-means clustering splits up the two clusters.

To run the kmeans() function in R with multiple initial cluster assignments, we use the nstart argument. If a value of nstart greater than one is used, then K-means clustering will be performed using multiple random assignments in Step 1 of Algorithm 10.1, and the kmeans() function will report only the best results. Here we compare using nstart=1 to nstart=20.

```
> set.seed(3)
> km.out=kmeans(x,3,nstart=1)
> km.out$tot.withinss
[1] 104.3319
> km.out=kmeans(x,3,nstart=20)
> km.out$tot.withinss
[1] 97.9793
```

Note that km.out$tot.withinss is the total within-cluster sum of squares, which we seek to minimize by performing K-means clustering (Equation 10.11). The individual within-cluster sum-of-squares are contained in the vector km.out$withinss.

We *strongly* recommend always running K-means clustering with a large value of nstart, such as 20 or 50, since otherwise an undesirable local optimum may be obtained.

When performing K-means clustering, in addition to using multiple initial cluster assignments, it is also important to set a random seed using the set.seed() function. This way, the initial cluster assignments in Step 1 can be replicated, and the K-means output will be fully reproducible.

10.5.2 Hierarchical Clustering

The hclust() function implements hierarchical clustering in R. In the following example we use the data from Section 10.5.1 to plot the hierarchical
clustering dendrogram using complete, single, and average linkage clustering, with Euclidean distance as the dissimilarity measure. We begin by
clustering observations using complete linkage. The dist() function is used
to compute the 50×50 inter-observation Euclidean distance matrix.

hclust()

dist()

```
> hc.complete=hclust(dist(x), method="complete")
```

We could just as easily perform hierarchical clustering with average or
single linkage instead:

```
> hc.average=hclust(dist(x), method="average")
> hc.single=hclust(dist(x), method="single")
```

We can now plot the dendrograms obtained using the usual plot() function.
The numbers at the bottom of the plot identify each observation.

```
> par(mfrow=c(1,3))
> plot(hc.complete,main="Complete Linkage", xlab="", sub="",
    cex=.9)
> plot(hc.average , main="Average Linkage", xlab="", sub="",
    cex=.9)
> plot(hc.single , main="Single Linkage", xlab="", sub="",
    cex=.9)
```

To determine the cluster labels for each observation associated with a
given cut of the dendrogram, we can use the cutree() function:

cutree()

```
> cutree(hc.complete, 2)
 [1] 1 1 1 1 1 1 1 1 1 1 1 1 1 1 1 1 1 1 1 1 1 1 1 1 1 2 2 2 2
[30] 2 2 2 2 2 2 2 2 2 2 2 2 2 2 2 2 2 2 2 2 2
> cutree(hc.average, 2)
 [1] 1 1 1 1 1 1 1 1 1 1 1 1 1 1 1 1 1 1 1 1 1 1 1 1 1 2 2 2 2
[30] 2 2 2 1 2 2 2 2 2 2 2 2 2 1 2 1 2 2 2 2
> cutree(hc.single, 2)
 [1] 1 1 1 1 1 1 1 1 1 1 1 1 1 1 1 2 1 1 1 1 1 1 1 1 1 1 1 1 1
[30] 1 1 1 1 1 1 1 1 1 1 1 1 1 1 1 1 1 1 1 1 1
```

For this data, complete and average linkage generally separate the observations into their correct groups. However, single linkage identifies one point
as belonging to its own cluster. A more sensible answer is obtained when
four clusters are selected, although there are still two singletons.

```
> cutree(hc.single, 4)
 [1] 1 1 1 1 1 1 1 1 1 1 1 1 1 1 1 2 1 1 1 1 1 1 1 1 1 3 3 3 3
[30] 3 3 3 3 3 3 3 3 3 3 3 4 3 3 3 3 3 3 3 3
```

To scale the variables before performing hierarchical clustering of the
observations, we use the scale() function:

scale()

```
> xsc=scale(x)
> plot(hclust(dist(xsc), method="complete"), main="Hierarchical
    Clustering with Scaled Features")
```

Correlation-based distance can be computed using the as.dist() function, which converts an arbitrary square symmetric matrix into a form that
the hclust() function recognizes as a distance matrix. However, this only
makes sense for data with at least three features since the absolute correlation between any two observations with measurements on two features is
always 1. Hence, we will cluster a three-dimensional data set.

as.dist()

```
> x=matrix(rnorm(30*3), ncol=3)
> dd=as.dist(1-cor(t(x)))
> plot(hclust(dd, method="complete"), main="Complete Linkage
    with Correlation-Based Distance", xlab="", sub="")
```

10.6   Lab 3: NCI60 Data Example

Unsupervised techniques are often used in the analysis of genomic data.
In particular, PCA and hierarchical clustering are popular tools. We illustrate these techniques on the NCI60 cancer cell line microarray data, which
consists of 6,830 gene expression measurements on 64 cancer cell lines.

```
> library(ISLR)
> nci.labs=NCI60$labs
> nci.data=NCI60$data
```

Each cell line is labeled with a cancer type. We do not make use of the
cancer types in performing PCA and clustering, as these are unsupervised
techniques. But after performing PCA and clustering, we will check to
see the extent to which these cancer types agree with the results of these
unsupervised techniques.

The data has 64 rows and 6,830 columns.

```
> dim(nci.data)
[1]   64 6830
```

We begin by examining the cancer types for the cell lines.

```
> nci.labs[1:4]
[1] "CNS"   "CNS"    "CNS"    "RENAL"
> table(nci.labs)
nci.labs
    BREAST            CNS         COLON K562A-repro K562B-repro
      7              5             7           1           1
  LEUKEMIA MCF7A-repro MCF7D-repro    MELANOMA        NSCLC
      6              1             1           8           9
   OVARIAN    PROSTATE        RENAL     UNKNOWN
      6              2             9           1
```

10.6.1   PCA on the NCI60 Data

We first perform PCA on the data after scaling the variables (genes) to have standard deviation one, although one could reasonably argue that it is better not to scale the genes.

```
> pr.out=prcomp(nci.data, scale=TRUE)
```

We now plot the first few principal component score vectors, in order to visualize the data. The observations (cell lines) corresponding to a given cancer type will be plotted in the same color, so that we can see to what extent the observations within a cancer type are similar to each other. We first create a simple function that assigns a distinct color to each element of a numeric vector. The function will be used to assign a color to each of the 64 cell lines, based on the cancer type to which it corresponds.

```
Cols=function(vec){
+      cols=rainbow(length(unique(vec)))
+      return(cols[as.numeric(as.factor(vec))])
+    }
```

Note that the `rainbow()` function takes as its argument a positive integer, and returns a vector containing that number of distinct colors. We now can plot the principal component score vectors.   `rainbow()`

```
> par(mfrow=c(1,2))
> plot(pr.out$x[,1:2], col=Cols(nci.labs), pch=19,
     xlab="Z1",ylab="Z2")
> plot(pr.out$x[,c(1,3)], col=Cols(nci.labs), pch=19,
     xlab="Z1",ylab="Z3")
```

The resulting plots are shown in Figure 10.15. On the whole, cell lines corresponding to a single cancer type do tend to have similar values on the first few principal component score vectors. This indicates that cell lines from the same cancer type tend to have pretty similar gene expression levels.

We can obtain a summary of the proportion of variance explained (PVE) of the first few principal components using the `summary()` method for a prcomp object (we have truncated the printout):

```
> summary(pr.out)
Importance of components:
                            PC1     PC2     PC3     PC4     PC5
Standard deviation       27.853  21.4814 19.8205 17.0326 15.9718
Proportion of Variance    0.114   0.0676  0.0575  0.0425  0.0374
Cumulative Proportion     0.114   0.1812  0.2387  0.2812  0.3185
```

Using the `plot()` function, we can also plot the variance explained by the first few principal components.

```
> plot(pr.out)
```

Note that the height of each bar in the bar plot is given by squaring the corresponding element of `pr.out$sdev`. However, it is more informative to

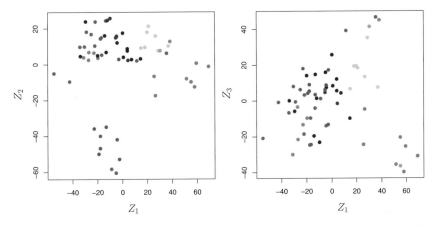

FIGURE 10.15. *Projections of the* NCI60 *cancer cell lines onto the first three principal components (in other words, the scores for the first three principal components). On the whole, observations belonging to a single cancer type tend to lie near each other in this low-dimensional space. It would not have been possible to visualize the data without using a dimension reduction method such as PCA, since based on the full data set there are $\binom{6,830}{2}$ possible scatterplots, none of which would have been particularly informative.*

plot the PVE of each principal component (i.e. a scree plot) and the cumulative PVE of each principal component. This can be done with just a little work.

```
> pve=100*pr.out$sdev^2/sum(pr.out$sdev^2)
> par(mfrow=c(1,2))
> plot(pve, type="o", ylab="PVE", xlab="Principal Component",
    col="blue")
> plot(cumsum(pve), type="o", ylab="Cumulative PVE", xlab="
    Principal Component", col="brown3")
```

(Note that the elements of pve can also be computed directly from the summary, summary(pr.out)$importance[2,], and the elements of cumsum(pve) are given by summary(pr.out)$importance[3,].) The resulting plots are shown in Figure 10.16. We see that together, the first seven principal components explain around 40 % of the variance in the data. This is not a huge amount of the variance. However, looking at the scree plot, we see that while each of the first seven principal components explain a substantial amount of variance, there is a marked decrease in the variance explained by further principal components. That is, there is an *elbow* in the plot after approximately the seventh principal component. This suggests that there may be little benefit to examining more than seven or so principal components (though even examining seven principal components may be difficult).

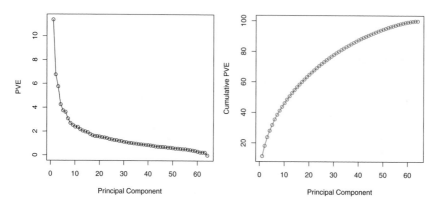

FIGURE 10.16. *The PVE of the principal components of the* NCI60 *cancer cell line microarray data set. Left: the PVE of each principal component is shown. Right: the cumulative PVE of the principal components is shown. Together, all principal components explain 100 % of the variance.*

10.6.2 Clustering the Observations of the NCI60 Data

We now proceed to hierarchically cluster the cell lines in the NCI60 data, with the goal of finding out whether or not the observations cluster into distinct types of cancer. To begin, we standardize the variables to have mean zero and standard deviation one. As mentioned earlier, this step is optional and should be performed only if we want each gene to be on the same *scale*.

```
> sd.data=scale(nci.data)
```

We now perform hierarchical clustering of the observations using complete, single, and average linkage. Euclidean distance is used as the dissimilarity measure.

```
> par(mfrow=c(1,3))
> data.dist=dist(sd.data)
> plot(hclust(data.dist), labels=nci.labs, main="Complete
    Linkage", xlab="", sub="",ylab="")
> plot(hclust(data.dist, method="average"), labels=nci.labs,
    main="Average Linkage", xlab="", sub="",ylab="")
> plot(hclust(data.dist, method="single"), labels=nci.labs,
    main="Single Linkage", xlab="", sub="",ylab="")
```

The results are shown in Figure 10.17. We see that the choice of linkage certainly does affect the results obtained. Typically, single linkage will tend to yield *trailing* clusters: very large clusters onto which individual observations attach one-by-one. On the other hand, complete and average linkage tend to yield more balanced, attractive clusters. For this reason, complete and average linkage are generally preferred to single linkage. Clearly cell lines within a single cancer type do tend to cluster together, although the

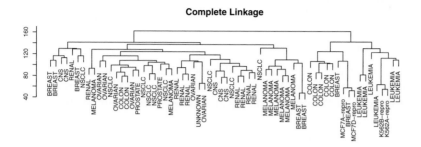

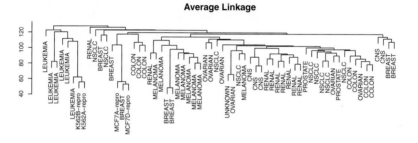

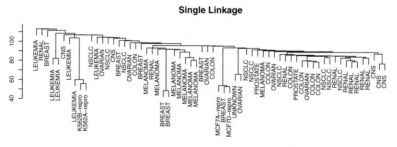

FIGURE 10.17. *The* NCI60 *cancer cell line microarray data, clustered with average, complete, and single linkage, and using Euclidean distance as the dissimilarity measure. Complete and average linkage tend to yield evenly sized clusters whereas single linkage tends to yield extended clusters to which single leaves are fused one by one.*

clustering is not perfect. We will use complete linkage hierarchical clustering for the analysis that follows.

We can cut the dendrogram at the height that will yield a particular
number of clusters, say four:

```
> hc.out=hclust(dist(sd.data))
> hc.clusters=cutree(hc.out,4)
> table(hc.clusters,nci.labs)
```

There are some clear patterns. All the leukemia cell lines fall in cluster 3,
while the breast cancer cell lines are spread out over three different clusters.
We can plot the cut on the dendrogram that produces these four clusters:

```
> par(mfrow=c(1,1))
> plot(hc.out, labels=nci.labs)
> abline(h=139, col="red")
```

The abline() function draws a straight line on top of any existing plot
in R. The argument h=139 plots a horizontal line at height 139 on the dendrogram; this is the height that results in four distinct clusters. It is easy
to verify that the resulting clusters are the same as the ones we obtained
using cutree(hc.out,4).

Printing the output of hclust gives a useful brief summary of the object:

```
> hc.out

Call:
hclust(d = dist(dat))

Cluster method   : complete
Distance         : euclidean
Number of objects: 64
```

We claimed earlier in Section 10.3.2 that K-means clustering and hierarchical clustering with the dendrogram cut to obtain the same number
of clusters can yield very different results. How do these NCI60 hierarchical
clustering results compare to what we get if we perform K-means clustering
with $K = 4$?

```
> set.seed(2)
> km.out=kmeans(sd.data, 4, nstart=20)
> km.clusters=km.out$cluster
> table(km.clusters,hc.clusters)
            hc.clusters
km.clusters  1  2  3  4
          1 11  0  0  9
          2  0  0  8  0
          3  9  0  0  0
          4 20  7  0  0
```

We see that the four clusters obtained using hierarchical clustering and K-
means clustering are somewhat different. Cluster 2 in K-means clustering is
identical to cluster 3 in hierarchical clustering. However, the other clusters

differ: for instance, cluster 4 in K-means clustering contains a portion of the observations assigned to cluster 1 by hierarchical clustering, as well as all of the observations assigned to cluster 2 by hierarchical clustering.

Rather than performing hierarchical clustering on the entire data matrix, we can simply perform hierarchical clustering on the first few principal component score vectors, as follows:

```
> hc.out=hclust(dist(pr.out$x[,1:5]))
> plot(hc.out, labels=nci.labs, main="Hier. Clust. on First
    Five Score Vectors")
> table(cutree(hc.out,4), nci.labs)
```

Not surprisingly, these results are different from the ones that we obtained when we performed hierarchical clustering on the full data set. Sometimes performing clustering on the first few principal component score vectors can give better results than performing clustering on the full data. In this situation, we might view the principal component step as one of denoising the data. We could also perform K-means clustering on the first few principal component score vectors rather than the full data set.

10.7   Exercises

Conceptual

1. This problem involves the K-means clustering algorithm.
   (a) Prove (10.12).

   (b) On the basis of this identity, argue that the K-means clustering algorithm (Algorithm 10.1) decreases the objective (10.11) at each iteration.

2. Suppose that we have four observations, for which we compute a dissimilarity matrix, given by

$$\begin{bmatrix} & 0.3 & 0.4 & 0.7 \\ 0.3 & & 0.5 & 0.8 \\ 0.4 & 0.5 & & 0.45 \\ 0.7 & 0.8 & 0.45 & \end{bmatrix}.$$

   For instance, the dissimilarity between the first and second observations is 0.3, and the dissimilarity between the second and fourth observations is 0.8.

   (a) On the basis of this dissimilarity matrix, sketch the dendrogram that results from hierarchically clustering these four observations using complete linkage. Be sure to indicate on the plot the height at which each fusion occurs, as well as the observations corresponding to each leaf in the dendrogram.

(b) Repeat (a), this time using single linkage clustering.

(c) Suppose that we cut the dendogram obtained in (a) such that two clusters result. Which observations are in each cluster?

(d) Suppose that we cut the dendogram obtained in (b) such that two clusters result. Which observations are in each cluster?

(e) It is mentioned in the chapter that at each fusion in the dendrogram, the position of the two clusters being fused can be swapped without changing the meaning of the dendrogram. Draw a dendrogram that is equivalent to the dendrogram in (a), for which two or more of the leaves are repositioned, but for which the meaning of the dendrogram is the same.

3. In this problem, you will perform K-means clustering manually, with $K = 2$, on a small example with $n = 6$ observations and $p = 2$ features. The observations are as follows.

Obs.	X_1	X_2
1	1	4
2	1	3
3	0	4
4	5	1
5	6	2
6	4	0

(a) Plot the observations.

(b) Randomly assign a cluster label to each observation. You can use the sample() command in R to do this. Report the cluster labels for each observation.

(c) Compute the centroid for each cluster.

(d) Assign each observation to the centroid to which it is closest, in terms of Euclidean distance. Report the cluster labels for each observation.

(e) Repeat (c) and (d) until the answers obtained stop changing.

(f) In your plot from (a), color the observations according to the cluster labels obtained.

4. Suppose that for a particular data set, we perform hierarchical clustering using single linkage and using complete linkage. We obtain two dendrograms.

(a) At a certain point on the single linkage dendrogram, the clusters $\{1, 2, 3\}$ and $\{4, 5\}$ fuse. On the complete linkage dendrogram, the clusters $\{1, 2, 3\}$ and $\{4, 5\}$ also fuse at a certain point. Which fusion will occur higher on the tree, or will they fuse at the same height, or is there not enough information to tell?

(b) At a certain point on the single linkage dendrogram, the clusters {5} and {6} fuse. On the complete linkage dendrogram, the clusters {5} and {6} also fuse at a certain point. Which fusion will occur higher on the tree, or will they fuse at the same height, or is there not enough information to tell?

5. In words, describe the results that you would expect if you performed K-means clustering of the eight shoppers in Figure 10.14, on the basis of their sock and computer purchases, with $K = 2$. Give three answers, one for each of the variable scalings displayed. Explain.

6. A researcher collects expression measurements for 1,000 genes in 100 tissue samples. The data can be written as a $1,000 \times 100$ matrix, which we call $\mathbf{X}$, in which each row represents a gene and each column a tissue sample. Each tissue sample was processed on a different day, and the columns of $\mathbf{X}$ are ordered so that the samples that were processed earliest are on the left, and the samples that were processed later are on the right. The tissue samples belong to two groups: control (C) and treatment (T). The C and T samples were processed in a random order across the days. The researcher wishes to determine whether each gene's expression measurements differ between the treatment and control groups.

As a pre-analysis (before comparing T versus C), the researcher performs a principal component analysis of the data, and finds that the first principal component (a vector of length 100) has a strong linear trend from left to right, and explains 10 % of the variation. The researcher now remembers that each patient sample was run on one of two machines, A and B, and machine A was used more often in the earlier times while B was used more often later. The researcher has a record of which sample was run on which machine.

(a) Explain what it means that the first principal component "explains 10 % of the variation".

(b) The researcher decides to replace the (j, i)th element of $\mathbf{X}$ with

$$x_{ji} - \phi_{j1} z_{i1}$$

where z_{i1} is the ith score, and ϕ_{j1} is the jth loading, for the first principal component. He will then perform a two-sample t-test on each gene in this new data set in order to determine whether its expression differs between the two conditions. Critique this idea, and suggest a better approach. (The principal component analysis is performed on $\mathbf{X}^T$).

(c) Design and run a small simulation experiment to demonstrate the superiority of your idea.

Applied

7. In the chapter, we mentioned the use of correlation-based distance and Euclidean distance as dissimilarity measures for hierarchical clustering. It turns out that these two measures are almost equivalent: if each observation has been centered to have mean zero and standard deviation one, and if we let r_{ij} denote the correlation between the ith and jth observations, then the quantity $1 - r_{ij}$ is proportional to the squared Euclidean distance between the ith and jth observations.

   On the USArrests data, show that this proportionality holds.

   Hint: The Euclidean distance can be calculated using the dist() *function, and correlations can be calculated using the* cor() *function.*

8. In Section 10.2.3, a formula for calculating PVE was given in Equation 10.8. We also saw that the PVE can be obtained using the sdev output of the prcomp() function.

   On the USArrests data, calculate PVE in two ways:

   (a) Using the sdev output of the prcomp() function, as was done in Section 10.2.3.

   (b) By applying Equation 10.8 directly. That is, use the prcomp() function to compute the principal component loadings. Then, use those loadings in Equation 10.8 to obtain the PVE.

   These two approaches should give the same results.

   Hint: You will only obtain the same results in (a) and (b) if the same data is used in both cases. For instance, if in (a) you performed prcomp() *using centered and scaled variables, then you must center and scale the variables before applying Equation 10.3 in (b).*

9. Consider the USArrests data. We will now perform hierarchical clustering on the states.

   (a) Using hierarchical clustering with complete linkage and Euclidean distance, cluster the states.

   (b) Cut the dendrogram at a height that results in three distinct clusters. Which states belong to which clusters?

   (c) Hierarchically cluster the states using complete linkage and Euclidean distance, *after scaling the variables to have standard deviation one.*

   (d) What effect does scaling the variables have on the hierarchical clustering obtained? In your opinion, should the variables be scaled before the inter-observation dissimilarities are computed? Provide a justification for your answer.

10. In this problem, you will generate simulated data, and then perform PCA and K-means clustering on the data.

   (a) Generate a simulated data set with 20 observations in each of three classes (i.e. 60 observations total), and 50 variables.

   Hint: There are a number of functions in R that you can use to generate data. One example is the rnorm() *function;* runif() *is another option. Be sure to add a mean shift to the observations in each class so that there are three distinct classes.*

   (b) Perform PCA on the 60 observations and plot the first two principal component score vectors. Use a different color to indicate the observations in each of the three classes. If the three classes appear separated in this plot, then continue on to part (c). If not, then return to part (a) and modify the simulation so that there is greater separation between the three classes. Do not continue to part (c) until the three classes show at least some separation in the first two principal component score vectors.

   (c) Perform K-means clustering of the observations with $K = 3$. How well do the clusters that you obtained in K-means clustering compare to the true class labels?

   Hint: You can use the table() *function in R to compare the true class labels to the class labels obtained by clustering. Be careful how you interpret the results: K-means clustering will arbitrarily number the clusters, so you cannot simply check whether the true class labels and clustering labels are the same.*

   (d) Perform K-means clustering with $K = 2$. Describe your results.

   (e) Now perform K-means clustering with $K = 4$, and describe your results.

   (f) Now perform K-means clustering with $K = 3$ on the first two principal component score vectors, rather than on the raw data. That is, perform K-means clustering on the 60×2 matrix of which the first column is the first principal component score vector, and the second column is the second principal component score vector. Comment on the results.

   (g) Using the scale() function, perform K-means clustering with $K = 3$ on the data *after scaling each variable to have standard deviation one.* How do these results compare to those obtained in (b)? Explain.

11. On the book website, www.StatLearning.com, there is a gene expression data set (Ch10Ex11.csv) that consists of 40 tissue samples with measurements on 1,000 genes. The first 20 samples are from healthy patients, while the second 20 are from a diseased group.

(a) Load in the data using `read.csv()`. You will need to select `header=F`.

(b) Apply hierarchical clustering to the samples using correlation-based distance, and plot the dendrogram. Do the genes separate the samples into the two groups? Do your results depend on the type of linkage used?

(c) Your collaborator wants to know which genes differ the most across the two groups. Suggest a way to answer this question, and apply it here.

Index

G. James et al., *An Introduction to Statistical Learning: with Applications in R*,   419
Springer Texts in Statistics, DOI 10.1007/978-1-4614-7138-7,
© Springer Science+Business Media New York 2013